Green Analytical Chemistry

Green Analytical Chemistry

Mihkel Koel and Mihkel Kaljurand
Institute of Chemistry, Tallinn University of Technology, Tallinn, Estonia

RSC Publishing

ISBN: 978-1-84755-872-5

A catalogue record for this book is available from the British Library

Published by The Royal Society of Chemistry,
Thomas Graham House, Science Park, Milton Road,
Cambridge CB4 0WF, UK

Registered Charity Number 207890

For further information see our web site at www.rsc.org

Foreword

The pursuit of Green Analytical Chemistry is an emerging field of endeavor that has already shown great achievements and even greater promise and potential for the future. As we consider that the word *analysis* means: "the resolution of anything complex into simple elements," it reminds us that to affect any complex system we must first understand it. Essential to our understanding is analysis. This is particularly important in the area of Green Chemistry.

Green Chemistry is centered on the concept of design. The definition of *Green Chemistry* itself is: "the *design* of chemical products and processes that reduce or eliminate the use and generation of chemical products and processes that reduce or eliminate the use and generation of hazardous substances." In order to design something you have to deeply understand it. You cannot design something by accident. If it were an accident, it wasn't design. Analysis is essential therefore to achieving the depth of understanding needed for design; design of products, processes, and systems.

Green Analytical Chemistry not only informs design by providing insight, it is currently reinventing the field of chemical analysis itself. It was not so long ago that there were studies done to identify some of the major significant waste generators in various industrial sectors. Surprisingly on the list were the environmental analytical laboratories. The same laboratories that were so essential in monitoring, measuring, and characterizing environmental problems, were themselves generating environmental problems. This was largely due to the quantity of solvents that were used at the time of the studies. Upon seeing these studies, I posited the mostly humorous, Anastas Environmental Analysis Principle (with apologies to Heisenberg) – "One cannot measure an environmental problem without causing another environmental problem." The great potential of Green Analytical Chemistry is to continue to make this "principle" antiquated and increasingly false.

One of the Twelve Principles of Green Chemistry deals exclusively with analytical chemistry in pursuit of real-time, in-process, in-field, non-destructive, non-materially or energy-intensive analysis. This book provides important insights into the various methodologies and techniques that are at the cutting edge of this rapidly growing field. The contributors to this volume are to be commended for their vision and insight. Readers of this volume, especially students, will find fertile research areas to grow new projects and technologies.

There is a story of a sign that supposedly hung in Einstein's office that read, "Not everything that can be counted counts, and not everything that counts can be counted." The astounding success of chemical analysis over the past century and even more so in recent decades, has made it clear that our ability to analyze needs to be coupled with our ability to characterize. Our ability to generate data is increasing at an astounding rate. Our ability to transform the data into information must strive to keep pace. The Green Analytical Chemistry challenge is to couple these abilities with the perspective that the ultimate goal is to generate knowledge and perhaps even wisdom. Only in this way will we have the ability to move the planet and civilization toward a sustainable trajectory. In the assessment of analytical instruments, the compass is more important than the speedometer. While it is often good to go faster and faster, it is more important to know that you are heading in the right direction. This book is an important step in the right direction.

Paul T. Anastas
New Haven, CT
USA

Preface

The principles of Green Chemistry are reaching into all the chemical disciplines. There is increasing demand for chemical analysis and the development of analytical chemistry continues at a steady rate. Every new discovery in chemistry, physics, molecular biology and material science has an application in analytical chemistry as well.

The use of toxic compounds and solvents in chemical analysis is an extremely pressing issue that makes Green Analytical Chemistry an emerging hot topic in industrial and governmental laboratories as well as in academia.

However, the relationship between Green Chemistry and Analytical Chemistry is especially close because analytical chemistry provides the means of evaluating and justifying Green Chemistry and is an efficient tool for determining the "greenness" of a chemical product or technology.

On the other hand, methods of chemical analysis cannot avoid the use of solvents, reagents and energy, and thus generate waste. The application of a Green Chemistry perspective in the assessment of analytical methods should be a natural development in chemistry and should coincide with its basic policy. Being green is a "must" in contemporary chemical analysis, *i.e.* the true cost of resources and generation of waste must be included in the design of every new method and in every comparison of procedures. It can be said that the goal of Green Analytical Chemistry is to employ analytical procedures that generate less hazardous waste, are safer to use and more benign to the environment.

This book is a modest attempt to portray the changing situation in analytical chemistry with regard to adopting the principles of Green

Green Analytical Chemistry
By Mihkel Koel and Mihkel Kaljurand

Published by the Royal Society of Chemistry, www.rsc.org

Chemistry. There are many other textbooks on different aspects of Green Chemistry, but none on Green Analytical Chemistry. We hope to fill this gap to some extent. The rationale for writing this book on Green Chemistry is to describe the current application of green principles in analytical chemistry and to comment on what more needs to be done. The main emphasis in analytical chemistry is on the metrological quality of the data, and only recently has attention been directed toward the environmental aspects of the way the data are obtained. In view of the theoretical as well as the practical importance of this subject, we urge the scientific community to exert itself in this respect.

The first two chapters are devoted to the general aspects of Green Chemistry and trends in analytical chemistry. An overview is provided of existing techniques for sample preparation and instrumental analysis in the context of solvent usage and safer chemicals. The next chapters review the current knowledge and efforts in this area, and cover diverse fields of instrumental analytical chemistry, such as separation science, optical and mass spectrometry, and analytical electrochemistry, that could guide others to new ideas and discoveries. The book endorses some prospective methods such as capillary electrophoresis and chemometrics that have been somewhat neglected in the literature on Green Chemistry. Microfluidic technologies have enabled the miniaturisation of established analytical techniques and enhanced performance. The expression "small is beautiful" is more valid in Green Analytical Chemistry than in many other situations. Micronisation can also be exploited to develop completely new approaches to chemical and biological processing.

The book describes efforts to make analytical chemistry greener: the application of Green Chemistry ideas and concepts in chemical analysis, an evaluation of the performance of current analytical methodologies from the perspective of Green Chemistry, and a discussion of the concept of green profiles of methods. We must emphasise that every step toward greening a particular method must respect the main analytical parameters such as selectivity, sensitivity, reliability, analysis time and cost.

This book has been written from an academic point of view; it is meant for senior scientists as well as novices who are just entering the field. The prospective audience for this book is likely to be managers of analytical research laboratories, but the book will also be of interest to teachers of analytical chemistry and even to green politicians.

We obtained most of the information for the book from the publications of scientists in this field. However, we did not attempt to provide a complete review of the literature on this topic. We tried to identify the general trends and the most promising applications that would profit from more attention for further development. There were also time constraints that required us to finish the work by mid-2009. We take responsibility for the choices made in selecting the topics and would be grateful for readers' comments on errors, oversights and other useful data on the systems discussed.

Our sincere thanks are devoted to Dolores Talpt Lindsay for her fruitful cooperation and serious efforts to improve our language and Jekaterina Mazina for her help with the illustrations. We are also grateful to the Tallinn University of Technology for supporting the preparation of this book.

Mihkel Koel and Mihkel Kaljurand
Tallinn University of Technology, Tallinn, Estonia

Contents

Green Analytical Chemistry
By Mihkel Koel and Mihkel Kaljurand

Published by the Royal Society of Chemistry, www.rsc.org

CHAPTER 1

Introduction to Green Chemistry

Science and technology are the cornerstones of the development of human society. Science provides humans with very powerful tools, the careless use of which endangers their world. The result could well be an environment inhospitable to human habitation. It is conceivable that without serious effort, this scenario will prevail and the further development of society could be jeopardized.

Chemistry is a very old discipline, with references to chemical transformations and debate about the nature of matter dating back to the times of the ancient Egyptians and Greeks. Modern chemistry began to emerge from alchemy in the seventeenth and eighteenth centuries, thanks to scholars such as Robert Boyle and Antoine Lavoisier, and made rapid advances in the following two centuries.

At the present time, the central role of chemistry in science is synthesis. The structure and bonding of molecules is at the core of the discipline, especially in organic chemistry, and using weaker intermolecular forces to assemble supra-molecules is a field with much still to explore. Understanding phenomena at the molecular level is vital to future innovation and invention.

The chemical sciences continue to be at the heart of multidisciplinary initiatives. Today, chemical tools — and the analytical tools chemists use — are especially pertinent to research in biology. For example, much of gene technology is chemistry. The chemical sciences provide the wide expertise for most scientific and technological developments and thus continue to make enormous contributions to social, cultural, economic and intellectual advances. Chemistry contributes to our well being, long life expectancy and economic prosperity. It satisfies the desire for better materials for everyday life and accommodation, for drugs to cure

Green Analytical Chemistry
By Mihkel Koel and Mihkel Kaljurand

Published by the Royal Society of Chemistry, www.rsc.org

illnesses and improve health, for pure water, and a host of other human activities.

For this reason, energy and sustainable chemistry are key themes in current discussions about the future. Chemicals are present in all spheres of human life. Among other things, they are used as pharmaceuticals, pesticides, detergents, fertilizers, dyes, paints, preservatives and food additives. The first and most influential description of the dangers related to chemicals in the environment is found in Rachel Carson's *Silent Spring*, published in 1962.[1]

Synthetic chemicals end up in the environment in many different ways. The chemical industry is a point source of emissions which create changes around that point. In everyday life the constituents and ingredients of consumer or household products and other open applications emit chemicals into the environment *via* non-point sources. Chemicals and their compositions do degrade and break down into water, carbon dioxide and inorganic salts, but very often the degradation is incomplete. Unknown transformation products can result from such biological and chemical processes as hydrolysis, redox reactions or photolysis. These unknown chemical entities remain in the environment and can be toxic to humans and environmental organisms. The latter situation is more serious as there is usually much less knowledge about the longevity and effects on the environment of the final transformation products than there is about the parent compounds. Even if there is some degree of degradation, the parent compounds will nevertheless remain at constant levels in the environment if the input rate is higher than their rate of degradation or mineralization. This situation has to do with the persistency of chemicals. Persistency is one of the most important criteria in the environmental assessment of chemicals.

Polychlorinated biphenyls (PCBs) are a classic example of persistent pollutants. PCBs were synthesized for the first time in 1877, and as early as 1899 severe health problems (chloracne) associated with the handling of PCBs were reported. Since then, the poisoning of rice oil by these compounds and their neurotoxic effects and carcinogenicity have been described in detail. Despite this knowledge, it was not until 1999 that PCBs were completely banned within the EU – 100 years after the first reports of their severe toxicity. This example clearly demonstrates that it is not only the time lag of the impact of the chemicals on environmental processes, but also the time lag of economic and political systems which has a significant effect.

In 1995, the Governing Council of the United Nations Environment Programme (UNEP) called for global action to assess the effects of POPs (persistent organic pollutants). The twelve worst offenders are known as the "dirty dozen" and include eight organo-chlorine pesticides: aldrin, chlordane, DDT, dieldrin, endrin, heptachlor, mirex and toxaphene; two industrial chemicals: hexachlorobenzene (HCB) and the polychlorinated biphenyl (PCB) group; and two groups of industrial by-products: dioxins and furans. It became clear that these POPs were deadly and that urgent global action was needed. The result was the Stockholm Convention on Persistent Organic Pollutants in May 2001.[2] The Convention outlaws the dirty dozen and also establishes a system to track additional substances that can be classified as POPs, to prevent the development of new problem chemicals.

According to the Stockholm convention, a half-life of more than 50 days in water is set as a criterion for POPs. Recent research has demonstrated that chemicals which are less persistent and have a higher polarity than PCBs are distributed globally as well, and can also accumulate in humans.

An example of pollutants with a much shorter history are the inert organohalogen compounds, known as freons. They were initially developed in the early 20th century as an alternative to the toxic gases that were used as refrigerants, such as ammonia, chloromethane and sulfur dioxide. These compounds, which contain only chlorine, fluorine, and carbon, are called chlorofluorocarbons (CFCs). Each freon product is designated by a number. For instance, Freon-11 is trichlorofluoromethane and Freon-12 is dichlorodifluoromethane. The most common is Freon-113, trichlorotrifluoroethane, used as a cleaning agent.

In the 1970s, scientific evidence showed that human-produced chemicals are responsible for observed depletions of the ozone layer. In 1978, the United States, Canada and Norway enacted bans on aerosol sprays containing CFC that are thought to damage the ozone layer. After the discovery of the Antarctic ozone hole in 1985, CFC production was sharply limited as of 1987 and was phased out completely by 1996 according to an international treaty. The Montreal Protocol on Substances that Deplete the Ozone Layer classifies Freon-11 and Freon-12 as Annex A substances, and bans their production and consumption.[3] It has now been confirmed that the upper atmosphere ozone depletion rate has slowed significantly during the past decade.

The interim replacements for CFCs are hydrochlorofluorocarbons (HCFCs), which contain chlorine that depletes stratospheric ozone, but to a much lesser extent than CFCs. Ultimately, hydrofluorocarbons (HFCs) will replace HCFCs with essentially no ozone destruction, although all three groups of halocarbons are powerful greenhouse gases.

The Montreal Protocol on Substances That Deplete the Ozone Layer is a landmark international agreement designed to protect the stratospheric ozone layer. The treaty was originally signed in 1987 and substantially amended in 1990 and 1992. The Montreal Protocol stipulates that the production and consumption of compounds that deplete ozone in the stratosphere – chlorofluorocarbons (CFCs), halons, carbon tetrachloride and methyl chloroform – are to be phased out by 2000 (2005 for methyl chloroform).

Chemistry, and the application of chemical technology, can be practised and useful materials synthesised in many ways, some of which are safer for the environment and human health than others. Safety for humans and the environment, as well as the economical use of resources, must be of utmost consideration in the development of new chemical processes. In other words, the true cost of resources and waste must be included in every calculation and comparison. The mind-set in chemistry must be changed from waste treatment to the prevention of waste generation.

New environmental regulations, and a growing social consciousness with regard to the protection of nature, have nudged the chemical sciences and industry toward a new framework in which pollution prevention is the central consideration. This fact presents an enormous technological and scientific challenge that requires the development of innovative products and tools to minimize the impact on the environment. Nevertheless, the same standards of quality and efficiency need to be retained in order to continue scientific advancement within a sustainable framework.

Concerned individuals have raised these questions and citizens are becoming aware of the critical situation. This is also true of chemical scientists and engineers and the chemical industry where a new wind is blowing. A comprehensive and accurate understanding of what constitutes the sustainable development of human civilization is required, taking into account the ecological, economic and social dimensions. This is simple in principle but much more difficult in practice.

In the long term, chemicals end up in the environment. Therefore, an important requirement is to focus not only on the optimization of the chemical synthesis of a molecule and its use, but also to take into account the molecules themselves and their effect on the environment. To quote Anastas and Warner in *Green Chemistry: Theory and Practice*: "Chemical products should be designed so that at the end of their function they do not persist in the environment and do break down into innocuous products".[4] Designing chemicals to meet both the requirements of their application and environmental considerations, throughout their life cycle, is ambitious and quite new. It must, however, become an essential part of the core of chemistry and pharmacy.

Agenda 21, Chapter 35 of *Science for Sustainable Development*,[5] focuses on the role and the use of the sciences in supporting the prudent management of the environment and development for the daily survival and future development of humanity. "The sciences should continue to play an increasing role in providing for an improvement in the efficiency of resource utilization and in finding new development practices, resources and alternatives. There is a need for the sciences to constantly reassess and promote less intensive trends in resource utilization, including less intensive utilization of energy in industry, agriculture and transportation. Thus, the sciences are increasingly being understood as an essential component in the search for feasible pathways towards sustainable development".

Scientists must develop their basic and applied research topics in accordance with the sustainable development of society. This also applies to chemistry and chemical research topics. Indeed, chemistry – expressed as the courage and curiosity to discover and formulate new materials and compounds to improve human life – plays the most important role in this process. However, chemistry is changing very slowly. Obviously, chemical companies are accustomed to petrochemicals and are reluctant to use alternative renewable feedstocks which may not be well suited to the usual petrochemical processing.

A general process scheme is presented in Figure 1.1 which illustrates the complexity and importance of factors other than the product outcome.

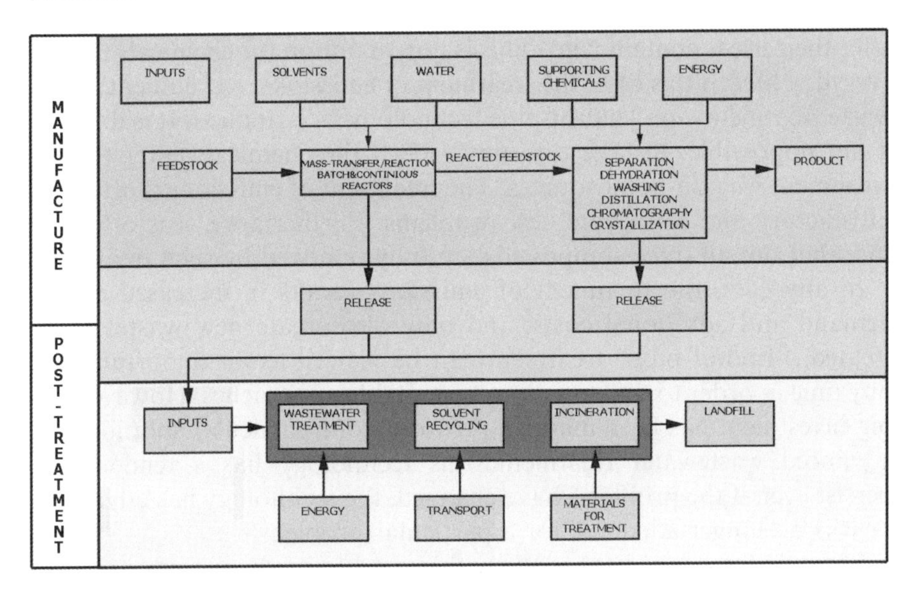

Figure 1.1 General process scheme.

The following describes a pharmaceutical industry manufacturing process based on this scheme. The normal manufacturing yield from a single stage ranges between 35 and 95%, with an average yield of 86% The typical primary manufacturing process involves six stages with an overall yield of 30–40%. Overall yield does not capture the use of reagents, solvents and catalysts. If these are included, the average total materials use is 16 kg/kg of stage product (intermediate). Even with a 100% yield at each stage, a 16 kg/kg ratio of materials use would result in an overall Mass Productivity of about 1%.[6]

These figures demonstrate that a large amount of the materials are not part of the product and could possibly end up in the environment as a result of the process. Reducing the release of these chemicals into the environment has become a crucial issue.

There are at least three options for decreasing the input of chemicals into the environment:

- the technical approach: advanced treatment (short- to medium-term);
- the education and training of users, *e.g.* retailers and consumers (medium-term);
- the substitution of critical compounds with benign ones (long-term).

The traditional short- to medium-term approach for the prevention and reduction of the input of chemicals into the environment during and after their use is containment. This is not an option for chemicals used in everyday life. In this case, the treatment of emissions, *e.g.* effluent, air or waste treatment – or "end-of-pipe technology" – is applied. It is difficult, if not impossible, to take this approach if the chemicals enter the environment *via* non-point sources. The treatment of emissions is often not satisfactory and can create new problems. Furthermore, it is often the case that not all trace compounds are fully removed by such treatment.

In any case, the treatment of emissions results in increased energy demand and additional costs, and may also create new wastes to be treated. "End-of-pipe" treatment can be viewed as an opportunity to buy time in order to develop more sustainable approaches. However, if a big investment has been made in providing one particular solution, *e.g.* advanced wastewater treatment, this technology has a tendency to persist even if the problems have changed, the technology has advanced, or it is no longer adequate for a particular problem.

Some progress has been made with regard to medium-term strategies and containment at source. Responsible care and product stewardship have been created and implemented within a number of industries, which has contributed to the reduction of emissions. In the case of chemicals used in open systems, this approach has its limitations and is not always very efficient.

The third and long-term approach is the objective of Green Chemistry – to make chemistry itself more sustainable, and the key to sustainability is the design of chemicals and pharmaceuticals which are not harmful to the environment. One approach is to design chemicals in such a way that they are readily degradable after their use. This means that the functionality of a chemical consists of more than just the properties required for successful applications. Functionality in a broader and sustainable sense would also include the rapid and complete degradability of a molecule after its use, for example in traditional sewage treatment. According to this assumption, when new chemical entities are created, it is necessary to take into account both the properties required for successful applications as well as what happens to them afterwards.

Compounds with few or no side effects or low toxicity can be called inherently safe. Such chemicals need little or no safety measures or special knowledge on the part of the user or applicant. A simpler expression of this concept is that chemicals should have the lowest possible impact on humans handling and ingesting them and also on the environment. If the environment is exposed to such compounds, no effects have to be considered. This calls for a different understanding of the functionality of a chemical: its manufacturing and use, as well as its fate after use. Such chemicals are benign by design. A life cycle assessment and optimization is conducted according to the specific conditions present at the different stages of a chemical before its synthesis and introduction into the market.

These steps include:

- raw materials
- synthesis
- production
- use
- fate after use

Such an approach requires the chemist performing the synthesis to have a different mind-set. He has to take an interdisciplinary approach and consider the world outside his laboratory. He is not only a craftsman who assembles molecules, but also an architect who designs them. This entails accounting for not only the functionalities of a molecule that are necessary for its application, but also for those functionalities which are important throughout the entire life cycle of the molecule. It is this aspect of sustainability which is specific to chemistry and at the core of a new understanding of the science. It includes the sustainable use of resources, as well as improved efficiency and efficacy.

At first glance, it does not seem that this third approach is feasible, nor that it has much to contribute to the goals of a Green and Sustainable Chemistry. However, a review of the chemical literature pertaining to stability, as described above, indicates that there are a considerable number of examples that demonstrate the feasibility of the approach.

Green Chemistry, which is still in its early stages of development, is commonly defined as chemical research aiming at the optimization of chemical processes and products with respect to material consumption and generation of waste, inherent safety, toxicity, environmental degradability, and so on.

Two important approaches to chemistry are pertinent to any discussion of sustainable development.

The first is the chemical and biochemical conversion of renewable resources, which is rapidly becoming a major theme for a science which is emphasizing the need for efficient catalysts and catalytic processes to set up truly green synthetic routes starting from renewable feedstock.

The second approach is to increase energy efficiency while increasing energy production from renewable/non-fossil fuel sources.

Both approaches are key to preventing the depletion of resources, which was first identified in the report for the Club of Rome on the *Limits of Growth* in 1972 and later updated in *Beyond the Limits* in 1993.[7]

"Green" in the context of chemistry has an ethical value and describes the relationship of chemical scientists and engineers with the environment. The growing trend to develop new catalysts, environmentally benign solvents and wasteless reactions, and the use of renewable source materials is giving rise to a Green Chemistry community. Scientists have started formulating the principles which can guide choices for better solutions for further developments in chemistry and chemical technology. It is evident that these principles must include important topics such as: (1) the prevention of waste production in the various steps of processes and process control; (2) the use of safer compounds or processes; (3) methods to guarantee sustainability and efficiency.

An initial and successful attempt was made by Paul Anastas and John Warner in 1998 with their book *Green Chemistry: Theory and Practice*.[4]

It is clear that the Twelve Principles of Green Chemistry presented in this book have now become an essential part of chemistry.

Integral to these principles is the reduction or elimination of wastes, and the reduction of the use or generation of hazardous substances into the design, manufacture and application of chemical products:

Prevent Waste: it is better to prevent waste than to treat or clean up waste after it has been created, and redesigning chemical transformations to minimize the generation of hazardous waste is an important first step in pollution prevention. By preventing waste generation, hazards associated with waste storage, transportation and treatment will be minimized.
Maximize Atom Economy: synthetic methods should be designed to maximize the incorporation of all materials used in the process into the final product. This Atom Economy concept, developed by Barry Trost,[8] evaluates the efficiency of a chemical transformation. Similar to a yield calculation, atom economy is a ratio of the total mass of atoms in the desired product to the total mass of atoms in the reactants. One way to minimize waste is to design chemical transformations that maximize the incorporation of all materials used in the process into the final product, resulting in few, if any, wasted atoms. Choosing transformations that incorporate most of the starting materials into the product is more efficient and minimizes waste.
Design Less Hazardous Chemical Synthesis: wherever practicable, synthetic methodologies should be designed to use and generate substances that possess little or no toxicity to human health and the environment. The goal is to use less hazardous reagents whenever possible and design processes that do not produce hazardous by-products. Often a choice of reagents exists for a particular transformation. This principle involves choosing reagents that pose the least risk and generate only benign by-products.
Design Safer Chemicals and Products: chemical products should be designed to effect their desired function while minimizing their toxicity. Toxicity and eco-toxicity are properties of the product. New products can be designed that are inherently safer, while highly effective for the target application. In academic labs, this principle should influence the design of synthetic targets and new products.
Use Safer Solvents/Reaction Conditions and Auxiliaries: the use of auxiliary substances (*e.g.*, solvents, separation agents, *etc.*) should be eliminated wherever possible and should be innocuous when used. The use of solvents leads to considerable waste. Reduction of solvent volume or complete elimination of the solvent is often possible. In cases where

solvent is needed, less hazardous replacements should be employed. Purification steps also generate large amounts of solvent and other wastes, *e.g.* chromatography supports. Purification should be avoided if possible and the use of auxiliary substances when needed should be minimized.

Increase Energy Efficiency: the energy requirements of chemical processes should be evaluated for their environmental and economic impacts and should be decreased. If possible, synthetic and purification methods should be designed for ambient temperature and pressure, so that energy costs associated with extremes of temperature and pressure are minimized.

Use Renewable Feedstock: a raw material or feedstock should be renewable rather than depleting, whenever technically and economically practicable; therefore, chemical transformations should be designed to utilize raw materials that are renewable. Examples of renewable feed-stocks include agricultural products or the wastes of other processes. Examples of depleting feed-stocks include raw materials that are mined or generated from fossil fuels (petroleum, natural gas or coal).

Avoid Chemical Derivatives: unnecessary derivatization (use of blocking groups, protection/de-protection, temporary modification of physical/chemical processes) should be minimized or avoided if possible, because such steps require additional reagents and can generate waste. Synthetic transformations that are more selective will eliminate or reduce the need for protecting groups. In addition, alternative synthetic sequences may eliminate the need to transform functional groups in the presence of other sensitive functionalities.

Use Catalysts: catalytic reagents that are as selective as possible are superior to stoichiometric reagents and can serve several functions during a transformation. Catalysts can enhance the selectivity of a reaction, reduce the temperature of a transformation, enhance the extent of conversion to products and reduce reagent-based waste (since they are not consumed during the reaction). By reducing the temperature, one can save energy and potentially avoid undesirable side reactions.

Design for Degradation: chemical products should be designed so that they break down into innocuous degradation products and do not persist in the environment after they have been used. Efforts related to this principle focus on using molecular-level design to develop products that will degrade into hazardless substances when they are released into the environment.

Analyze in Real Time to Prevent Pollution: it is always important to monitor the progress of a reaction to know when the reaction is

complete or to detect the emergence of any undesirable by-products. Whenever possible, analytical methodologies should be developed and used to allow for real-time, in-process monitoring and control, to minimize the formation of hazardous substances.
Minimize the Potential for Accidents: one way to reduce the incidence of chemical accidents is to choose reagents and solvents that minimize the potential for explosions, fires and accidental release. Risks associated with these types of accidents can sometimes be reduced by altering the form (solid, liquid or gas) or composition of the reagents.

Another important originator of Green Chemistry concepts is Roger A. Sheldon. He has defined Green Chemistry as follows: "Green Chemistry efficiently utilises (preferably renewable) raw materials, eliminates waste and avoids the use of toxic and/or hazardous reagents and solvents in the manufacture and application of chemical products".[9]
The two most important concepts are:

- the efficient utilization of raw materials and the elimination of waste,
- the health, safety and environmental aspects of chemicals and their manufacturing processes.

This compact definition emphasizes that Green Chemistry addresses both chemical products and the processes by which they are manufactured. The main thrust clearly must be directed toward product life cycle analysis and the design of greener products and processes.

Economic incentives are driving global changes and the principles of Green Chemistry have the potential to reshape our way of doing business. Increases in the cost of waste disposal and the price of petrochemicals, plus the discovery of toxic chemicals in the proximate environment, have brought Green Chemistry considerations to the forefront. The application of Green Chemistry principles throughout the life cycles of compounds is resulting in support for new synthetic initiatives and the invention of new approaches to the use of existing compounds.

It has reached the point where the industry has realized the benefits of preventing waste through careful synthetic route selection. Processing that uses fewer unit operations and is performed under more concentrated conditions can markedly reduce waste, cycle times and labour costs, as well as the impact on the environment. Unfortunately, the twelve principles do not explicitly include a number of important concepts which are highly relevant to environmental impact, such as the inherency of a product or process, the need for life cycle assessment, the possibility of heat recovery from an exothermic reaction, *etc.*

For that reason, Paul T. Anastas and Julie B. Zimmerman added an engineering perspective to the concepts of Green Chemistry, and outlined the principles for Green Engineering: "Green Engineering is the development and commercialization of industrial processes that are economically feasible and reduce the risk to human health and the environment".[10]

Inherent rather than Circumstantial: designers must strive to ensure that all materials and energy inputs and outputs are as inherently non-hazardous as possible as a first step toward a sustainable product, process, or system.
Prevention instead of Treatment: it is better to prevent waste than to treat or clean up waste after it has been produced. Regardless of its nature, the generation and handling of waste consumes time, effort and money.
Design for Separation: separation and purification operations should be designed to minimize energy consumption and materials use. Up front consideration with regard to separation and purification, avoids the need to expend materials and energy to harvest the desired output across all design scales and throughout the life cycle.
Maximize Mass, Energy, Space and Time Efficiency: products, processes and systems should be designed to maximize mass, energy, space and time efficiency. If a system is designed, used, or applied at less than maximum efficiency, resources are being wasted throughout the lifecycle.
Output-Pulled *versus* Input-Pushed: products, processes and systems should be "output-pulled" rather than "input-pushed" through the use of energy and materials. Designing according to Le Chatelier's principle minimizes the amount of resources consumed to transform inputs into desired outputs. For example, reactions can be driven by pulling out products rather than increasing inputs, such as additional starting material, heat or pressure.
Conserve Complexity: embedded entropy and complexity must be viewed as an investment when making design choices with regard to recycling, reuse or beneficial disposal. End-of-life design decisions on recycling, reuse or beneficial disposal should be based on the material and energy invested, and on the subsequent complexity across all design scales. Natural systems should also be recognized as having complexity advantages that should not be needlessly sacrificed in manufacturing transformation or processing.

Durability rather than Immortality: targeted durability, not immortality, should be a design goal. However, this strategy must be balanced with the design of products that are durable enough to withstand anticipated operating conditions for their expected lifetime, to avoid premature failure and subsequent disposal.
Meet Need, Minimize Excess: this strategy can be applied across design scales to limit the expenditure of underutilised and unnecessary materials and energy. Technologies that target the specific needs and demands of end users offer an alternative to devising solutions for extreme or unrealistic conditions.
Minimize Material Diversity: material diversity in multi-component products should be minimized to promote disassembly and value retention. Diversity becomes an issue when considering end-of-useful-life decisions, which determine the ease of disassembly for reuse and recycling.
Integrate Local Material and Energy Flows: the design of products, processes, and systems must include integration and interconnectivity with available energy and material flows. By taking advantage of existing energy and material flows, the need to generate energy and/or acquire and process raw materials is minimized. At the process scale, this strategy can be employed to use the heat generated by exothermic reactions to drive other reactions with high activation energies.
Design for Commercial "Afterlife": products, processes and systems should be designed for functionality in a commercial "afterlife". To reduce waste, components that remain functional and valuable can be recovered for reuse and/or reconfiguration. The next-generation design of products, processes or systems must consider the use of recovered components with known properties at their highest value level as functional components.
Renewable rather than Depleting: material and energy inputs should be renewable rather than depleting. Whether the material and energy inputs are renewable or depleting has a major influence on the sustainability of products, processes and systems.

Martyn Poliakoff, a great advocate of Green Chemistry,[11,12] has expressed these principles for chemistry and engineering in mnemonics which emphasize the results which would be obtained from applying them – "Improvements Productively" (see Table 1.1).

These principles illustrate the areas where the most intensive study is needed: catalysis and catalysts; prevention of chemical derivatization; replacement of solvents and auxiliary substances; use of renewable resources; and saving energy. The influence of these principles can be

Table 1.1 Mnemonics proposed by M. Poliakoff "Improvements Productively" for Green Chemistry and Green Engineering principles.[12] (Reproduced with kind permission of The Royal Society of Chemistry).

	Green Chemistry		*Green Engineering*
P	Prevent wastes	I	Inherently non-hazardous and safe
R	Renewable materials	M	Minimize material diversity
O	Omit derivatization steps	P	Prevention instead of treatment
D	Degradable chemical products	R	Renewable material and energy inputs
U	Use safe synthetic methods	O	Output-led design
C	Catalytic reagents	V	Very simple
T	Temperature, Pressure ambient	E	Efficient use of mass, energy, space and time
I	In-Process monitoring	M	Meet the need
V	Very few auxiliary substances	E	Easy to separate by design
E	E-factor, maximize feed in product	N	Networks for exchange of local mass and energy
L	Low toxicity of chemical products	T	Test the life cycle of the design
Y	Yes, it is safe	S	Sustainability throughout product life cycle

illustrated best by the number of publications referring to Green Chemistry (see Figure 1.2).

Not only are there papers in scientific journals, but numerous books have been published on Green Chemistry as well, which contain comprehensive descriptions of various aspects of Green Chemistry. Moreover, the specialized journal *Green Chemistry*, published by the Royal Society of Chemistry in the UK, has had a significant impact.

It has been said that: "Green Chemistry has gone from blackboard conjecture to a multimillion-dollar business".[13] Indeed, Green Chemistry has "arrived". Academic chemical societies are supporting research programs and academic networks. There are research centres at colleges ranging from the University of Oregon in the USA to the University of York in the UK.

There are concerns that a narrow interpretation of Green Chemistry as "the use of chemistry for pollution prevention", might lead to doing the same things better, rather than rethinking solutions altogether. For example, a less-polluting alternative to a synthetic organic pesticide could be organic farming, which may obviate the need for the pesticide completely. Today's chemistry is already able to provide many similar alternative materials and processes, and it is necessary to work actively with the public in order to introduce these practices into everyday life.

Research and investment in nanoscience and nanotechnology is providing new properties and capabilities of materials based on their size and geometry. This is different from the traditional practice of chemistry – to control material properties through composition. An

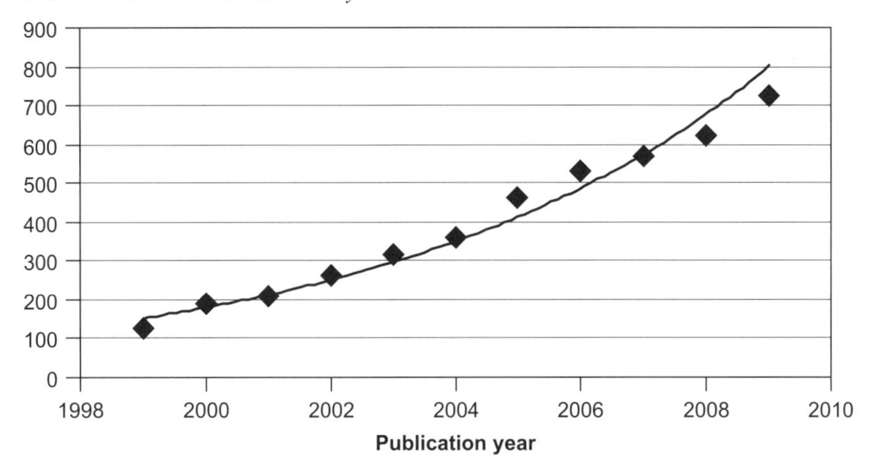

Figure 1.2 Number of publications containing the keyword "Green Chemistry" by ISI Web of Knowledge (the solid line indicates exponential growth).

increasing amount of scientific research and analysis explores the potential effects of these materials on human health and the environment. Nevertheless, there is still a large gap between studying the potential environmental and human health implications associated with new materials or technologies, and beginning to design materials and technologies in such a way that potentially harmful effects are inherently reduced or eliminated.

Although the new materials field is developing rapidly, consideration of environmental and human health impacts at the design phase is essential to its success and long-term acceptance. Many substances currently used to make nanomaterials are questionable and their synthetic processes are often quite inefficient, leading to substantial waste generation, energy consumption and water use. Furthermore, the toxicity and environmental impacts of the new materials themselves are not well understood, and may in some cases prove to be significant. Just as the principles of Green Chemistry and Green Engineering have been successfully applied in many sectors for environmental and economic benefits, they can also be used in the design of emerging materials and technologies.

For example, the design of nanoparticles with benign toxicological effects on humans should follow approximately the same framework as that for designing safer chemicals, including the following:[14]

- modification or termination of biological pathways of action;
- alteration of reactive functional groups;

- reduction or elimination of bioavailability;
- reduction or elimination of the need for associated hazardous substances;
- design for end-of-useful-product-life (or complete biodegradation on environmental trigger).

This is true not only for nano-scale substances, but also for the design of new materials and processes through Green Chemistry and Green Engineering. The result is the development of enhanced performance without adverse consequences to humans and the biosphere, while preserving the properties minimizing or eliminating the negative implications of materials. In addition to research on physical and chemical properties, intensive studies must be started simultaneously on the influence of new materials and technologies on environmental and human health. Very limited toxicological information is available initially on these new compounds and extra resources need to be dedicated quickly to assess comprehensively their potential risks to human and ecosystem health.

REFERENCES

1. R. Carson, *Silent Spring*, Houghton Mifflin, Boston, 1962.
2. *The Stockholm Convention on Persistent Organic Pollutants*, Stockholm, 2001; http://chm.pops.int/
3. *The Montreal Protocol on Substances that Deplete the Ozone Layer as either adjusted and/or amended in London 1990, Copenhagen 1992, Vienna 1995, Montreal 1997*, Beijing 1999, UNEP, 2000; http://www.unep.org/ozone
4. P. T. Anastas and J. C. Warner, *Green Chemistry: Theory and Practice*, Oxford University Press, N. Y., 1998.
5. http://www.un.org/esa/sustdev/documents/agenda21/
6. S. W. Austin, *Environmental Decision Making and Metrics*, presented at the 12th Annual Green Chemistry and Engineering Conference, Washington DC, 2008.
7. D. H. Meadows, D. L. Meadows, J. Randers and W. W. Behrens III, *The Limits to Growth, A Report to The Club of Rome*, Universe Books, 1972.
8. B. M. Trost, *Science*, 1991, **254**, 1471–1477.
9. R. A. Sheldon, *Green Chem.*, 2007, **9**, 1273–1283.
10. P. T. Anastas and J. B. Zimmerman, *Environ. Sci. Technol.*, 2003, **37**, 94A–101A.

11. S. L. Y. Tang, R. L. Smith and M. Poliakoff, *Green Chem.*, 2005, ' 761–762.
12. S. L. Y. Tang, R. Bourne, R. L. Smith and M. Poliakoff, *Green Chem.*, 2008, **10**, 268–269.
13. *The USA Today*, November 21, 2004.
14. M. J. Eckelman, J. B. Zimmerman and P. T. Anastas, *J. Ind. Ecol.*, 2008, **12**, 3316–328.

CHAPTER 2

Concepts and Trends in Green Analytical Chemistry

It is impossible to draw a definitive line between analytical chemistry and other branches of chemistry. Analytical chemistry is greatly influenced by developments in other areas of chemistry, as well as in physics and mathematics. However, the relationship between Green Chemistry and analytical chemistry is especially close, because analytical chemistry provides the means whereby Green Chemistry can be evaluated and justified, as well as providing an efficient tool for determining the "greenness" of a particular chemical product or technology.

On the other hand, analytical methods require solvents, reagents and energy, and they generate waste. The previously-mentioned principles of Green Chemistry also apply to analytical chemistry, the most important of which are:

- prevention of waste generation;
- safer solvents and auxiliaries;
- design for energy efficiency;
- safer chemistry to minimize the potential for chemical accidents.

In addition, there are principles of Green Engineering related to analytical equipment and processes:

- ensure all inputs and outputs are inherently non-hazardous;
- maximize mass-, energy-, space- and time-efficiency;
- limit underutilized and unnecessary materials and energy;
- minimize diversity of materials;
- design for commercial "afterlife".

Green Analytical Chemistry
By Mihkel Koel and Mihkel Kaljurand

Published by the Royal Society of Chemistry, www.rsc.org

Thus, the principles of Green Chemistry can be applied to analytical chemistry as well as to other areas of chemistry and chemical technology.

The development of instrumental methods to replace "wet chemistry" in sample preparation and treatment is a general trend in analytical chemistry. The main result is increased reliability of the analysis, greater precision and time saving, which positively correlates with a substantial reduction in waste. In most cases, the result of instrumental analytical methods is a decrease in the volume of sample needed for analysis. Efforts to integrate microfluidics and processing on a microscale can substantially decrease the sample quantity and accompanying generation of waste.

Instrumentalisation is related to the development of direct techniques of analysis (*e.g.* various laser-spectroscopic methods) or solventless analytical processes. However, in most cases the samples under study are very complicated mixtures with interfering matrices, which require treatment of the sample with solvents before analysis and precludes the use of wasteless methods. In the search for alternative solvents the main objective is not just replacement, but the acquisition of additional advantages from different properties of the solvents to improve the selectivity, sensitivity and reliability of the analysis, as well as to reduce analysis time.

In general, the development of instrumental methods leads to an efficient use of energy, especially when the method is highly automated and uses a minimal amount of sample. The hyphenation of several methods for sample treatment and separation of components or the integration of separation with complex methods of detection, enables an efficient use of energy.

Additional energy saving is possible when microwave treatment or even microwave heating is incorporated into the process. Ultrasonic irradiation may also have a strong impact on saving time and energy in several sample treatment procedures. The development of photochemical methods in analytical chemistry is also widely supported.

Most of the procedures described also result in safer chemistry. However, in many cases of sample preparation and treatment, different chemical methods for the derivatization and chemical modification of samples are still used. The development of new methods aims to identify less toxic compounds and processes with reduced waste generation. These are common trends in analytical chemistry, and it is very often the case that "Green Chemistry" is not even cited in the related literature.[1] Most of the challenges presented in Table 2.1 are related to Green Chemistry.

Table 2.1 Fields of cutting edge research in analytical chemistry.[1]

Microsystems	Miniaturization: "lab-on-a-chip" and total analysis systems; Microfluidics, "chip-to-world" interfacing; Automated sampling strategies; Remote (wireless) control and data acquisition.
New materials	Composites, alloys, biomaterials; Polymers, molecularly imprinted materials; Nanostructured materials.
Information processing	Multiple, localized simultaneous measurements to yield: – a spectrum of information – information on different species/processes – analysis in the frequency domain; Chemometrics and multivariate analysis techniques.
Sensors	Wide selection of sensing principles; Highly selective, sensitive sensors; biosensors; Stability, no need for calibration; Real time: "instantaneous" response, no kinetic effects.

It can be stated that the goal of Green Analytical Chemistry is to employ analytical procedures that generate less hazardous waste, and that are safer to use and more benign to the environment. This goal can be achieved by developing new analytical methodologies, or by at least modifying traditional methods to incorporate procedures that use less hazardous chemicals if safer chemical substitutions have not yet been discovered.

The path to greening analytical methodologies includes incremental improvements in established methods, as well as completely rethinking analytical approaches. Strategies include changing or modifying reagents and solvents; reducing the use of chemicals through automation and advanced flow techniques; miniaturization; and even eliminating sampling for laboratory testing by measuring analytes *in situ* in the field. The ideal green analysis would take place *in situ*, without sampling or adding reagents, and would respect all of the Green Chemistry principles related to analytical methods. It must be emphasised, however, that the main analytical parameters, such as selectivity, sensitivity, reliability and analysis time, must satisfy all requirements at an acceptable price.

2.1 THE ROLE OF ANALYTICAL CHEMISTRY

The main role of analytical chemistry is to serve as a controlling tool or set of tools. This role supports the development of chemical engineering and technology for the production of chemicals. As a tool, analytical procedures provide information about chemical substances, their occurrence in organisms and the environment. The analysis of the behaviour of chemical products and their interactions with the environment could not be possible without the chemical analysis of components and degradation products.

This leads to the second important role of analytical chemistry, in providing a source of information about the environment – hydrosphere, atmosphere or land (see Figure 2.1). Analytical chemistry is used to assess the quality of our food, drinking water, air, soil and oceans.

The third important role of analytical chemistry is related to the production of materials and goods, in the control of the process and quality of the production. Analytical chemistry is an informational tool for balancing economic and environmental pressures. It provides information about the sustainability of our world under those pressures.

The development of standards and specifications used in industry and agriculture is strongly based on the evidence obtained through analytical chemistry, and monitored by chemical evaluation methods. Analytical chemistry is the only way to determine the environmental friendliness of new methods, processes and products.

In cooperation with producers and users, analytical chemistry has gained considerable influence on socio-political decisions as well. Careful and thorough chemical analysis promotes the use of new technologies and logistical solutions, not only for chemicals, but for all other products as well. Analytical chemistry provides the data necessary for making decisions about human and environmental health. The results of chemical analysis are the strongest evidence for the need to establish new laws and administrative prescriptions. The continuous development of

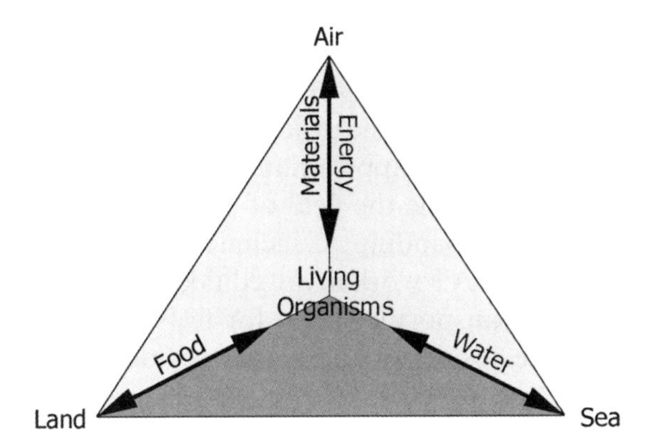

Figure 2.1 The ecosystem.

analytical methods has provided the capability to measure very small quantities of substances, which has been the basis for setting restrictions and enacting legislation, such as the regulations on the emission of volatile organic compounds (VOC): Vienna Convention (1985); Montreal Protocol (1987: ref. 2); USA Clean Air Act (1990), and the ban on persistent organic pollutants (POP): Stockholm Convention (2001).[3]

Obtaining fast, precise and accurate results will always be the primary business of an analytical chemist, and for that, reliable systems must be established. Every system of chemical analysis for monitoring or analyzing processes using an instrument or sensor can be divided into two broad categories: the pre-treatment steps (including digestion, extraction, drying and concentration) as well as the separation step, and the generation of the analytical signal category.

Analysis starts with treatment of the sample and its preparation for further separation into components. The components have to be detected in a way that allows for their individual quantification, a description of their characteristics, and the generation of potentially unique data for the identification of a substance.

Decisions have to be made at various points along the path:

- Is sample treatment or conversion needed?
- Is separation into components needed?
- Is calibration needed?

The signal obtained must be identified (correctly assigned to the analyte under study) and finally the information obtained must be interpreted objectively in order to make a decision and take corrective action if needed. The related schematic is presented in Figure 2.2.

Analytical chemistry has a special relationship with the chemical industry. Process analytical chemistry (PAC) or process analytical technology (PAT) has a history of more than 70 years; for most of that history it was used mainly for problem solving and determining the composition of the products of the process. The term "process analytical technology" (PAT) is a more appropriate term than "process analytical chemistry" (PAC) to describe the field of process analysis, as measurement technologies are expanding to include many physical characterization tools. The early PAT work involved taking samples from various process streams and transporting them for tests in a central analytical laboratory. Often the results of these analyses did not yield complete information about the process, as the sample properties were often altered by shipping and conditioning. Eventually it was realized that real-time measurements almost always result in more valuable data

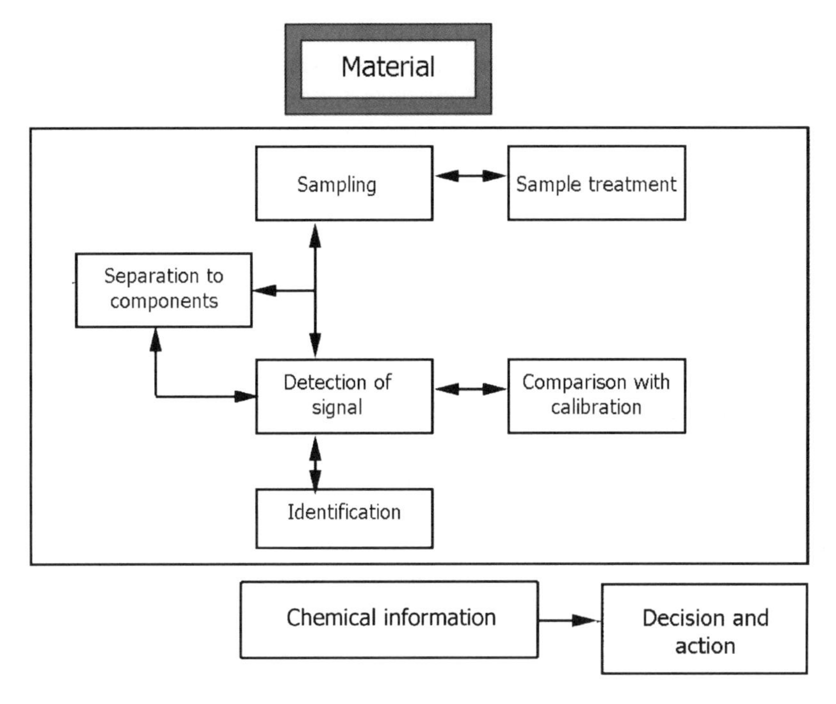

Figure 2.2 A general scheme of the analytical system.

about a process. The first real-time measurements in a production environment (other than physical measurements of temperature, pH and flow) were often made with modified laboratory instrumentation. Many industrial processes have benefited from applying these laboratory-based technologies in manufacturing. Currently PAT is mostly comprised of analytical technologies applicable to process with remote, real-time and high-throughput measurements. The technology is rapidly advancing in analytical chemistry as well, and the field is being redefined by miniaturization, sampling systems, high-throughput requirements, and increased sophistication in data modelling and processing. It is becoming common to use instrumentation that allows for a process to be studied and critical process parameters to be determined on site and when it takes place. Once these parameters are identified and monitored, process models can be developed to substantially improve process control.

Requirements for real-time, portable and highly sensitive analytical instruments capable of detecting and quantifying biological or chemical hazards are creating new challenges for technological advances in the field. Research is focused on improvements in real-time measurement for both chemical and biological phenomena. This involves an expanding range of techniques including vibrational and scattering spectroscopy, chromatography, mass spectrometry, acoustics, chemical imaging, light-induced fluorescence and light scattering. Measurement science continues to build on advances in technology found within the computing and communication industries, as well as sensor-related research conducted in university and research laboratories. These developments in the areas of miniaturization, electro-optics, new materials and information/data handling are revealing the potential for measurement advances using emerging nanomaterials, photonics, improved bioassay techniques, and the like. The field of nanoscience is already having an effect, as nanomaterials are making significant improvements in separation science and sensing, as well as providing potential building material for new sensors. The future of PAT will be influenced significantly by developments in micro-instrumentation, and versatile micro-analytical systems will provide the modularity and flexibility needed to improve laboratory and process control operations.

High-throughput experimentation and process intensification are fields where large numbers of small-volume samples are being created for product development and process optimization studies. However, advances in micro-hardware devices (microreactors, extractors, *etc.*) rarely address how these various microunit operations will be monitored and controlled. The demand is growing to provide miniaturized measurement devices, scanning sensor devices, and rapid data handling with high-throughput and intensive experimentation platforms. The objectives are changing from simply proving the feasibility of a chemical reaction to more in-depth scientific studies and industrial piloting. This is having a positive impact on PAT.

Process intensification as a way to reduce capital and operating expenditures is based on an understanding of the underlying fundamentals of the process. High-throughput experimentation is an essential step in estimating optimal parameters. The miniaturization of production unit operations is also a cost-effective way to gather data for process engineering. As a result of real-time process analysis and control, reductions in resource use and waste generation are achieved while simultaneously enhancing productivity and the quality of the product.

Most of these analytical procedures are involved in quality control systems as well. In addition to the steps already described, there must always be a definition of the problem, a method of obtaining representative samples, specification of the required level of accuracy and acceptable uncertainty (perhaps from current legislation), a method of data treatment and communication of results. The main objective is to ensure that specified levels of product quality are attained, however, the concept has definite "green" implications, since it is used simultaneously to measure impurity levels, toxicity, waste, *etc.*

Current legislation deals with both product quality and the green implications. It is clear that the analytical devices and sensors that are being developed will be applicable to more complex and potentially difficult situations, such as online or periodic monitoring of diverse industrial processes or in the field as a pollution diagnostic, which coincide with the principles of Green Chemistry. The new green challenge is to meet the informational needs of chemists, industry and society, while reducing the human and environmental impact of the analyses. The nature of the analytes, the matrices and the methods of analytical signal generation have considerable bearing on the likelihood of creating a green analytical method, and some analytical processes have a greenness advantage.

Many analytical procedures require hazardous chemicals as part of sample preservation and preparation, quality control, and equipment calibration and cleaning – effectively creating wastes in larger quantities and with greater toxicity than that of the sample being analyzed. For all of these reasons, a green approach is becoming increasingly important in analytical chemistry.

2.1.1 Field Analysis

In traditional analytical chemistry, sample analysis begins with collecting a sample from a site and transporting it to a laboratory. The sample is stored under specified conditions before it is analyzed in the laboratory. Usually, preparation of the sample, sometimes with a separation and/or a pre-concentration step, is required prior to the actual analytical measurements. Although they provide precise and accurate analytical results, laboratory-based analytical methods are time-consuming and costly. Depending on the organisation of the work in the laboratory, up to a few days may be needed before the final analytical results of the samples are available. Often, this leads to delays in decisions that are based on the analytical data. In order to prevent further problems, especially in matters involving human safety and/or product reliability, some of those decisions need to be made in real-time.

Especially important measurements are related to environmental pollutants, following their accidental release where immediate action is required. Problems relating to pollutants are many-fold depending on the way in which they are introduced into the environment, their chemical properties and their biochemical effects (*e.g.* toxicity). Common contaminants include: pesticides, hydrocarbons, oil-production chemicals, toxins, flame-retardants and other industrial or domestic waste products.

The demand for advanced analytical methods for field use in order to reduce the cost and length of time of analyses, has been the driving force behind an increasing interest in field analytical chemistry (FAC) in recent years. Wider interest and the management of natural systems have generated a requirement for data to be collected with greater frequency and spatial resolution than was deemed necessary in the past. Therefore, the analysis of a greater number of samples to better describe variability and change over time is needed, and the instruments must have the capability to gather data over longer periods of time. Moreover, one needs to keep in mind that quality assurance and quality control are as important for field measurements as for analytical measurements in the laboratory.

Unlike traditional laboratory-based methods, field measurement is accomplished on-site where the analyte/sample of interest is located and from that fact the general requirements for an ideal FAC method can be derived:

- fast instrument response times for the acquisition of necessary information on a real-time or near real-time basis;
- *in situ*/on-site analyses with communication links to a central location;
- little or no requirement for sample preparation;
- large area placement capability – portable instrumentation;
- minimum requirement for power, consumables *e.g.,* gases and solvents, clean space for handling samples;
- cost-effective analysis.

To better satisfy these demands, field studies are now increasingly focused on the development of monitoring networks based on on-site sensors and biosensors. With the use of remote control, these sensors provide different levels of information in order to monitor ecosystems both on land and underwater, and to predict the effects of increasing numbers of pollutants on these ecosystems.

Recent developments in miniaturizing electronics and wireless-communication technology have led to the emergence of environmental sensor networks, which will greatly enhance monitoring of the natural environment and, in some cases, facilitate new measurement techniques or the previously impossible deployment of sensors. Especially promising are biosensors which offer prospects for combining the recognition of biological events with electronic signal transduction, and for designing a new generation of bio-electronic devices that perform novel functions. Some examples of the successful development of field instruments are provided in Chapter 4.4.1.

The general acceptance of these techniques in the field will depend on solutions to the problems of inherent high cost, complexity, size, weight, and power requirements. However, the implementation of environmental monitoring programs still has two parts:

- screening methods based on high-throughput analysis capable of continuous on-line, on-site, low-cost monitoring;
- confirmatory laboratory analysis of positive samples using validation techniques.

These are necessary to guarantee high quality and trustworthy results.

Analytical measurement in the field of chemical warfare is another area where field analysis is important, or even obligatory. Biological and chemical warfare agents present a broad spectrum of threats. Traditional chemical agents – nerve, vesicant and blood agents – have acute toxicities in the range of 10^{-3} g person^{-1} and are relatively easy to detect. The chemical structure of the newer chemical agents (toxic industrial chemicals and aerosols) and bioregulators (neuropeptides and psychoactive compounds) is more complex, requiring more sophisticated analytical methods for identification and detection. The most difficult chemical agents to detect are the cytotoxins and neurotoxins with chronic toxicities as low as 10^{-10} g person^{-1}. However, the primary reason that analyses are not routinely performed on-site is the lack of approved equipment that can be transported conveniently to the test site. To date, there are only a few instruments approved for the field verification of chemical warfare agents, the most familiar of which is a gas chromatograph coupled with a mass spectrometer (GC-MS). While GC-MS is one of the most powerful analytical instruments available, it is also one of the most complex and difficult to maintain and operate on a routine basis in the field. Another instrument on this list is the ion mobility spectrometer (IMS), which is used as a routine tool for field detection of explosives and chemical weapons.

The following approaches may be proposed where an increased sophistication of field analysis is required:

- One-dimensional sensors or selective detectors which rely on the selective detection of the analyte. This is the simplest approach for chemical detection of a specific compound in the presence of other background compounds and interferences;

- Two-dimensional sensors, in which a separation step is coupled with selective detection to overcome the interference caused by the complex nature of the agents and their matrices, and to provide a more specific response and a broader range of application;
- Three-dimensional instruments, in which a two-dimensional separation step is coupled with selective detection to improve the analysis of unknown analytes, where absolute identification is recommended. Because of the complexity of these instruments, they are difficult to operate in the field without the support of a full mobile laboratory. These kinds of instruments are often referred to as hyphenated instrumentation, which often can provide structural information as well.[4] However, they should not be regarded as a panacea for every analytical problem, whether in the laboratory or in the field;
- More complicated four-dimensional instruments, in which a three-dimensional separation step is coupled with selective detection. In this case, hyphenation is developed to an even higher level.

Four-dimensional instruments result when separation techniques such as chromatographic separations that occur in the order of minutes, ion mobility separations (IMS) that occur in the order of milliseconds, and mass separations that occur in the order of microseconds, are coupled in series. Moreover, with some mass spectrometers it is possible to obtain MS-MS spectra or even MS^n spectra. Thus, it might be possible to have more than four-dimensional sensors, such as GC-IMS-MS-MS-MS. Increasing the dimensions of the space in which a compound is identified decreases the possibility of false responses, although it correspondingly increases the complexity of the instrument. These kinds of hyphenated instruments can hardly be considered viable in the field, as at the present time they are too fragile even for mobile laboratories.

Other examples of complex systems are related to the measurement and monitoring of oceans and seas that possess different types of ecosystems and a variety of interactions between them. The ability to distinguish between natural variability and anthropogenic change is an important issue, and requires exact observations to be made over long periods of time, with high temporal and spatial resolution. In addition, a range of chemical and physical measurements must be made simultaneously. The requirements in this area are even more complex than measurements on land because of the very different physical environment. Depending on the information required, the measurements can vary from fish stock surveys to single chemical analyses; from processes such as denitrification to the taxonomy of toxic algae or the detection of pathogens. Pollutants can enter the marine system from known or unknown point sources, as well as through diffuse inputs, for example, *via* the atmosphere. Their concentrations have to be studied to identify sources, environmental pathways, persistence or uptake into biota, and degradation or biotransformation. Monitoring of "hot spots" by remote sensing to predict damage to or death of a species in the sea, as well as eutrophication over the longer term, requires technology capable of functioning in field conditions.

In these studies, the target analytes vary as do the matrices in which the analysis is carried out. In the simplest case it is seawater, but other matrices, such as sediments or biota, are also frequently analyzed; the same variability occurs in measurements on land. Both the measurement strategy and the instrumentation must be carefully chosen to meet the requirements of the data being sought. In the case of seas and oceans, the contaminant concentrations in watercourses are dynamic, changing both as a result of varying inputs and changes in water flow (affected by rainfall). Monthly sampling and analysis makes it is extremely unlikely that the maximum concentration for that period will be detected and the calculated average is based on a limited number of observations.

This highlights the importance of an early warning system that could be part of a system to track trends or for precise concentration quantification to ensure compliance, for example, with discharge licence conditions or environmental standards. This is a type of process analytical technology, but the processes in this case are environmental.

Clearly, the requirements placed on field instruments and sensors are demanding because they have to withstand harsh environmental conditions and must operate for extended periods of time, often without maintenance.

The requirements for data collection instrumentation at remote locations are as follows:

- robustness (mechanical and operational);
- self-calibration;
- sensitivity, with a broad detection range;
- reliable and flexible data logging regimes;
- easy interface with electronic logging and telemetry;
- resistance to varying environmental conditions;
- low power consumption;
- long service intervals;
- user friendliness (preferably "fool-proof");
- low cost.

As pointed out many times previously, the data quality assurance for instrumentation deployed remotely and for long periods of time is a major issue and must be addressed.

Molecular and biological events can now be monitored and translated into digital signals by means of sensors and this creates new possibilities for analytical devices.

The above-mentioned downscaling and integration of chemical assays to provide analytical microsystems can be considered a positive step towards the development of Green Analytical Chemistry screening tools, and holds considerable promise for faster and simpler on-site monitoring of priority pollutants. For every type of on-site environmental or industrial application, reduced dimensions and a high degree of automation, minimal solvent/reagent consumption and waste production, are very attractive in terms of efficiency and speed. The amount of waste generated by microsystems is reduced by *ca.* 4 to 5 orders of magnitude, in comparison, for example, to conventional liquid chromatographic assays (*e.g.*, 10 μL *vs.* 1 L per daily use). Such a significant improvement in the rate of waste generation and material consumption has enormous implications for Green Chemistry (see Chapter 5.2.1.1).

Microfluidic analytical devices are more functionalised microsystems that integrate multiple sample-handling processes with the actual measurement step on a microchip platform. For obvious reasons, such devices are referred to as "labs-on-a-chip". Complete assays involving sample pre-treatment (*e.g.*, pre-concentration/extraction), chemical/biochemical derivatization reactions, electrophoretic separations and detection, can be performed on single microchip platforms. Such analytical microsystems rely on electrokinetic fluid "pumping" and obviate the need for pumps or valves. Highly effective separations combined with short assay times have been achieved by combining long separation channels and high electric fields. The enhanced performance and advantages of microfluidic devices relative to conventional (bench top) analytical systems have been demonstrated both in theory and through experiment.[5]

The high degree of integration offered by "lab-on-a-chip" devices implies that the principles of Green Chemistry can be applied to all steps in the analytical process. The interface of microfluidic devices with the external environment is problematic, but there are already developments allowing the continuous introduction of actual samples into micrometer channels which would make "lab-on-a-cable" devices compatible with real life monitoring applications.[6]

Underwater monitoring and analysis is technically more complicated. For this purpose, the "lab-on-a-cable" has been proposed.[7] As opposed to *in situ* sensors (that lack the sample preparation steps essential for optimal analytical performance), the "lab-on-a-cable" incorporates several functions into a single sealed, submersible package. This first generation of submersible microanalyzers integrates *in situ* microdialysis sampling with reservoirs for reagent, waste and calibration/standard solutions, along with a micropump and fluidic network on a cable platform. The sample and reagent are thus brought together and allowed to react in a reproducible manner. The internal buffer solution ensures independence of sample conditions such as pH or ionic strength. The on-cable measurement of metal-complexation or enzyme-inhibition reactions has thus been achieved. Other groups of researchers have developed submersible reagent-based flow spectroscopic analyzers for long-term oceanographic monitoring or process control.[8] These examples illustrate how miniaturization results in economy and efficiency, which is consistent with Green Analytical Chemistry. However, the cost of reliable autonomous sensing is still far too high for widespread deployment.

Ongoing efforts toward commercialisation, coupled with regulatory compliance, as well as acceptance by plant and environmental engineers, should lead to the translation of this research activity into large-scale environmental and industrial applications.

Data from analyzers (sensors or "labs-on-a-chip") must be collected and processed. Improvements in wireless communication should, however, provide the advancements necessary. A wireless-sensor network is comprised of spatially-distributed autonomous devices, using sensors or biosensors linked to monitor physical or chemical conditions at different locations. This technique could, in the near future, be one of the main advances in continuous environmental monitoring, especially for environments which are difficult to monitor.

Even in the early days of the Internet it was said that the future world would operate on the basis of complex interlocking control loops, ranging from localized sensor-actuator systems to platforms that aggregate information from multiple heterogeneous sources. In recent years, there have been significant advances in hardware and software for building wireless-sensor networks. However, to ensure effective data gathering by sensor networks monitoring remote external environments, some problems must be solved:

- the ability of the network to react rapidly and accurately to its environment, and to provide relevant data to users;
- the ability of network nodes to function properly in harsh and variable environments;
- the lifetime of the network must be sufficiently long to deliver data from the nodes on the events under study.

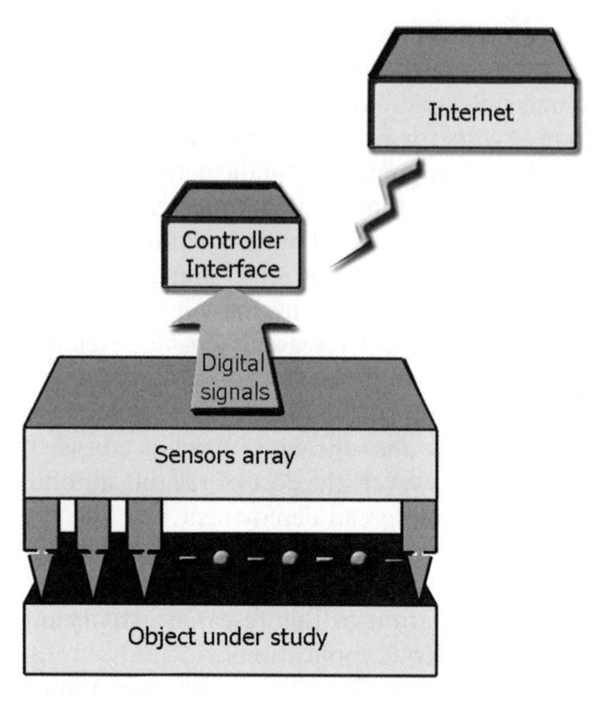

Figure 2.3 Wireless sensor networks.

A general scheme of a sensor network is presented in Figure 2.3. The first step in the development of wireless-sensor networks is creating autonomous blocks of sensing elements (nodes) to form an array of sensors. A node is the smallest component of a sensor network that has integrated sensing and communication capabilities: a basic capability to form digital signals, some data-storage capacity and a microcontroller that performs basic processing operations for the sensor. Usually several sensor nodes are integrated for different parameters, forming an array of sensors equipped with a controller board that enables interfacing with other sensors (provided that the signal is transmitted in the appropriate form for the controller). It also includes a power supply, usually provided by an on-board battery. The next step is providing a controller which collects and organises the necessary data from nodes, and relays it to a central station through wireless communication. Typically this communication is two-way and instructions can be sent to the sensor array from a central station.

The concept of analytical monitoring in the field is important and a great deal of work has been done in laboratories with very promising results. However, the transition from laboratory to wireless-sensor networks to their application in the real world is not totally straightforward, and there are several barriers involved:

- the lack of stability, availability and the high price of sensing elements (especially biological ones);
- the slow progress from proof-of-concept to application to actual samples and environmental scenarios (while meeting the outlined instrumentation requirements);
- the slow acceptance of new sensor-based methods to gain accreditation and status as a recommended method of analysis in a regulatory context;
- a global market which is still emerging and highly specialised, and in which manufacturers and distributors have little interest in getting involved.

The motivation to overcome these problems and to find solutions is strong, and Green Chemistry lends confidence to positive developments towards widespread and industrial applications.

2.1.2 Screening

The previous chapter discussed field analysis in which various considerations were emphasized: the observation of objects or the monitoring of a situation over time and *in situ*.

This chapter focuses on chemical screening, which is by definition the examination of a large number of objects, such as people, in order to identify those with a particular problem or feature, in this case, features related to chemistry.

The simplest example is at an airport, where many bags are examined to detect those which may contain weapons or explosives. People are also screened by passing through a metal detector. From this example, some requirements of the screening process and its related instrumentation become clear:

- the sufficient sensitivity and selectivity of analysis;
- a non-detrimental effect on the object examined;
- a capacity to provide many analyses of the same level of quality over a long period of time.

Screening can be conducted under stationary conditions, where samples are transported to a laboratory and measurements are performed on stationary instruments, or where the instrumentation is mobile/portable and measurements are performed on-site. Field screening tests can be qualitative, *e.g.* chemical spot tests, semi-quantitative or even quantitative, *e.g.* portable X-ray fluorescence or portable anodic stripping voltammetric measurements. A qualitative test consists of a screening analysis which returns either a positive result (indicating the presence of the analyte of interest) or a negative response (indicating the absence of the analyte). A quantitative method, when used for screening analysis, can be treated in a qualitative manner by converting quantitatively measured results to positives or negatives by comparison with a pre-specified threshold value. A semi-quantitative test consists of a screening analysis, wherein a value is recorded that is an estimate of the concentration of analyte present in the test sample, but the confidence interval for the measured value is greater than it is in a corresponding quantitative analysis. For on-site analysis, accurate quantitative tests for field measurements may not be available, depending on the analyte(s) or specific field situation.

On-site screening tests are often used for applications in the environmental and industrial hygiene fields. Screening measurements are performed in the field in order to estimate human exposure to toxic chemicals, or to estimate the content(s) of toxic chemicals in materials or on contaminated surfaces, so that elevated exposure to toxic substances can be avoided. The principal aim in conducting a screening analysis may not be to quantify the level of a particular toxic chemical, but rather to determine if the chemical in question is present above or below a regulatory or recommended standard value or action level. Apart from standards and action levels established by governmental agencies, many other agencies have developed regulatory standards which may be more stringent than those set by national and international standard-setting organisations.

Thus, as an alternative to more thorough testing methods, screening tests which are based on qualitative or semi-quantitative methods are often used for making immediate decisions in the field, *e.g.* for compliance or risk assessment. To ensure the quality of these screening tests and the decisions that are made based on their results, screening methods need to be evaluated with sufficient data and should meet basic performance criteria prior to their being used for decision-making purposes. Although quantitative, semi-quantitative and qualitative methods demonstrate different characteristics, it is desirable that the performance criteria for all three categories of method be consistent.

A minimum requirement of a test method is related to rates of false positive and false negative results. A test result is a false negative if a negative result is obtained, but the true value is above the standard value or action level.

Conversely, a result is a false positive if a positive result is obtained when the true value is below the threshold. Obviously, the objective is to minimise the rates of both false positives and false negatives as much as possible. An ideal qualitative test method would have a zero false negative and a zero false positive rate. In other words, the ideal method would always yield a positive result when the true concentration is above the standard, and a negative result when the true concentration is below the threshold value. However, such performance is clearly unrealistic. A method can conceivably have either a zero false negative rate or a zero false positive rate, but it is practically impossible to obtain both, since the positive or negative response rate cannot, in practice, be immediately changed from 0 to 100% at a single (the standard, or threshold) concentration level. Thus, an uncertainty region around the standard value should be allowed to accommodate the change in response rate according to the change in analyte concentration, which is not instantaneous. Also, a qualitative method usually demonstrates a small, but non-zero, false positive and/or false negative response rate. Hence a realistic requirement for a qualitative screening method should allow for limited false negative and false positive response rates, along with a specified uncertainty region around the standard value.

Frequently, methods that yield a numerical value rather than a yes/no (positive/negative) response are employed for screening analysis in field measurement. When a semi-quantitative or quantitative method is used in screening tests, a threshold value may be established to convert each measurement to a qualitative outcome. The performance of a measurement method that returns a numerical value is characterised by its bias and precision, or its measurement distribution. If measurements are normally distributed, the distribution of results at each concentration level is determined by its mean and standard deviation. As a qualitative test procedure, the performance function depends on the threshold value and its distribution of numerical measurements (which would be converted to positive or negative responses). Such quantitative methods for screening tests can have low precision or high bias, resulting in poorer overall accuracy than definitive quantitative methods used for fixed-site laboratory analysis. Of course, quantitative methods that meet more stringent performance requirements for accuracy may also be used for screening purposes, but whether the measured analyte concentration is above or below the standard or action level of interest must be decided beforehand for methods that are used for making decisions.

A statistical treatment is required that allows the examination of the performance criteria and the performance characteristics of field-screening test methods.[9] In order to unify the performance criteria for different types of method, the same performance function used for quantitative methods is used to characterise both qualitative and semi-quantitative methods. False negative rates, false positive rates, sensitivity and specificity are key characteristics of screening methods that can be determined from pertinent performance curves. The performance characteristics of each method are related to the uncertainty region that is associated with that method, and the applicable uncertainty regions can be gleaned from performance curves. In certain conditions, multiple test results can be used to improve decisions based on those results.

Advanced medical care systems in developed countries have problems with regard to the pollution of the evnironment with drug residues and their related degradation products. Nowadays, it is a well-established fact that pharmaceutical drugs used during medical treatment may be partially excreted in unmetabolized form, enter municipal sewage systems and even survive the passage through sewage treatment plants. Therefore, sewage treatment plant effluents are a major source of the introduction of pharmaceuticals into the aquatic environment.

Another, even bigger, source of pharmaceuticals in the environment is agriculture. In modern agricultural practice, veterinary drugs are being used on a large scale, administered for treating infection or to prevent infection.[10] In order to prevent the outbreak of diseases and also in cases of disease, veterinary drugs are in some instances administered as feed additives or *via* drinking water. Furthermore, pharmaceuticals employed in veterinary medicine may be introduced into the soil (and eventually into the water) *via* manure, or may find a direct way into the aquatic system when used in fish farms. In this way, waste water may reveal the use and misuse of pharmaceuticals in our society.

Residues of pharmaceuticals have most likely been present in our environment for quite a long time however, it is only recently that progress in instrumental analytical chemistry has allowed the reliable measurement of low ng L^{-1} concentrations of pharmaceuticals in various types of water sample. The limits of detection of analytical methods may decrease even further over the next few years.

It is evident that every synthetic chemical used in daily life can show up in measurable concentrations in the environment and this is true for all kinds of drugs. Unfortunately, the consequences for the ecosystem of the ubiquitous presence of low concentrations of pharmaceuticals are still not fully known. It is quite clear that environmental risk assessments must be based on reliable data about the actual concentrations of pharmaceuticals in aquatic systems. Cooperation between analytical chemists and toxicologists is needed in order to answer the question whether low concentrations of xenobiotics have an impact on living organisms.[11]

Residues of pharmaceuticals in aquatic systems are not yet included in regular monitoring programs. The identification of a limited set of pharmaceuticals that are representative of toxic effects may be advantageous (but a final selection of such a set has not yet been made).

The danger is recognised and there are steps being taken toward legislation for monitoring drug residues (including antimicrobials), and studies are underway to estimate the maximum concentration of residue resulting from the use of a medicinal product (expressed in $mg\,kg^{-1}$ or $\mu g\,kg^{-1}$ on a fresh weight basis) that may be legally permitted or recognized as acceptable in or on food.

The high costs of instrumental analysis may be prohibitive to more extended studies. The screening of these residues is not an easy task, because the analytical methods must be capable of achieving extremely low limits of detection (LOD). It is also a complex task because while low LOD values can be achieved, they vary widely depending on the analyte being determined, the sample preparation, the technique used and the sample matrix, which is especially important for food samples. Another issue is that it can be difficult to reach the required sensitivity levels for all the analytes with one pre-treatment method. There are many different compounds and matrices that legislation requires to be monitored. Despite the activity in this area of research, there are still many gaps for certain matrices and species that residue laboratories are required to monitor in their national residue plans. For this reason, multi-residue "catch-all" methods, or even combinations of methods for different drug residues using definitive techniques such as LC-MS, are highly appealing in terms of fulfilling the legislated requirements, as well as for their high throughput and sensitivity. However, they are expensive, require solvents and chemicals for sample pre-treatment, and qualified specialists to operate the laboratories.

Time and cost issues are more important for screening methods than obtaining a complete analysis, therefore, a compromise must be reached, for example, on the removal of matrix interferences. A simple extraction system might be more suitable than a more complex one with higher recoveries, or a set of selective detectors might be used instead of a universal one. Seeking economic alternatives would be beneficial for monitoring programs, and effective solutions can make the monitoring process more environmentally benign. A prospective means of green screening for developing countries – paper-based microfluidic sensors – is described in Chapter 5.2.3.2.

In general, flow-based automatic methods offer several features such as simplicity, versatility, low cost and high-sample throughput, that are advantageous for rapid and reliable determination of different parameters in both pure compounds and complex matrices. In addition, reaction/determination takes place in a contained environment, minimizing operator exposure to the organic solvents used. Furthermore, the reaction conditions (time, mixing and pH) are strictly controlled, improving the repeatability of results. One of the main advantages associated with flow-based methods for parameter assessment is the significant decrease in analysis time compared to a batch procedure, as reaction completeness is not expected. However, when results comparable to those obtained from absolute methods are sought, the direct translation of batch methods to flow systems is not feasible, especially for samples containing slow-reacting analytes. Computer-controlled flow techniques offer the flexibility of flow management and allow the implementation of two different assays in the same manifold, which can be performed sequentially. In this way, the sample is processed by both methods at the same time, avoiding discrepancies that may arise due to modification of the sample over time, and providing a more reliable comparison of methods.

Flow-based systems are easily applied to food samples such as wine, beer, coffee, fruit juices, tea, olive oil, honey and plant extracts. The majority of these samples are liquid and can be directly introduced into the systems. When necessary, in-line dilution can be performed automatically by means of a dilution coil or a mixing chamber.

Immunoassays have attractive features for organic trace analysis due to the fact that they require little sample pre-treatment, exhibit high sensitivity, and are inexpensive compared to the instrumental analysis described previously. A considerable number of immunoassays have been developed and used for residue analysis of pesticides in water samples, but immunoassays for pharmaceuticals in the aquatic environment are quite rare. Although test kits for pharmaceuticals are commercially available, these kits are, in most cases, optimised for samples like blood, urine, or food; the applicability of these kits to environmental samples has not been investigated in the majority of cases. Unfortunately, immunoassays are not well suited to the simultaneous determination of analytes of differing chemical structures.

With fluorescence- and chemiluminescence-based methods, detection limits comparable to LC-MS measurements can be reached even without sample pre-treatment (except filtration). There are fully automated immunosensor systems based on total internal reflection fluorescence.[12] This system can be used for monitoring pharmaceuticals, hormones, endocrine-disrupting chemicals and pesticides in water. It consists of a laser diode, the light of which is directed onto the bevelled edge of a bulk optical glass slide. The laser beam is guided by total internal reflection along the glass, which is coated on one side with the analyte molecules. This side is part of a flow cell. Samples are mixed with an appropriate antibody that is labelled with a fluorescent tag and injected into the system. Due to the competition between the analyte molecules in the sample and those immobilised on the glass surface, a sample-dependent amount of antibody gets bound to the glass surface and is excited by the guided laser beam. The emitted light is collected by optical fibres and detected by photodiodes. The range of applications depends on the availability of appropriate antibodies. Detection limits of approximately $1\,ng\,L^{-1}$ are possible with this device.

Instead of measuring a set of selected pollutants, an alternative approach would be to measure the biological effects of substances in water samples directly. Biosensors, based on natural receptors, bacteria or cells, may be appropriate for such purposes and indeed have been developed in recent years.

Rapid progress in the development of analytical procedures for the residue analysis of pharmaceutical drugs has been facilitated by the existence of considerable expertise in the area of pesticide residue analysis. Strategies successfully used in the routine analysis of trace pesticides have been directly applied to the analysis of pharmaceutical residues.

Analytical methods for the analysis of polychlorinated biphenyls (PCBs) and organochlorine pesticides (OCPs) are widely available, and result from the vast amount of work on the development of environmental analytical methods and research that has taken place on persistent organic pollutants (POPs) over the past 30–40 years.[13]

Biological samples (fish, aquatic and terrestrial mammals, and birds) as well as soils and sediments, are important objects of study for the analysis of PCB/OCPs. These matrices and areas of the environment have higher concentrations of PCBs and most OCPs than water or air, making them more suitable for routine monitoring and more relevant in the context of human and wildlife exposure. Milk and blood are important matrices for monitoring POPs in humans, as are live specimens of marine mammals and birds.

The collection, preparation and storage of samples, as well as specific quality control and reporting criteria are critical to the successful application of new approaches for extraction, isolation, identification and quantification of individual congeners/isomers of PCBs and OCPs. However, the establishment of an analytical laboratory, and the application of this methodology to currently acceptable international standards is a relatively expensive undertaking. Furthermore, the current trend to use isotope-labelled analytical standards and high-resolution mass spectrometry for routine POP analysis is particularly expensive. With the signing of the Stockholm Convention on POPs and the development of global monitoring programs, there is an increased need for laboratories in developing countries to detect PCBs and OCPs, but the high costs of contemporary laboratories limit the participation of scientists, especially from developing countries. Thus, a major priority is the need for low-cost methods that can be easily implemented in developing countries.

Access to modern capillary gas chromatography (GC) equipment, with either electron capture (ECD) or low-resolution mass spectrometry (MS) detection to separate and quantify OCP/PCBs, is essential and mandatory in order to make regional and international comparisons.

Existing analytical methods for PCB/OCPs can detect over 100 individual components at low ($ng\,g^{-1}$) concentrations in many environmental media using high-resolution capillary GC-ECD. However, the number of certified values for OCP/PCB congeners in certified reference materials is more limited (approximately 23 PCB congeners and 15 OCPs in NIST 1588a cod liver). At a minimum, the OCPs/PCBs for which there are certified values in readily available reference materials (CRM) should be specified (approximately 38). With this number of analytes, the information would be useful for regulatory action, as well as for source identification using multivariate analysis or other "fingerprinting" methods.

Inter-laboratory comparisons of POP analysis over the past ten years have shown that the availability of accurate analytical standards is a fundamental requirement of an analytical program designed to quantify trace organic contaminants such as POPs. The detection of PCBs/OCPs requires the analysis of blank samples, because of the ubiquitous nature of these contaminants. If the results for blank samples are significant (for example, averaging more than 10% of the average level of total PCBs) then a blank correction should be carried out.

Quality assurance (QA) programs are critically important for verifying the performance of analytical methods for POPs within a laboratory and between laboratories. QA requirements for PCB/OCP analysis are well known and include the use of certified reference materials, field and laboratory blanks, quality control charts to monitor long-term laboratory performance, participation in inter-laboratory studies, and guidelines for sampling and analysis.

Detection limits depend not only on the analytical method used, but also on the sample size and QA considerations, *e.g.* on information available from blank or control samples and recovery studies. A comparison of the detection limits for POPs achieved using a wide variety of instrumentation, with action limits for POPs in food and tissue residues which correspond to guidelines, suggests that the current GC-ECD and GC-MS methodologies for PCBs/OCPs not only meet, but can exceed these limits (in some cases by orders of magnitude). However, screening of samples, especially in areas where there is a known usage of OCPs or PCBs, could also be accomplished with bioanalytical methods, such as specific, commercially available, enzyme-linked immunosorbent assays (ELISA).

In general, ELISAs are very useful tools for the rapid assessment of PCB/OCP contamination, especially in areas of former heavy use. They are particularly well suited to laboratories in developing countries that may have access to spectrophotometric equipment, but not to GC instrumentation.

New analytical techniques, such as two-dimensional GC (2D-GC) and "fast GC" using GC-ECD, may be well suited for wider use in routine PCB/OCP analysis in the near future given their relatively low cost and ability to provide high-resolution separations of PCBs/OCPs. Procedures with low environmental impact, (solid-phase microextration, micro-LC, *etc.*) are increasingly being used and may be particularly appropriate for developing countries. This type of system is presented in Figure 2.4, where SPE is coupled to a GC-MS system using several valves, and the GC is equipped with additional columns for solvent removal.

One very important area of screening is that of human bodily fluids by non-invasive methods for detecting biomarkers. It has been found that putative biological marker(s) correlate with disease states and, as a consequence, clinical researchers are currently investigating techniques for the accurate and reproducible measurement of such biomarkers. In recent years, an accumulation of evidence has indicated that urine is the ideal biological fluid amongst those available (*e.g.* blood, urine, saliva or cerebrospinal fluid) for routine disease screening, particularly where large numbers of people are concerned. In fact, urine is extremely easy to collect, can be transported in an intact state simply by using preservatives and freezing, and, in many cases, it can be used directly without any pre-treatment.

In addition, the screening of urine can yield reliable information on a large series of specific biochemical constituents, including inorganic ions, organic acids, purines, pyrimidines, amino acids, peptides and

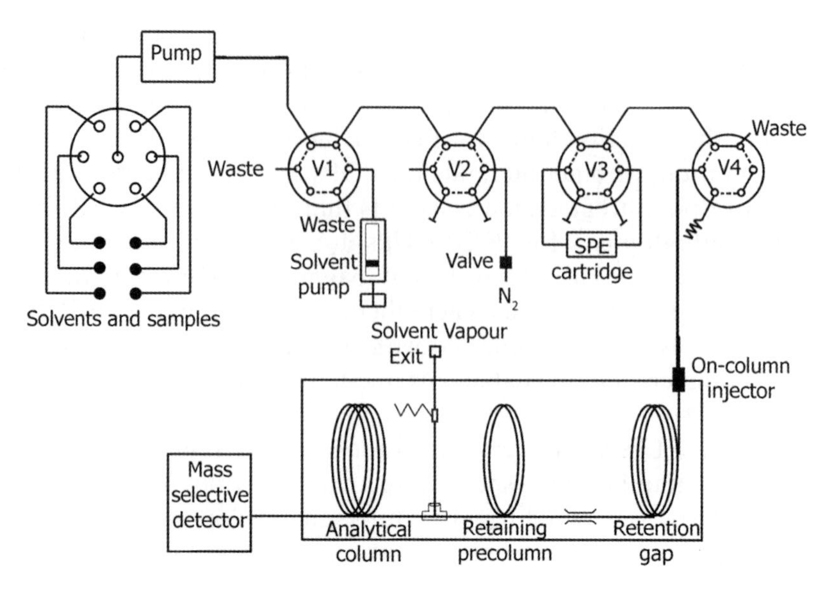

Figure 2.4 Scheme of an on-line SPE–GC–MS system.[47] (Reproduced with kind permission of Elsevier).

proteins, that are expected to change in response to a particular disease state. Moreover, the fact that some analytes become significantly more concentrated in urine than in other fluids, making measurement easier, supports the firm belief that this is an important matrix for performing multi-component profiling analyses. Urinary compounds may either be individual markers of disease states or, more interestingly, provide detailed information about the progression of a pathological process. Since changes in their profile directly reflect changes occurring in cellular homeostasis, their accurate quantification allows a systemic understanding of the dynamic molecular processes operating in a particular disease state.

However, the need for accurate identification of minute amounts of a specific biomarker present in such a complex matrix as urine, requires sensitive detection and, in general, complex and timely sample pretreatment, which results in a high cost of analysis. The availability of fast, reproducible and easy-to-apply analytical techniques that would allow the identification of a large number of these analytes is thus highly desirable since they could provide detailed information about the progression of a pathological process.

To date, numerous methods have been developed for the urinary screening of biomarkers. Most of these are based on techniques which use the same sophisticated instrumentation as employed in the screening of pollutants and drug residues, as described previously: gas chromatography mainly combined with mass spectrometry (GC-MS), high-performance liquid chromatography mainly combined with MS (LC-MS) and even nuclear magnetic resonance (NMR). The only techniques for creating urine profiles which can be regarded as simpler and cheaper are paper chromatography and ion-exchange chromatography.

A new and viable analytical alternative to other instrumental methods in this area is capillary electrophoresis (CE). In particular, the high resolution of capillary zone electrophoresis (CZE) allows it to be considered as one of the best separation techniques for the analysis of charged analytes. Additionally, the development of micellar electrokinetic chromatography (MEKC) has provided an even more promising method, since this can be used for the selective separation of both neutral and ionic compounds. In fact, the partition of solutes between the micelle and the surrounding aqueous medium provides a sophisticated means for achieving a very high selectivity of analytes with closely related structures. The limit of detection (LOD) of an analyte often determines the applicability of a CE technique, in which only very small sample volumes (a few nL) can be introduced. To overcome this important drawback and to increase the sensitivity of a method, fluorescence detectors may be used.

The use of bioanalytical methods based on sensor techniques and procedures with low environmental impact (SPME, microscale, low solvent use, *etc.*) for this purpose are anticipated to become routine practice.

Another area of bioanalysis which has become increasingly important is the development of reliable methods for the detection, identification, screening and quantification of genetically modified organisms (GMO). The recent advances in molecular techniques for GMO screening reveal the following trends.[14] Exponential amplification by the polymerase chain reaction (PCR) is still a fundamental step in molecular methods of GMO detection and quantification. For GMO screening, the classical agarose electrophoretic techniques are being replaced by capillary electrophoresis (CE), which provides shorter separation times, automation, lower detection limits and an extended linear dynamic range, through the use of laser-induced fluorescence detection.[15] CE allows for the discrimination of target and internal standard amplified fragments differing by only 10 ppb, thereby reducing the effect of the size difference on the amplification efficiency. CE chips are particularly promising in this area because agarose electrophoretic techniques do not provide sequence confirmation. Microtiter, well-based hybridization assays offer high-sample throughput for a relatively small number of GMO-related sequences. In order to meet the challenge posed by the continuously increasing number of GMOs, various multiplex assays have been developed for the simultaneous amplification and/or detection of several GMOs. It is expected that biosensors, including surface plasmon-resonance sensors, quartz crystal microbalance piezoelectric sensors, thin-film optical sensors, dry reagent dipstick-type sensors and electrochemical sensors, will play a leading role in the low-cost, rapid and simple screening of GMO in cases where high throughput and automation are not required. Disposable biosensors that provide visual detection of GMO without instrumentation are particularly attractive. Biosensor development is targeted at making molecular tests simpler, faster and less costly. DNA biosensors based on optical, electrochemical and piezoelectric transducers have been developed for the detection of amplified GMO-related sequences. GMO quantification requires the determination of two target DNA sequences, *i.e.* a GMO-specific sequence and a plant-specific reference gene. The reference gene permits compensation for differences in the amount and integrity of isolated genomic DNA between samples. The ratio of the copies of the two sequences, expressed as a percentage, gives the relative GMO content of the sample. In regard to GMO quantification, there are a steadily growing number of validated endogenous reference genes. Plasmid DNA constructs are becoming the material of choice for constructing calibration graphs. Real-time PCR, based on homogeneous fluorometric hybridization assays, is the most widely used technique. However, the latest advances in quantitative competitive PCR have allowed high throughput and multiplexing ability. Rapid, simple methods that can be automated are still in high demand for DNA extraction from raw materials and food products.

2.1.3 Monitoring

In the previous chapter, analytical procedures related to the screening of chemicals in different situations were discussed. Process monitoring – in which the state of an environment (in its wider meaning) is monitored and observations of the situation for any chemical changes which may occur over time are made using chemical measuring devices or analyzers – is similar. The development of monitoring has to be related to the principles of Green Chemistry as well. It provides a rapid return of chemical information (with an alarm in case of a sudden discharge), while minimizing errors and cost. Methodologies need to be developed for real-time, in-process monitoring and control prior to the formation of hazardous substances.

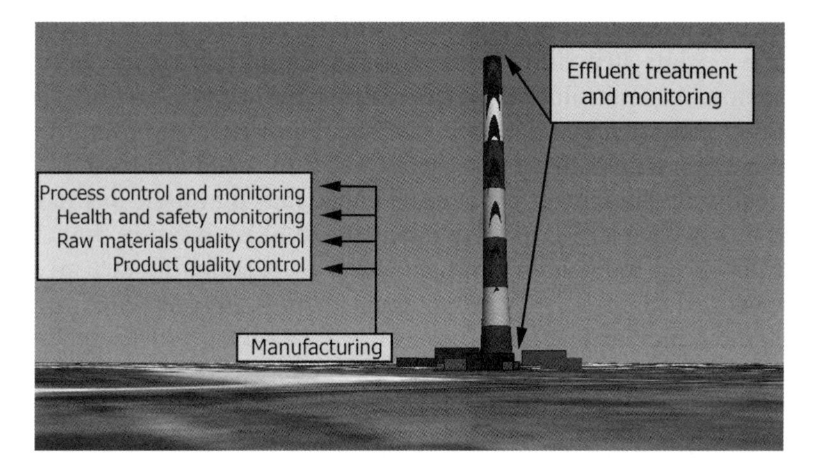

Figure 2.5 The processes in industry that require monitoring.

The first link with monitoring is related to industrial processes and the environment around the industry, in cases where the analytical data is obtained close to the production operation. The industrial application of analytical chemistry is extensive and has various functions that are independent of the size of an enterprise (see Figure 2.5). Even where processes are automated as much as possible, or where processes can reasonably avoid the production of toxic waste, they need to be monitored.

Monitoring allows the use of reagents to be optimised and permits the composition of waste and effluents to be determined. This demonstrates the dual role of the analyst – to improve and optimise the process being analysed, and to ensure that effective and environmentally benign methods are used. Similarly, a real-time field measurement capability for continuous environmental monitoring is preferable to the traditional approach of sample collection and transport to a central laboratory.

Process monitoring is a developing area because of the rising needs related to advancing technologies and the use of new materials, and also because of the challenges associated with new regulation devices and requirements for reducing pollution. To be efficient and effective, analysis should be carried out in real time, *i.e.* the result of the analysis should be obtained "instantaneously". This implies that samples to be analysed undergo no pre-treatment or digestion, and if a chemical reaction has to be carried out to prepare the sample, then the process is fast and does not lead to any time limitations. Such an approach precludes most derivatization reactions, that are used to increase the selectivity and sensitivity of the analytical procedure, and also assumes that appropriate steps have been taken to minimise interference.

Additionally, instruments or sensors should ideally not need to be calibrated, should require only a small amount of sample, and should provide accurate results with a level of certainty sufficient for the purpose required. Therefore, the general challenges that arise in the monitoring of any chemical process or environmental analysis are the same as those that need to be addressed for green or sustainable measurement processes.

Possible candidates for the necessary flexibility could be (chemical) sensors. Continuous and on-line flow and injection methodologies working with sensors that possess the required sensitivity and selectivity are especially important.

There are two approaches to process monitoring which may conflict in some cases, but will generally be complimentary (see Figure 2.6). Off-line analysis can provide the benefit of knowledge gained from previous monitoring of chemical processes, with the higher sensitivity that stationary laboratory instruments provide. In this way, correct points for sensor placement can be identified in the design of new plants and modified processes as a quality and environmental control measure.

On-line monitoring allows for the continuous optimization and efficient use of reagents, and the determination of the composition of waste and effluent, as well as their variation over time, which are necessary for a sustainable development scenario. A faster analysis time enables a greater throughput of samples and the portability of devices enables the generation of large amounts of temporal and spatial information, thus eliminating the problem of unrepresentative sampling.The flow systems are compatible with industrial processes where flow-through reactors are used, and a small portion of the reactor contents can be sampled for analysis purposes. Flow analysis was recognized many decades ago as an effective approach to improve the response time in analytical

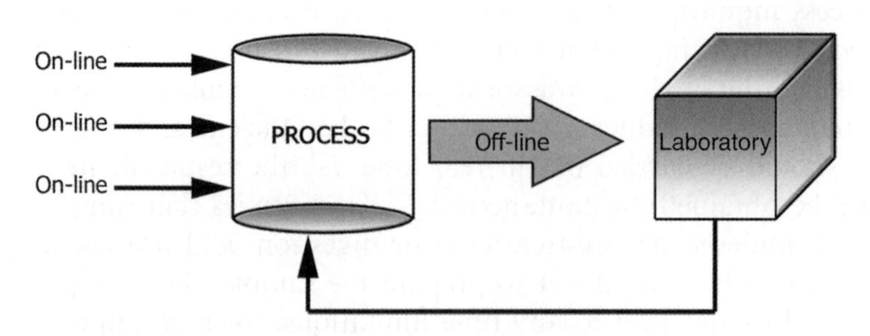

Figure 2.6 Activities related to process monitoring.

experiments; convection leads to an increase in the sensitivity of the detector, a decrease in the detection limit and an increase in reproducibility. As an alternative to continuous flow, the concept of flow-injection analysis (FIA) was introduced in 1975,[16] and has since had a widespread effect on how chemical analyses are undertaken.[17]

The current trend in analyzers, especially for on-line use, is toward the use of portable sensors that are amenable to miniaturization. A chemical sensor can be defined as a tiny device that, as a result of a process of chemical interaction, transforms chemical or biochemical information into a signal. It consists of three parts: a recognition element, a transducer and a signal processor.[18] The function of the transducer is to transform the signal obtained by the sensor element into an electrical signal.

Progress in electronics and instrumentation means that extremely low electrical signals can now be measured. With wireless technology, responses from individual sensors can be received automatically, the signals transformed into analytical information, and changes over time assessed from a remote location. A particular challenge is that of periodic calibration, or the complete removal of the need for calibration. If it can be guaranteed that all sensors within a batch are exactly equal within stipulated limits of uncertainty, then only one of them needs to be calibrated. A second challenge is a real-time response without kinetic effects, which is one of the hurdles in using flow systems.

A list of possible sensors is presented in Table 2.2. More detail is devoted to sensors in Chapter 4.3.

Better instrumentation and improvements in the fabrication of sensors have provided a substantial increase in sensitivity and the signal-to-noise ratio, which has led to a decrease in detection limits. This points the way

Table 2.2 Classification of types of sensors used in the detection and quantification of chemical species.[1]

Type of sensor	*Examples*
Calorimetric	Thermistor
Electrochemical	Potentiometric: solid state, glass membrane, dissolved gas, polymer membrane
	Voltammetric sensors and biosensors, ISFET,[a] ENFET[b]
Gas and solid electrolyte	Oxygen
Mass	Piezoelectric, surface acoustic wave
Optical	Spectrophotometric, colorimetric, fiberoptic
Surface plasmon resonance	Interfacial "concentrations" of biomolecules

[a]Ion-selective field effect transistors.
[b]Enzyme field effect transistors.

towards the miniaturization of detectors. At the same time, physical characterisation techniques, particularly those based on local probe microscopies, now permit the routine examination of solids and surfaces at the nanometer level. This allows the nanoscale aspects of materials to be understood and exploited, and new nanostructured materials to be developed. Nanosensors require nanomaterials such as nanotubes, nanoparticles and nanowires, which are the focus of much research.[19] Nanosensors are already used in environmental analysis in which sensors of different structural types are fabricated, including those formed by self-assembly, that can be adapted for monitoring chemical processes.[20]

Of particular interest, as an extension of this concept, is molecular recognition *via* molecular imprinted polymers (MIP). MIPs are also useful in sample pre-treatment since they can be tailored to target specific analytes. This provides a means of solid-phase extraction, and consequently pre-concentration, which circumvents many of the problems which can arise in trace analysis in complex matrices.

Thus, miniaturization, together with simplification and automation, has been the focus of increasing research in recent years. The concept of miniaturized total analysis systems (μTAS) was invented in the 1990s,[21] partly in response to the fact that a laboratory fully equipped to deal with a range of analytical problems requires a large investment, large amounts of reagents, the probable need for access to other physical techniques, and usually the instruments can only be used in the laboratory. Miniaturization, implied by the term "lab-on-a-chip", leads to the consumption of fewer reagents (less than in FIA approaches) and often to the ability to use the system outside the laboratory.

The goal of "lab-on-a-chip" is that all steps of the process are carried out using a microfluidic system on the same device – integration of pre-treatment, separation and detection. The investigation of different ways to construct microfluidic systems, fabricated in silica, glass, and polymers, and incorporating multiphase flows and membrane functionalities, has undergone an explosive increase over the last decade. Since the channel dimensions are in the order of micrometers, instead of millimetres as in FIA where the flow is pump-driven, the consumption of analyte and reagent is correspondingly reduced. Special challenges arise with respect to their fabrication and use – their accuracy, reproducibility, calibration, *etc.* – that cannot be solved in the same way as for larger systems. The miniaturized systems, therefore, must be robust, preferably disposable and require only a minimum of direct operator intervention, particularly for environmental monitoring.[22]

However, the introduction of miniaturized systems into process monitoring has not been successful, despite promising results in university laboratories. In many industrial applications, the analyte must be detected in the presence of many interfering substances that may be present in large quantities. Since applications and their associated matrices vary widely, the sensing principle must be incorporated into a miniaturized total analysis system (μTAS), that will perform sample pre-treatment, protect the sensor from aggressive matrices and even provide a preliminary separation. This modular system is configured according to its exact application.[23] There are still problems with the technology, arising from sample introduction, small sample volumes and the low limits of detection (LOD). There is also a need for a good interface between the environment and the microfluidic device. The general principles of preparing and using μTAS are described in more detail in Chapter 5.2.3.

As mentioned previously, much research has focused on methods of flow-injection in solution. A small volume of sample is injected into a carrier stream where, in the course of the flow (that is imposed mechanically, gravitationally or by pressure difference), a sample is diluted by dispersion and may be processed on-line on its way to a detector. The principal advantages are that the various processes with the injected sample are carried out mechanically, with high analysis efficiency and a reduction in contamination of the measuring system (especially important in the case of trace analytes). In FIA with electrochemical detection, dispersion of the sample plug occurs within an electrolyte carrier stream, so that the electrolyte is effectively added to and dilutes the sample before reaching the detector electrode.

Finally, it is possible to carry out some on-line sample processing in the measuring system and thereby reduce matrix effects by packing a small amount of solid sorbent in the micropipette tip. The sample is aspirated and then washed, and the species of interest is retained on the sorbent. It is then released by addition of an appropriate eluent. The process can be easily automated, is relatively simple, and is a miniaturized system, since very little sorbent or sample solution is needed.

An important research topic in the field of chemical sensors is the sufficient miniaturization of the sensor so that it can be placed in different key locations, is portable, reduces poisoning problems and is sufficiently inexpensive to be disposable if necessary, as discussed above. This represents a movement toward the development of new materials that are robust and provide the necessary performance for use in sensors and biosensors. Additional criteria, such as low cost and ease of use, have also been taken into account. One example of these materials involves the use of carbon-film electrodes, made from carbon-film electrical resistors, which are sufficiently inexpensive to be able to be used as disposable electrodes if required. As electrical resistors, each end is covered with a metal cap plus external connecting wire. To make electrodes, one of the contacts is removed and the other end, including the wire, is covered with a plastic tube and epoxy resin. The procedure is quick and effective. Examination of these electrodes by electrochemical techniques such as cyclic voltammetry and electrochemical impedance spectroscopy shows that they are at least as good, and often better, than other forms of carbon as an electrode material.[24] Owing to their small size, they can be inserted in places where larger "normal" electrodes cannot, and they are more robust than microelectrodes. It is encouraging that even with inexpensive means, sensitive electrochemical detection can be obtained that requires a small amount of sample and generates a small amount of waste.

In traditional process control and environmental monitoring, samples are collected and transported to a central laboratory to determine their chemical composition. The use of sensors and detectors to continuously measure important chemical properties has significant technical and cost advantages over traditional sampling and analysis. By providing a fast return of analytical information in a timely, safe and cost-effective fashion, such devices offer a direct and reliable assessment of the production process or the effects and gradient of contaminants on site, while greatly reducing analytical costs.

Electrochemical methodologies in analysis are powerful and portable enough to allow monitoring measurements to be carried out in the field. The advantages of electrochemical systems include high sensitivity and selectivity, a wide linear range, minimal space and power requirements, and low-cost instrumentation. Therefore, electrochemical devices offer unique opportunities for addressing the challenges of Green Analytical Chemistry by providing effective process monitoring while minimizing the environmental impact.

Devices based on potentiometry (*e.g.* pH), conductivity and amperometry (*e.g.* oxygen) have all been used for many years.[25] The past two decades have seen enormous advances in electro-analytical chemistry, including the development of ultramicro-electrodes, the design of tailored interfaces, the coupling of biological components with electrical transducers, the micro-fabrication of molecular devices, and the introduction of "smart" sensors and sensor arrays. Advances in controlled-potential techniques have greatly increased the ratio between analytical and background currents, to allow quantisation down to the parts per trillion range. Similarly, recent developments in polymer membrane ion-selective electrodes have pushed the detection limits of these potentiometric devices to a similar level. An example is the measurement of trace metal ions that are common contaminants in environmental matrices, such as in waters used in industrial processes, *etc*. Since problems can arise from the complexity of the matrices, coating the electrodes with an inert polymer is particularly useful. The application of a nanoporous ion-exchange coating prevents the passage of organic and biological molecules, whilst permitting the trace metal ions to pass through. This strategy allows the measurement of trace metals in environmental samples, even those which contain surfactants.

These developments have led to a substantial increase in the popularity of electroanalysis, with relatively fast, small, easy-to-use, "smart" and "environmentally friendly" electrochemical systems. A vast array of devices for on-site and *in situ* environmental and industrial monitoring has been developed in recent years.[26]

Electrochemistry provides a good example of the elimination of hazardous materials from the analytical protocol. Electrochemical measurements are strongly influenced by the material of the working electrode. For many years, mercury was the first choice for electrode material due to its very attractive behaviour, high reproducibility, renewability and smooth surface. The unique properties of mercury drop electrodes led to the 1958 Nobel Prize in Chemistry. Both dropping and hanging drop electrodes have been widely used in connection with various polarographic and voltammetric techniques. Routine applications of these mercury-based procedures can result in the daily generation of a few grams of mercury waste. However, because of the toxicity of mercury, alternative (greener) electrode materials are highly desirable for both centralised and field applications. Various non-mercury electrodes have been examined and their replacement proposed – one possibility is to use bismuth-film electrodes, which offer high-quality, trace metal measurements that compare favourably with those of mercury electrodes. Bismuth can be considered a "green" element, with very low toxicity and widespread pharmaceutical use.

A more widespread "environmentally friendly" material for electrodes is carbon, which has been widely used in electroanalysis for more than three decades, due to its low background current, wide potential window, chemical inertness, low cost, and suitability for various sensing and detection applications. The analytical power and scope of carbon (and other solid) electrodes can be greatly enhanced through a deliberate modification of their surfaces. The resulting chemically modified electrodes can be advantageous in monitoring applications through a preferential accumulation of target contaminants, the exclusion of unwanted materials (*e.g.*, surface-active macromolecules), or acceleration of desired electron transfer reactions. The miniaturization of solid electrodes offers several fundamental and practical advantages, including a dramatic reduction in sample consumption. The significant reduction of resistance (ohmic drop) effects at these tiny electrodes greatly facilitates voltammetric measurements in low-ionic-strength water samples.

The advantages of using membranes in the electrochemical analysis of metals are mentioned above, but electrochemistry does not replace all the atomic spectroscopic techniques that are commonly used for measuring trace metals in central laboratories, and which are unsuitable for on-site analyses. Electrochemical stripping analysis has always been recognized as a powerful tool for measuring trace metals. Its remarkable sensitivity is attributed to the "built-in" pre-concentration step, during which the target metals are collected on the working electrode. The portable instrumentation and low power demands of stripping analysis satisfy many of the requirements for on-site and *in situ* measurement of trace metals. Automated stripping flow analyzers were developed for continuous, on-line monitoring of trace metals in the mid-1970s. Remote/submersible probes circumvent the need for solution pumping and offer greater simplification and miniaturization. This type of remote metal monitoring has been achieved by eliminating the need for mercury electrodes, oxygen removal, forced convection or supporting electrolytes (which previously prevented the direct immersion of stripping electrodes into sample streams).

For a voltammetric, microelectrode array for real-time monitoring of the fluxes of metals across the sediment–water interface, a protective agar coating is proposed. Such surface protection is crucial for achieving the robustness required for long-term *in situ* monitoring in the presence of coexisting surface-active constituents.

The interest in biosensors, in which highly specific biological recognition elements are coupled with electrical transducer technology, has been driven by the needs of environmental and industrial applications for faster, simpler, cheaper and better monitoring tools.[27] The remarkable selectivity of the biological recognition process allows for the direct measurement of target analytes in actual untreated samples. Electrochemical biosensors are the most promising of the bioprobes currently available and have great potential for environmental and industrial monitoring. A variety of schemes for implementing electrochemical biosensing, based on different combinations of biocomponents and electrode transducers, have been suggested. These rely on the immobilisation of enzymes, antibodies, nucleic acids or whole cells onto amperometric or potentiometric electrode transducers. These transducers convert the biological recognition process into a usable electrical current or potential signal.

The integration of electrochemical biosensors with remotely deployed probes still has some specific problems with the *in situ* monitoring of pollutants. When using enzyme electrodes in a submersible operation, one must consider the influence of actual field conditions (pH, salinity, temperature) on biocatalytic activity. Furthermore, the lack of long-term enzymatic stability can limit the prolonged operation of such biocatalytic probes.

In research laboratories, electrochemical biosensors have been developed for monitoring organic streams and harsh environments common to industrial process control. In particular, organic-phase flow detectors could lead to a wide range of previously inaccessible industrial processes.[28]

To address the needs of decentralised (field) testing, it is necessary to move away from the cumbersome electrodes and cells commonly used in research laboratories. The exploitation of advanced microfabrication techniques allows the replacement of traditional ("beaker-type") electrochemical cells and bulky electrodes with easy-to-use sensor strips as described in Chapter 4.3.3. Both thick-film (screen-printing) and thin-film (lithographic) fabrication processes have been used for the high-volume production of highly reproducible, effective and inexpensive sensors strips. Such strips rely on planar working and reference electrodes on a plastic or silicon substrate. These strips become self-contained electrochemical cells onto which the sample droplet is placed. The thin-film fabrication method also facilitates the development of cross-reactive electrode arrays[29] that are useful for multiparameter process or pollution control (in connection with advanced signal-processing algorithms).

Despite major advances, there are still many challenges related to the achievement of stable and reliable monitoring. These include long-term stability (of both the recognition and transducer elements), related baseline drift, matrix effects, reversibility and *in situ* calibration. For identification purposes, the comparison of the physico–chemical parameters and the known spectra is still a basic approach. The growing amount of data produced by automated instrumentation and high-throughput technologies highlights the importance of data processing, where chemometric methods provide the main tools for further development.

2.2 THE ROLE OF INSTRUMENTAL METHODS AND AUTOMATION IN THE GREENING OF ANALYSIS

The need for chemical analysis is growing rapidly, especially for environmental purposes. A real-time field measurement capability is needed for continuous environmental monitoring on land, and even more so in seas and oceans. Another rapidly expanding field is process analytical chemistry, for obtaining analytical data on large-scale production operations.[30,31] This capability provides improved process control and minimises environmental impact. These requirements are putting pressure on laboratories to increase their throughput and shorten response times to produce data for decision-making. At the same time, the quality of the data must be maintained or improved. To achieve the highest possible gains in productivity, automation has to be focused on the laboratory operations where it can have the most impact. This requires carefully studying the process flow of an organization and identifying the points where automation could have a measurable impact. The solution is to use sensitive instruments with high productivity, in order to implement automation to improve the overall productivity of the laboratory. The elimination of defects and errors in laboratory processes is closely related to automation, and laboratory automation can often be a way to minimise human errors.

Many changes in analytical chemistry have been driven by automation, miniaturization and system integration with high throughput for multiple tasks. The fundamental, high-level architecture of laboratory automation has always involved applying technology to the transport of samples through various experimental conditioning, processing or treatment stages, followed by gathering, storing and analyzing data resulting from the experiment. Laboratory automation technology is generally adopted from other fields which have stronger market forces driving technological research. A major step has been taken in equipping complicated and expensive instruments with autosamplers, and independent overnight runs are becoming common practice in control laboratories. Another form of automation is developing laboratory robots for situations where more process flexibility is needed. These robots can easily be reprogrammed when changes are needed, or more complicated movement sequences are not possible with fixed autosamplers.

A more general approach is to automate an entire laboratory procedure, using a variety of available system architectures, all of which are similar at a high level and designed to automate one or more of three categories of operations: sample transport, sample processing, and data collection and handling. A sequence of laboratory steps or functions that when combined become a "unit" operation is referred to as a laboratory unit operation (LUO),[32] a concept that facilitates breaking into components, understanding and evaluating laboratory automation architectures. These become the building blocks for all laboratory procedures. The following operations are divided into three main categories with numerous subcategories:

(1) ***Sample Transport***: the movement of a sample from one stage of the experimental process to another. Common transport mechanisms are:
- manual transport;
- linear transport devices (*e.g.* conveyors, belts);
- positioning stages (*e.g.* turntables, cartesian platforms);
- robotics;
- fluidic flow.

(2) ***Sample Processing***: the experimental treatment of a sample. Common processing steps include:
- manipulation: the physical handling of laboratory materials. This includes capping, crimping, labelling, sealing, reagent addition, the handling of consumables and the movement of objects such as doors and lids of instruments;
- conditioning: modifying or controlling the sample environment. Includes mixing, shaking, vortexing, heating, cooling and atmospheric blanketing;
- grinding: reduction of sample particulate size;
- weighing: quantitative measurement of sample mass, including taring the sample container, opening/closing the balance door;
- separation: includes solid-phase extraction, liquid–liquid extractions, precipitation, filtration and centrifugation.

(3) ***Data Collection & Handling***: recording, analyzing and storing the data from the experiment:
- measurement: direct measurement of physical properties, usually *via* an electronic detection device;
- data acquisition: recording raw direct measurement data *via* transfer of information from the detection device/instrument. This can be done *via* electronic transfer or *via* electronic display and manual recording;
- data processing: conversion of raw measurement data to useful information within the context of the experiment;
- data storage: creating records of raw measurement and processed data;
- documentation: creating records of metadata related to the experiment such as day, time, operator ID, sample ID, experimental conditions, reagent lot numbers, temperature, *etc*.

All laboratory procedures involve one or more operational units. Even a simple procedure, such as weighing a sample, involves several LUOs: transport of the sample to and from the balance, either manually or *via* an automated sample transport mechanism; measurement, in this case measuring a tare weight and the weight of the sample, and data acquisition, *via* visual display or electronic transmission to a database.

Sample Processing. All laboratory procedures are candidates for automation. To achieve high sample throughput, a laboratory must determine which LUOs in a given procedure are good candidates for automation, and which would be of most benefit to the laboratory if automated. Automating a single step is usually not very effective – the problem must be solved holistically.

Sampling and sample pre-treatment typically account for more than 60–70% of the total analysis time, and the quality of these steps largely determines the success of an analysis from a complex matrix. Therefore, streamlining sample processing is the top priority for increasing the productivity of the laboratory. The main focus in the analysis process should be on sample pre-treatment – conditioning and manipulating the sample. Sample treatment can be done separately and then introduced for measurement – off-line processing – or directly connected with the measurement instrument – on-line or at-line treatment. In the case of off-line processing, an additional step to transport the sample to the instrument is needed. When automated sample preparation takes place in a closed system many problems are prevented, such as human error, contamination, sample loss and degradation of the analytes due to air or moisture. The automation of sample processing by at-line or on-line coupling of the extraction and clean-up steps with the analysis system is an important goal in the development of effective, economical and environmentally friendly systems that satisfy many current demands. In most cases these systems relate to fluidic samples and their transport. In a typical on-line system, the moving carrier stream containing liquid samples (and reagents if necessary) is transported and/or combined in a tubular channel, and passed through the tubular channel from one sample-processing device to another, with each device performing discrete operations, which could be rather complicated. The completed reaction mixture then passes through a detection device – usually a chromatograph or spectrometer – and subsequently a signal is recorded.

A wide selection of extraction techniques are used in on-line systems in laboratories, such as thermal extraction, solid-phase extraction (SPE), membrane-based sample pre-treatment, supercritical fluid extraction (SFE) and, to a lesser extent, pressurised liquid extraction (PLE), and microwave- and sonication-assisted liquid extraction (MAE and SAE). The selection of a technique is dependent on the type of sample and the analytes of interest. On-line combinations of extraction and chromatographic methods have been used to analyse many types of complex samples, including food, fossil fuel, agricultural samples, soils, sediments, particulates, water industrial products, plant and animal tissues, blood, serum and urine. However, only a few techniques have been commercialised and used routinely in control laboratories. Most are related to the analysis of volatile compounds by means of vapour-phase extraction and gas chromatography.

The basic steps of on-line coupled vapour-phase extraction-GC systems are described as follows; there are only slight differences for liquid samples and liquid chromatography (LC):

- extraction of the analytes: vapour-phase extraction or liquid/fluid-assisted extraction;
- trapping of the analytes onto an intermediate SPE, membrane or GC injector;
- Focusing of the transferred analytes in front of a GC column: cold-trap, solvent effects, *etc*;
- GC separation of the target analytes.

An intermediate trap is not needed in many applications; it is mainly used in on-line combinations for the liquid/fluid-assisted extraction of solid samples. It is also used for clean-up and/or re-concentration of the extract and for replacing the extraction solvent with a non-polar, volatile solvent. Transferring the analytes directly from the extraction unit or from the intermediate trap should be done efficiently, and is usually performed by thermal or liquid desorption. Refocusing is vital for the quality of the chromatographic peak shape. In vapour-phase extraction methods, such as dynamic headspace extraction and membrane-assisted thermal extraction, where the extracted analytes are in a gas phase, cold-trap is typically used for refocusing. Techniques developed for large-volume injection, such as solvent effects and stationary phase refocusing, are used in liquid-assisted extraction.

With regard to instruments, coupling is most commonly performed using multiport valves and one or more pumps for the dynamic extraction or transfer of the extract to the chromatographic system. In some cases, *e.g.* thermal extraction methods, pressurised gas can be used instead of pumps. Extraction can be performed in static, dynamic or combined mode in on-line systems, as long as the extraction system allows the on-line transfer of the extract. An additional clean-up step is often included in the on-line system. Using SPE between solvent extraction (SFE, PLE, MAE, SAE) and the chromatograph is commonly required if the solvent volume is too large for direct transfer, the type of solvent is not suitable for the chromatograph or the extract requires further clean-up. Techniques developed for large-volume injection, such as on-column, loop-type and vaporiser interfaces, are used to reduce the volume of the extract. The volume of the extract must be small because of the limited capacity of the chromatographic system. In practice, the volume of the extract must frequently be reduced before the transfer. The solvent (or fluid) must also be compatible with the chromatographic system.

The amount of sample is more critical in on-line than in off-line systems. The concentrations of target analytes and matrix compounds should both be considered because in an on-line configuration the whole extract is transferred to the chromatographic column, in contrast to traditional off-line techniques in which only a small part is injected. This makes the on-line method very sensitive, which can easily lead to overloading of the analytical column. Miniaturizing the extraction system is often required to prevent this. This is accomplished with small extraction columns (in SPE) or small extraction vessels, (*e.g.* in MAE, SAE, PLE). In practice, the volume of liquid sample typically varies from a few hundred microlitres to tens of millilitres. Up to 1000 mg is usually sufficient for trace level analysis of solid and semi-solid samples. Careful homogenisation of solid samples is crucial to ensure that they are representative.

Another parameter to be considered in an on-line system is cleaning of the extraction system to avoid memory effects. This is usually done by pumping suitable solvent(s) through the extraction and interface system during the chromatographic analysis. If the samples are very dirty, it may be necessary to change the SPE cartridge or membrane between each analysis. Some commercial systems automatically replace SPE cartridges. This additional step (use of extra solvent and cartridges) does make the method less green, and must be avoided if possible.

A wide variety of systems have been developed and several show promise for improving analysis. However, the full potential for combining on-line extraction techniques for analytical practice and routine research has not yet been fully exploited. One reason for this is the lack of commercial instruments for on-line coupling of solvent or fluid-based extraction techniques with a chromatograph. The acceptance of on-line systems also depends on their robustness and operational simplicity.

Flexible systems with simple optimization steps are required. More complex instruments are needed for solid and semi-solid samples, and they will require further development before routine applications are feasible. However, even the more complex instruments can be excellent tools for specific studies, in which very high sensitivity is needed or the analytes are labile.

Flow Systems. A good example of an on-line system in which sample preparation and measurement are combined is flow-injection in solution that is completely based on the fluidic transport of the sample. In this method the unit operations (sample transport, sample processing, and data collection and handling) are directly combined. A small volume of sample is injected into a carrier stream where, in the course of the flow (which is imposed mechanically, gravitationally, or by difference in pressure), it is diluted by dispersion and can be processed on-line on its way to a detector. The principal advantages of this method are: the ability to perform various processes mechanically, high analysis efficiency and reduced contamination of the measuring system, which is especially important for trace analytes.

Flow systems are compatible with industrial processes where flow-through reactors are used and a small portion of the reactor contents can be diverted for analysis. Flow analysis was recognized many decades ago as an effective way to improve the response time in analytical experiments, and convection leads to increased sensitivity at the detector, a decreased detection limit and increased reproducibility.[33]

Crucial factors in the correct functioning of such systems are:

- the injection or insertion of a discrete, well-defined volume of sample solution into a flowing carrier stream (inert or reagent, with additional reagents added subsequently);
- the precise and reproducible timing of the manipulation to which the injected sample zone is subjected in the system, from injection to detection (*i.e.*, controllable dispersion);
- the creation of a concentration gradient for the injected sample that provides a transient, but reproducible readout of the recorded signal.

This approach consumes less reagent than off-line methods and can be automated using computer-controlled valves and commutators.

FIA (flow injection analysis) is based on the continuous pumping of carrier and reagent solutions, is fully computer controlled, and incorporates the use of a multi-position valve from the ports of which individual, precisely metered zones of sample and reagent(s) are aspirated sequentially by means of a syringe pump and stacked in a holding coil. They are then, dispersed with each other, forwarded to a suitable detector. This approach is called sequential injection analysis (SIA). The SIA system yields substantial savings not only in consumables, but also in waste generation.

The direction of development of FIA is toward further downscaling – the so-called "lab-on-valve" (LOV) – and the introduction into the system of a low-pressure chromatographic separation step using monolithic columns: sequential injection chromatography (SIC). Different gradient methods, particularly the stopped-flow method, have proven to be very powerful tools in many contexts, especially in bioanalytical applications for the assay of substrates and the determination of enzymatic activities (the latter having been always very difficult to execute). Sensitive analytical procedures based on detection by bioluminescence and chemiluminescence, an important inherent property of which is the accurate timing of the flow injection manifold, are also being developed.

Kinetic discrimination schemes, in which even subtle differences in chemical reaction rates are exploited, have provided an extra degree of freedom in FIA, SIA and LOV systems to perform chemical assays, and have given rise to novel and unique applications. This versatility is primarily due to the capability of the systems to propel and aspirate liquid streams as well as solid materials such as beads, at a user-defined flow rate. Equally importantly, stopped flow can be performed on all unit operations as required. These characteristics are essential in order to take advantage of kinetics in performing on-line analytical measurements. FIA and related concepts are described in more detail in Chapter 4.2.

Micronization. Advances in microelectronics and microfluidics have made it possible to miniaturize analytical systems, which are then easily automated. This reduces energy requirements, sample volumes, reagent consumption and waste generation, and increases sample throughput. The current trend is to use portable sensors, that facilitate miniaturizing the analytical system. With the help of these devices, chemical and biochemical information is transformed into electrical signals that can easily be processed.[34] Innovations in electronics and instrumentation allow extremely small electrical signals to be measured. With wireless technology, responses from individual sensors can be received automatically, the signals transformed into analytical information, and longitudinal changes assessed from a remote location.

Automation and miniaturization have been the trend in recent years, and miniaturized total analysis systems (μTAS) have become common and economical replacements in labs, and in portable and mobile instruments as well.[35]

The principle of "lab-on-a-chip" in which all the steps of the process are carried out by means of a microfluidic system on one device, that integrates sample pre-treatment, separation and detection, is very useful for analytical chemistry. The miniaturized systems, that are robust, usually disposable, and require only a minimum of direct operator intervention, are well suited to (process) control and (environmental) monitoring.[36]

The high-level, microfluidic laboratory unit operation (LUO) is very similar to an automatic analyzer. They vary in specific details and the scale of technology used for each operation:

- sample transport takes place throughout the assay in microfabricated channels of a substrate (usually glass or polymer). Flows are in the range of nano or picoliters per minute through micrometer-diameter channels;
- micro- or nanoscale sample processing devices (mixing, dialysis, extraction, ion exchange, *etc.*) are created/integrated in the substrate;
- due to the low volume of sample, the completed reaction mixture is pumped through a sensitive detection device (spectroscopic or electrochemical sensor), and a signal is subsequently recorded electronically using a computer.

The microfluidics approach appears to be simply a scaled-down version of earlier fluid flow technologies, but in fact the factors that govern flow and mixing at the nano-scale are radically different from those of the macro world. Special challenges that arise with respect to the fabrication of microfluidic systems, accuracy, reproducibility and calibration, *etc.* cannot be solved in the same way as for large systems. It is also difficult to apply microfluidic devices to macro-scale assays. These systems are well suited to performing routine, well-defined and moderately complex chemical or biochemical processes. They are less suitable in situations that require a high level of flexibility and re-configurability, or for very complex processes. Microfluidic systems can be fabricated from silica, glass and polymers, and it is possible to incorporate multiphase flows and membrane functionalities. Their principal advantage is their very low consumption of solvents and chemicals, which makes them attractive as environmentally friendly applications.

It is necessary to scale down the detecting components for microfluidic devices, and the detectors used in conventional analyzers are usually replaced by sensors. Small and portable chemical sensors can be placed in various key locations. There are high expectations associated with advancements in optical transduction that afford robustness, versatility, ease of miniaturization, and compatibility with new technologies (*e.g.* nanotechnology). For example, some advances in electrochemical devices are a result of the introduction of new nano-technological materials (*e.g.* nanoparticles and carbon nanotubes), which provide improved responses. Validation, which requires reliable and robust solutions, is an important issue for integrating sensors into environmental-control programs. There are some critical aspects that must be emphasised, such as the small channel dimensions of these miniaturized systems. The solutions used must necessarily be pure, because the presence of even minute amounts of solid particles will cause clogging problems; small dimensions also result in short residence times, which require fast chemistries in sample processing.

Improved instrumentation and fabrication of sensors have allowed a substantial increase in sensitivity and signal-to-noise ratio, which has led to a decrease in detection limits. However, these sensor-based, micro- and especially nano-systems need further development to attract more attention for wider industrial or commercial use. Public advertisement could be useful for the propagation of industrial applications. In addition to advances in chemical sensors, there is intensive development in the field of biosensor technology, in which specific biological recognition elements are used in combination with a transducer for signal processing. Biosensors have been expected to play a significant analytical role in medicine, agriculture, food safety, homeland security, environmental and industrial monitoring. New sensing elements can improve affinity and specificity, and the mass production of molecular recognition components may ultimately dictate the success or failure of detection technologies. The possible contribution of biosensors in two main fields of genetic engineering – genetically-transformed cells and genetically-engineered receptor molecules – must be emphasized.

Biosensor technology is often designed to detect one single or a few target analytes. Effective biosensors must be versatile enough to support interchangeable biorecognition elements, and miniaturization must be feasible to permit automation for parallel sensing and ease of operation at a competitive cost. Large-scale biosensor arrays that are comprised of highly miniaturized signal-transducer elements, enable real-time parallel monitoring of many species and are an important driving force in biosensor research. However, the commercialisation of biosensor technology is not going as well as could be expected from the publications and patenting activities. The slow and limited transfer of technology could be attributed to cost considerations and technical barriers.

Mechanical Transport of Samples. In addition to closed on-line systems in which the fluidic transport of samples is used, there are other possible means of sample transport that can be used for automating laboratory processes. Automated mechanical transport systems are well known in industry, but in science laboratories, except for the autosamplers for large instruments, they are not common. Mechanical transport mechanisms need to be automated when the sample-containing vessels have to be physically moved to and from various workstations or devices, each of which performs one or more sample processing operations. These devices and workstations have to be designed to be accessible by the transport mechanism. Sample processing workstations vary in complexity, performing operations such as weighing, liquid or solid dispensing or transferring, capping/uncapping, centrifuging, filtering, or separating liquids or solids. The processed sample is then transported for laboratory analysis (spectrophotometry, chromatography, electrochemical analysis, *etc.*). When an analyzer is included in the sample processing workstation, the electronically collected data can be integrated into the workstation domain.

Robotic arms are the most versatile mechanical approach to sample transport, but they are usually expensive and not always the most appropriate solution. Early laboratory applications tasked the robotic arm with duties similar to those of the human arm, because of their inherent flexibility and programmability. This robot-centric approach proved to be quite inefficient, slow and unreliable because, although the robotic arm was an excellent general-purpose sample transport device, it was slow in performing sample processing operations. Robotic transport mechanisms are now used only to deliver and retrieve samples. They are highly flexible and reconfigurable, and they facilitate the physical transfer of samples to and from a variety of workstations, but robotic work envelopes are fixed and, in most cases, somewhat limited. Moreover, because complex sample processing procedures are performed in a robotic work envelope, this often results in the robotic movement of samples being the rate-limiting component of the system. The complex processes needed to prevent workstation time conflicts usually require the creation of sample processing schedules, which is a time-consuming task. For systems that may require periodic reconfiguring or process change, scheduling software is highly recommended. The integrated system, in which the robot arm is the transport device, is generally controlled by a system PC. Most robot arms will have their own controller that must be interfaced with the system controller (PC). This requires some degree of programming expertise.

Linear transport devices, *i.e.* belts or conveyors, have only relatively recently entered the laboratory automation environment. Flexible conveyors have the advantage of being fast and relatively inexpensive, with a work envelope that can be extended to almost any length, but, unlike robotic arms, they do not provide a "pick-and-place" capability for samples or vessels. That function must therefore be built into the workstations to and from which the conveyor transports samples, or an intermediate device must be used to transport the sample from the conveyor to the workstation and back. Considerable precision of movement can be obtained with linear transport devices, but high-precision devices are much more costly. The sequence coordination of linear systems can be fairly simple and does not require sophisticated system scheduling software. A special type of linear transport device is a positioning stage – a platform that can perform fast, reliable, repeatable and accurate positioning of loads. Positioning stages are often found in automated liquid handling workstations and in workstations that incorporate spectrophotometric measurements, for which the sample must be positioned accurately relative to a beam of light.

Data Processing. Data processing begins by obtaining an analytical signal from a measurement, and an electronic device usually presents the detector output. Electronic interfacing in laboratory automation spans a very wide range of sophistication and complexity, but does not tend to be on the cutting edge, simply because the demands placed on the interface are usually modest. The information that passes through an interface may be very simple, *e.g.* a two-state signal: high or low, open or closed. It might also be a complex, binary stream of serial or parallel data. Data rates can range from hundreds to giga-bits of information per second.

Ethernet is a physical and data link layer technology for local area networks (LANs) that is becoming more common in laboratory automation and instrumentation, although still not approaching the level of use of RS-232. Most laboratory instruments do not require the speed provided by Ethernet, but the ease of use and standardization of the modular connectors make the interface appealing. The cost of including an Ethernet connection in laboratory equipment has always been a barrier, especially when compared to RS-232, but widespread use of the network interface has significantly lowered the cost. Some instrument providers offer serial to network interfaces for their serial interface-equipped devices. Other laboratory equipment offers optional network capability.

The main tasks of data processing are recording, analyzing and storing data resulting from an experiment, but they also include data related to the experiment, such as day, time, operator ID, sample ID, experimental conditions, reagent lot numbers, temperature, *etc.* that are required for laboratory reports. Computers equipped with the relevant software are tools for laboratory automation in this regard. All contemporary instruments are computer-controlled, which means that recording of raw direct measurement data from the detection device, processing of measurement data in the required format, and storing of data are conducted in one location very close to the analytical instrument.

Data processing is not yet included among the principles of Green Chemistry, but because it helps to prevent human error and saves time, it can be considered a part of an environmentally friendly laboratory system. Furthermore, it is possible to extract useful information from raw measurement data using certain mathematical procedures, instead of further chemical processing. This method is called chemometrics and it is widely used in spectroscopic and chromatographic processing. It is an exceptionally green approach that reduces the consumption of chemicals and solvents, and the generation of waste.

2.3 ASSESSMENT OF ANALYTICAL METHODOLOGIES

Products and processes are often labelled "green" for advertising purposes without any solid evidence, because the term garners a positive public perception. This is a subjective motive, but there is also an objective reason, and that is the lack of measurable characteristics that can be applied to the greenness of a product or process. This is also the case for analytical chemistry, where the main emphasis is on the metrological quality of the data, and only recently has attention been directed toward the environmental friendliness of the way the data are obtained. This problem has been recognised by scientists and engineers and some useful concepts have already been developed.

2.3.1 Parameters for Assessment

The E-factor. The first parameter – the E-factor[37] – is in principle very simple and well suited to the characterization of chemical processes. Although it does not consider material life cycle stages apart from production, the E-factor is a measure of environmental impact and sustainability that is commonly employed by chemists. The E-factor consists of the ratio of product to the total inputs (or all materials used in the production process), and is expressed by the following equation:

$$E = \frac{\text{input material, kg}}{\text{product, kg}}$$

The E-factor considers all chemicals involved in production. Energy and water inputs are generally not included in E-factor calculations, nor are products of combustion, such as water vapour or carbon dioxide. Because of its simplicity and despite the difficulties of including recycled compounds in formulas, this parameter is receiving growing attention and is even being used to analyse complex processes.[38] However, this parameter is not directly applicable to analytical chemistry because the "product" of an analytical laboratory is not measurable in kilograms, and the equation cannot be used to calculate its E-factor. All chemicals and solvents could in principle become waste, even after careful recycling. However, the ratio of the requisite amount of chemicals and solvents to the amount of sample required to obtain a measurable analytical signal can be used to compare different analytical methods.

Atom Economy. Atom economy is another important metric of material efficiency in Green Chemistry.[39] Atom economy calculates the efficiency with which atoms that are used as feedstocks in chemical transformations are incorporated into the final product. Unlike the E-factor, which is based on the specific conditions and circumstances of a process, atom economy is an intrinsic metric that measures the theoretical efficiency of a process under perfect conditions. This metric is most frequently applied to chemical transformations in which substances of discrete molecular structure are transformed into new, homogeneous target products. To apply this approach to analytical measurements, one must be able to assess how many molecules of analyte will give a measurable analytical signal in cases where the theoretical limits of one molecule are known. According to this approach, the sensitivity of the method correlates with its greenness. Areas of analytical chemistry with highly sensitive methods are optical spectroscopy (especially fluorescence), electrochemistry and bioanalysis (due to the highly selective nature of the reactions involved).

Unfortunately, it is exceptionally rare that only a single analyte is analyzed without any interfering matrix; complex sample preparation is usually involved in analyses, for which the approach of atom economy cannot be used.

An assessment of the greenness of analytical methods is difficult because of their complexity and the existence of many parameters. Analytical method databases illustrate this situation well. One of the most comprehensive databases is the National Environmental Methods Index (NEMI),[40] which is a free, Internet-searchable database of analytical methods. This database continues to be expanded to include a growing number of methods and types of method. Within the database, methods can be easily searched, sorted and compared. The current search options include the analyte (name or CAS number), media type (water, air, soil/sediment, or tissue), instrument and detector (over 80 choices), method subcategory (biochemical, organic, inorganic, microbiological, physical, or radiochemical), and method source (USGS, EPA, ASTM, Standard Methods, and many others).

With ever-increasing information in analytical method databases, users have a greenness measure and supporting data to make more informed method selections and improve the environmental friendliness of analytical laboratory operations. NEMI is foremost in adding to the greenness profiles of methods listed in the database. However, the primary consideration in selecting a method is that it meets the performance requirements (*e.g.* detection level, acceptable bias and precision, *etc.*) of the user. It is hoped that users will demand "greener" methods.

Greenness Profiles of Analytical Methods. A more specific approach to the assessment of analytical methods is based on the fact that methods can be hazardous to the environment and to human health, and that greener analytical methods use fewer hazardous solvents, safer chemicals, prevent waste and conserve energy during sample preparation and analysis.

Four criteria have been proposed for creating greenness profiles: Persistent/Bioaccumulative/Toxic (PBT), Hazardous, Corrosive, and Waste.[41]

According to these criteria, a method is defined as less green if it is:

- PBT – a chemical used in the method is listed as persistent, bioaccumulative and toxic (PBT), as defined by the EPA's Toxic Release Inventory (TRI);

- hazardous – a chemical used in the method is listed on the TRI or on one of the RCRA's D, F, P or U hazardous waste lists;
- corrosive – the pH during analysis is less than 2 or greater than 12;
- waste – the amount of waste generated is greater than 50 g.

In order to create greenness profiles, acceptance criteria were developed and applied to methods. Acceptance criteria translate analytical method data (including chemicals used, pH, and waste generated) into a greenness profile.

The above-mentioned criteria were defined and developed collaboratively by a large number of experts in environmental methods. Their consensus was that these are the most important criteria from a regulatory viewpoint, as well as from the perspective of the twelve Principles of Green Chemistry referenced earlier. On the regulatory side, the PBT chemicals identified in the EPA's Toxic Release Inventory (TRI), the Resource Conservation and Recovery Act (RCRA)'s D, F, P and U hazardous waste lists, and the characteristics of hazardous wastes, such as the definition of corrosive, were referred to when developing the definitions of acceptance criteria.

A greenness profile symbol (see Figure 2.7) was developed to provide an easily recognizable summary of the greenness profile of the method. This four-quadrant circle with the quadrants labelled: PBT, Hazardous, Corrosive and Waste, represents the acceptance criteria. If a method is not identified as being "less green" as defined in the above profile criteria, the quadrant associated with that acceptance criterion is filled-in (in NEMI with the colour green). If a method is identified as being "less green" as defined in the above profile criteria, the quadrant associated with that acceptance criterion is not filled in. This generates a greenness profile symbol for every method. For example, if a method does not contain any

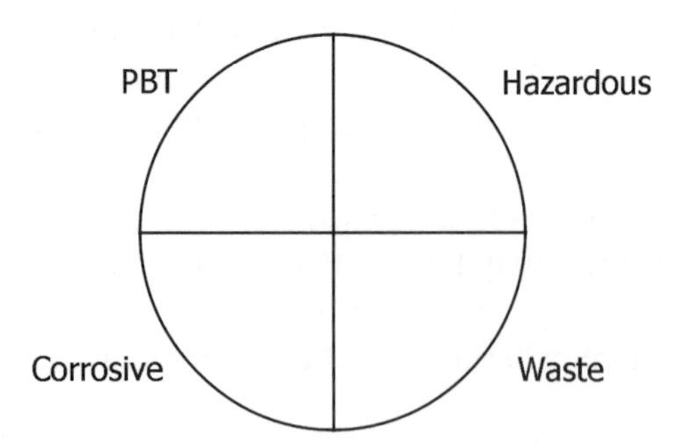

Figure 2.7 Greenness profile symbol.

chemicals on the TRI or RCRA lists, the Hazardous quadrant would be green. When an analyte search is conducted in NEMI, the greenness profile symbols appear in a column of the search results table. The greenness of a method is relative, and the boundary between "less green" and greener has been set by the acceptance criteria defined above.

Over two-thirds of the methods in NEMI have sufficient information to be evaluated according to the greenness profiles described above. The most common reasons for the inability to evaluate the greenness profiles of methods are summarized as follows:

- there is a lack of information on the sample size and/or chemicals;
- the method is not available electronically or was not accessible in a book, manual, or CD;
- parts of the analytical procedure are not documented in the method, but are referenced in another source.

Some trends have been identified from using this database to evaluate the greenness profiles of methods.

The most frequent cause of a method being "less green" was the failure to meet the waste criterion, *i.e.* the method generated greater than 50 g of waste. Two-thirds of the methods evaluated failed the waste criterion. The methods of testing for organic compounds frequently used large sample sizes and relatively large amounts of solvent for extraction. Inorganic methods also frequently failed to meet the waste criterion because strong mineral acids were added to samples for preservation or digestion.

The second most common reason for a method to be "less green" was the use of hazardous chemicals in the procedure. Approximately half of the methods evaluated failed to meet this criterion.

Approximately one-fifth of the methods evaluated were "less green" according to the corrosivity criterion because the sample pH was adjusted to either <2 or >12.

Only 5% of the NEMI methods failed to meet the PBT greenness criterion, and they also failed to meet the hazardous chemical criterion. The most commonly encountered PBTs in analytical methods were lead and mercury compounds.

This leads to the obvious conclusion that many methods generate waste because they use a large amount of solvent and hazardous chemicals. The criteria related to waste and hazardous chemicals must be taken seriously to obtain environmentally friendly methods.

A disadvantage of this approach is that an energy criterion is not included in the greenness profile. Assessment criteria are being developed in which energy considerations are included in the greenness profile.[42] This assessment has five potential risk categories – health, safety, environmental, energy and waste – based on toxicity, bioaccumulation, reactivity, waste generation, corrosivity, safety, energy consumption, and related factors. In addition, two different levels have been developed to score chemical methods from 1 to 3 on each attribute using readily available chemical data. The assessment tool allows individual researchers to make their own judgments about conflicting criteria. The visual presentation is in the form of a pentagram (see Figure 2.8).

The colours represent the levels of risk potential according to five criteria: Health, Safety, Environmental, Energy and Waste.

The energy score is calculated according to the total amount of energy in kWh required to analyse one sample: for $\leq$0.1 kWh, the score is 1 (green), for $>$0.1 to $\leq$1.5 kWh, the score is 2 (yellow), and for $>$1.5 kWh, the score is 3 (red). Energy consumed in the analytical process, including sample preparation and analysis, is accounted for in this score.

This three-level system gives more flexibility in assessing methods, but requires more data to create the profile. The number of parameters required to profile the greenness of a method can be debated, but it is certain that the availability of this kind of profiling helps in the selection of a method that performs best in terms of environmental friendliness and metrological quality.

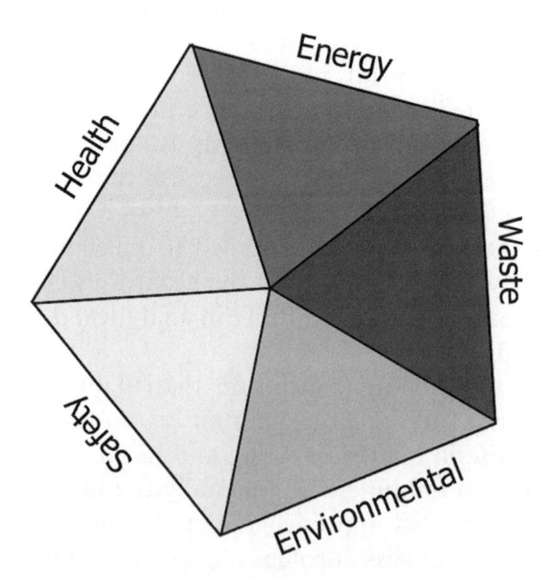

Figure 2.8 Modified greenness presentation symbol.

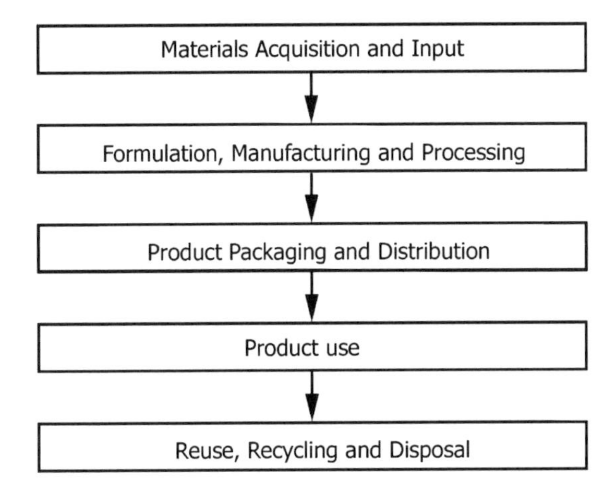

Figure 2.9 Steps in the product life cycle that must be considered in analysis.

2.3.2 Use of Life Cycle Analysis

A life cycle analysis (LCA) is a technique for quantifying and assessing the inputs and outputs affecting the environmental performance of a product throughout its life cycle, from production, through use and, finally, to disposal (see Figure 2.9). LCA includes all aspects of production such as capital and maintenance costs, efficiency, safety and hazard risk management, including waste management. The life cycle parameters of a product are associated with its flexibility of design and operation.

LCA can be a powerful tool to help in the systematic study of the environmental influence of a product and assist in identifying opportunities to improve its environmental performance. A careful review of the following input and output flows will assist in producing environmentally friendly results:[43]

- gain visibility – know exactly what you have;
- analyze inventory – rank products by hazard and cost;
- select products – eliminate costly, unsafe, redundant chemicals;
- source products – maintain a greener inventory for a more positive impact;
- follow performance metrics – track and report on progress.

Standardized life cycle assessment (LCA) protocols already exist, such as the ISO 14040 series, but standards and data for toxicity assessments are lacking; for that reason, few studies with comprehensive, quantitative analyses have been conducted.

LCA relies on information about resource use, emissions and impacts throughout the life cycle. For emerging technologies, such data are difficult to locate in an LCA database. Inventory data about fine and specialty chemicals are also difficult to find, even for industrial products. In most studies inventory data are estimated from laboratory experiments based on engineering heuristics and judgment. The emissions and downstream treatment are hard to estimate or model, and often are ignored or gauged *via* sensitivity analysis.

A standard impact analysis is usually used for analyzing inventory.[44] A life cycle study also requires upstream information. Because life cycle information is only available for common chemicals, and not for specialized or new chemicals and their precursors, it becomes necessary to create a life cycle tree that links materials for which an inventory is available to the final product. Un-reacted materials, by-products, co-products and lost solvents could potentially become pollutants if they are not collected, treated or disposed of carefully. Quantitative values for such pollutants resulting from processes are usually missing in the literature.

Some LCA chemical databases are available, but need more inputs.[45] It is the task of the analytical chemistry community to address this incompleteness of databases and lack of data, by providing reliable data about chemicals and their degradation products under different conditions. All of the inorganic chemicals, organic compounds, solvents, un-reacted materials, by-products, emissions and wastes that are related to a particular product must be confirmed by analytical chemistry.

However, there are mathematical approaches to estimate the chemical properties of compounds based on the small amount of data that does exist. The correlation of structure and properties is at the core of chemistry and chemical language. The systematic approach to this problem is called quantitative structure activity relationship (QSAR), and it also includes structure property relationships (SPRs) and others. Such approaches have been used in environmental science for the assessment of existing chemicals. In the development of new compounds, QSAR provides at least a correct orientation for the design of new chemical entities. For example, when a lead structure for a pharmaceutical has been discovered, variations of this structure are screened by *in silico* systems to find the most promising candidates in terms of activity and with the least undesirable side effects. The application of such methods to the design of new chemicals considers the functionalities of molecules over their entire life cycle and identifies degradability, such as hydrolysis, photolysis, biodegradation, and other elimination pathways such as sorption onto particulate matter and bioaccumulation in organisms under environmental conditions, as parameters correlating with chemical structure. This reduces costs, time and animal trials.

A comprehensive environmental analysis and evaluation of the greenness of any product or technology must also take into account the product life cycle, because the environmental impact may simply move to other stages of the life cycle. Replacing mature and high-yielding industrial production with newer solvent-based techniques may not be promising from an environmental life cycle viewpoint, despite their advantages over existing techniques. Furthermore, the process of producing new solvents poses significant environmental challenges that are more difficult to overcome than those related to common solvents. In general, the replacement of a conventional process by one that is based on new solvents may be environmentally attractive if the preparation of the solvent has a small life cycle impact, permits a high degree of reusability, and results in high yields and easy separation. Solvents such as ionic liquids that have a large life cycle impact per kilogram may only be attractive for the production of high-value-added materials such as fine chemicals and pharmaceuticals. Furthermore, with continued advances in the development of new functional materials and tunable solvents, along with guidance from life cycle research, it may be possible to develop truly green solvents that are suitable for a wide range of chemical processes.

The same is true for nanomaterials used in analytical methods. Most of them are so new and have been studied so little that there is not enough data for detailed LCAs or material flow analyses. Eventually, as the production capacity for nanomaterials increases, it is likely that comprehensive evaluations of their environmental impact will become more common.

LCA is a general approach that can greatly assist in estimating the greenness of product. Despite the uncertainties, lack of industrial data, and other challenges, it is important to include life cycle considerations in the core principles of Green Chemistry; LCA is most useful during the early stages in the development of a new technique, before a technology is widely adopted. It should also be supported with evidence of how the packaging and distribution of the product affect its environmental impact.[46]

Analytical instrumentation is an essential part of the analytical process and a thorough life cycle assessment of these instruments must consider the greenness of the analytical method. One design consideration that affects EOL (end-of-life) options is the operating life span of an instrument. As a general rule, an instrument that requires a large input of resources should also be durable and reusable. Equipment that has a short life span should be designed in such a way that the minimum amount of resources is expended to obtain results, especially if reuse or recycling is unfeasible. Chemical processes must be carefully designed to ensure that the use of resources is optimized.

In principle, chemical analysis is a chemical process, but its output is not a tangible product. LCA is not directly applicable to information obtained from the process of measurement, but it is definitely applicable to the components (chemicals, solvents, instrumentation, data processing equipment, *etc.*) used in the process. Therefore, LCA can be used to assess the environmental impact of analytical methods.

REFERENCES

1. Ch. M. A. Brett, *Pure Appl. Chem.*, 2007, **79**, 1969–1980.
2. *The Montreal Protocol on Substances that Deplete the Ozone Layer as either adjusted and/or amended in London 1990, Copenhagen 1992, Vienna 1995, Montreal 1997, Beijing 1999, UNEP 2000;* http://www.unep.org/ozone (last accessed 03/01/2010).
3. *The Stockholm Convention on Persistent Organic Pollutants,* Stockholm 2001; http://chm.pops.int/ (last accessed 03/01/2010).
4. W. H. McClennen, N. S. Arnold and H. L. C. Meuzelaar, *TrAC*, 1994, **13**, 286–293.
5. S. C. Jakeway, A. J. de Mello and E. L. Russell, *Fresenius J. Anal. Chem.*, 2000, **366**, 525–539.
6. S. Attiya, A. Jemere, T. Tang, G. Fitzpatrick, K. Seiler, N. Chiem and D. J. Harrison, *Electrophoresis*, 2001, **22**, 318–327.
7. J. Wang, J. Lu, T. Tian, S. Ly, M. Vuki, W. Adeniyi and R. Armennderiz, *Anal. Chem.*, 2000, **72**, 2659–2663.
8. L. Moore, D. Veltkamp, J. Cortina, Z. Lin and L. Burgess, *Sens. Actuators, B*, 1997, **38**, 130–135.
9. R. Song, P. C. Schlecht and K. Ashley, *J. Hazard. Mater.*, 2001, **83**, 29–39.
10. T. A. McGlinchey, P. A. Rafter, F. Regan and G. P. McMahon, *Anal. Chim. Acta*, 2008, **624**, 1–15.
11. W. W. Buchberger, *Anal. Chim. Acta*, 2007, **593**, 129–139.
12. J. Tschmelak, G. Proll and G. Gauglitz, *Talanta*, 2005, **65**, 313.
13. D. Muir and E. Sverko, *Anal. Bioanal. Chem.*, 2006, **386**, 769–789.
14. D. S. Elenis, D. P. Kalogianni, K. Glynou, P. C. Ioannou and Th. K. Christopoulos, *Anal. Bioanal. Chem.*, 2008, **392**, 347–354.

15. T. Levandi, C. Leon, M. Kaljurand, V. Garcia-Canas and A. Cifuentes, *Anal. Chem.*, 2008, **80**, 6329–6335.
16. J. Ruzicka and E. H. Hansen, *Anal. Chim. Acta*, 1975, **78**, 145.
17. E. H. Hansen and M. Miró, *TrAC*, 2007, **26**, 18.
18. R. W. Cattrall, *Chemical Sensors*, Oxford University Press, Oxford, 1997.
19. L. He and C.-S. Toh, *Anal. Chim. Acta*, 2006, **556**, 1.
20. J. Riu, A. Maroto and F. X. Rius, *Talanta*, 2006, **69**, 288.
21. A. Manz, N. Graber and H. M. Widmer, *Sens. Actuators, B Chem.*, 1990, **1**, 244–248.
22. L. Marle and G. M. Greenway, *TRAC*, 2005, **24**, 795–802.
23. N. Blom, J. C. Fettinger, J. Koch, H. Lüdi, A. Manz and H. M. Widmer, *Sens. Actuators, B*, 1991, **5**, 75–78.
24. C. Gouveia-Caridade and Ch. M. A. Brett, *Electroanalysis*, 2005, **17**, 549–555.
25. J. Wang, *Analytical Electrochemistry*, Wiley, New York, 2nd edn., 2000.
26. J. Wang, *TrAC*, 1997, **16**, 84–88.
27. K. R. Rogers and C. L. Gerlach, *Environ. Sci. Technol.*, 1996, **30**, 486A–491A.
28. J. Wang, *Talanta*, 1993, **40**, 1905–1909.
29. K. Albert, N. S. Lewis, C. Schauer, G. Soltzing, S. Stitzel, T. Vaid and D. R. Walt, *Chem. Rev*, 2000, **100**, 2595–2626.
30. J. Workman, K. E. Creasy, S. Dohetry, L. Bond, M. Koch, A. Ullman and D. Veltkamp, *Anal. Chem.*, 2001, **73**, 2705.
31. J. Workman Jr., M. Koch, B. Lavine and R. Chrisman, *Anal. Chem.*, 2009, **81**, 4623–4643.
32. J. Hurst and J. W. Mortimer, *Laboratory Robotics, a Guide to Planning, Programming and Applications*, VCH Publishers Inc., 1987.
33. J. Ruzicka and E. H. Hansen, *Anal. Chim. Acta*, 1975, **78**, 145.
34. R. W. Cattrall, *Chemical Sensors*, Oxford University Press, Oxford, 1997.
35. A. Manz, N. Graber and H. M. Widmer, *Sens. Actuators, B*, 1990, **1**, 244.
36. L. Marle and G. M. Greenway, *TrAC*, 2005, **24**, 795–802.
37. R. A. Sheldon, *Chem. Ind.*, 1992, 903–906.
38. R. A. Sheldon, *Green Chem.*, 2007, **9**, 1273–1283.
39. B. M. Trost, *Science*, 1991, **254**, 1471–1477.
40. http://www.nemi.gov (last accessed 03/01/2010).
41. L. H. Keith, L. U. Gron and J. L. Young, *Chem. Rev.*, 2007, **107**, 2695–2708.

42. D. Raynie and J. L. Driver, *Green Assessment of Chemical Methods*, presented at the 13th Green Chemistry and Engineering Conference, Maryland, USA, June 23–25, 2009
43. http://www.dolphinsafesource.com – Green product selector.
44. *Handbook on Life Cycle Assessment: Operational Guide to the ISO Standards*, ed. J. B. Guinée, Springer, New York, 2002. Database for impact assessment method available at: http://www.leidenuniv.nl/interfac/cml/pmo/index.html (last accessed 03/01/2010).
45. SimaPro 7 software, PRé Consultants, Amersfoort, The Netherlands, 2007; http://www.pre.nl/simapro/default.htm (last accessed 03/01/2010).
46. P. T. Anastas and R. L. Lankey, *Green Chem.*, 2000, **2**, 289–295.
47. T. H. Hankenmeier, S. P. J. van Leeuwen, J. J. Vreuls and U. A. Th. Brinkman, *J. Chromatogr., A*, 1998, **811**, 117–132.

CHAPTER 3

"Greening" Sample Treatment

Analytical chemistry is the measurement science that generates, processes, and evaluates signals about the composition and structure of samples. Analytical chemistry can thus be defined as the chemical discipline that deals with obtaining information about the chemical composition of samples (elements, ions, species, and their structure).[1] This approach provides the basis for using information theory to build a general theoretical basis of analytical chemistry[2] to describe and mathematically quantify the main parameters of the analytical process:

- sensitivity – the effects on the signal caused by given species and factors;
- specificity – the ability to detect or determine an individual analyte without interference from accompanying species, expressed qualitatively and quantitatively;
- selectivity – the ability to detect or determine a given species without interference, *i.e.* independent of and unaltered by each other and additional constituents of the sample:
- ruggedness (robustness) – the applicability of the method when small variations occur in method conditions, operator skill, and sample composition;
- precision (certainty of measurement) – the reproducibility of the entire analytical procedure from sampling to measurement; this is a primary characteristic of analytical methods.

Analytical methods are based on *chemical reactions* and *electrochemical processes*, as well as on *interactions* with all forms of energy (particularly radiation) that give unambiguous signals directly from the

Green Analytical Chemistry
By Mihkel Koel and Mihkel Kaljurand

Published by the Royal Society of Chemistry, www.rsc.org

site where something *significant to a chemist* is taking place – in the bulk or surface of a solid, liquid or gas. The analytical procedure to obtain data is comprised of several steps: field sampling and sample handling, laboratory sample preparation and separation, detection, quantization, and identification. This must all be done with high metrological quality, which means providing the above-mentioned parameters with measurement results. Whenever analytical methods are changed, enhanced or replaced, the aim should be to improve the metrological quality of the procedure. The application of Green Chemistry principles to analytical procedures almost always leads to an improvement in the quality of the method:

- the reduction of solvents and other compounds in the process decreases possible negative interference with the analyte;
- the reduction of steps in the sample preparation or separation process decreases sources of measurement errors and uncertainties;
- miniaturization and energy savings result in a more robust and simpler analytical process.

It must be emphasised that greening analytical chemistry should aim not only at yielding a more environmentally friendly process, but also at providing results of higher metrological quality.

Analytical signals are obtained as a result of interactions between relevant energy forms and levels of constituents of the sample (*e.g.* atoms, ions, molecules). Usually, to perform this interaction in the most optimal way, some external energy system (*e.g.* chemical energy, radiation, heat) has to be applied through some media (solvents).

The source of an analytical signal can be a chemical reaction, electrochemical or physical process (*e.g.* neutralization reaction, electrolysis, absorption and emission of radiation), and the signals are measured in physical quantities like volume, mass, electrical charges, differences in temperature, and of radiation energy, characterized qualitatively and quantitatively by wavelengths or frequencies, and intensities. Analytical instruments are required to receive analytical signals and present them in a usable form (*e.g.* data file of values, spectra, chromatograms or pictures) for further information processing. The principles of Green Engineering can be applied to equipment and instrumentation to obtain a green analytical procedure from the beginning to the final result. The most relevant principles are:

- ensuring all inputs and outputs are inherently non-hazardous;
- maximizing mass, energy, space and time efficiency;

- limiting underused and unnecessary materials and energy;
- minimizing material diversity;
- designing for a commercial "afterlife".

These principles relate to some developments in analytical chemistry that coincide with Green Chemistry and Engineering:

- replacing "wet" chemistry in sample preparation and treatment;
- searching for alternative solvents and using solventless techniques;
- reducing the need for derivatization with different detection methods and signal processing;
- treating samples efficiently by replacing thermal energy with alternatives (ultrasonic irradiation, microwaves, laser irradiation);
- using photochemical activation and mechanical activation;
- analyzing samples directly without treatment using spectroscopy, surface analysis, and laser-spectroscopic techniques;
- decreasing sample volume by miniaturization, using "lab-on-a-chip," biosensors or immunoassays;
- using automation or hyphenated techniques.

This list clearly indicates that current developments are directed towards increased sensitivity and specificity, and the optimal use of energy. This can be better achieved with a variety of specialized instruments rather than universal ones. Considering these principles is especially important when designing new methods and technologies; automation and miniaturization are not ends in themselves, but rather means to increase environmental friendliness and analytical metrological quality.

Sampling, and especially sample preparation, frequently involves the use of solvents and other chemicals that generate large amounts of waste, often followed by clean-up and pre-concentration steps. Therefore, sample preparation techniques that use a small amount of organic solvent, or none at all, are important in greening analytical chemistry. In the following sections we will discuss some of these options in more detail.

3.1 INTRODUCTION TO ENVIRONMENTALLY BENIGN SAMPLE TREATMENT

The procedure for generating an analytical signal has several steps: sampling and sample handling, sample treatment (separation, derivatization, *etc.*), and detection. Based on this signal, quantization and statistical evaluation are performed to obtain chemical information about the analyte. Each of these steps contributes to obtaining correct results, but sample treatment is a key component of the analytical process. The use of solvents and other chemicals usually results in the generation of large amounts of emissions and environmentally harmful waste.

Various sample treatment steps may be performed which use different solvents and reagents, but typically sample treatment methods involve liquid–liquid or liquid–solid extraction with an organic solvent, often followed by clean-up and pre-concentration steps. These methods are time-consuming, labour-intensive and costly, depending on the amount of solvent required. This indicates that sample treatment is an effective area in which to apply Green Chemistry principles because:

- the consumption of organic solvents can be significantly reduced;
- vapour emissions and the generation of liquid and solid waste can be reduced;
- reagents of high toxicity and/or ecotoxicity can be eliminated;
- labour and energy consumption can be reduced.

For these reasons, sample treatment techniques that use a small amount of organic solvent, or none at all, are important in greening analytical chemistry. Water is considered to be the ideal solvent, because it is non-toxic, cheap and readily available. Unfortunately, water as a solvent has some limitations, due to the insolubility of non-polar organic compounds and the instability of reactive reagents or substrates in this medium. However, water chemistry must always be considered for removing reactants during the workup procedure – a process that usually consumes large amounts of organic solvent and energy. The extraction volume of organic solvent can exceed the volume of water by factors of up to 30.

Even with the most advanced instrumental techniques for the separation, detection, identification and determination of analytes, sample treatment is often the time and efficiency bottleneck in many analytical procedures.[3] It is also the primary source of errors and discrepancies between laboratories. Thus, the quality of this step is a key factor in a successful analysis, and the judicious choice of an appropriate sample preparation procedure greatly influences the reliability and accuracy of an analysis. There is a wide range of techniques, many of which have changed little over the last 100 years.[4] For example, Soxhlet extraction, which is a well-established technique in common use today, was developed in 1879. All sample preparation techniques have the following objectives in common: to extract analytes from the matrix, to bring them to a specified level of concentration, to remove possible interferences (clean-up step) and, when required, to convert the analytes into a more suitable form for detection or separation. The solvents may be organic liquids, supercritical fluids, pressurized or superheated liquids.

The most common extraction techniques for solid and semisolid matrices, apart from Soxhlet extraction, are sonication-assisted extraction, supercritical fluid extraction (SFE), microwave-assisted extraction (MAE), pressurized liquid extraction (PLE) and matrix solid-phase dispersion (MSPD). Soxhlet extraction is the oldest technique used for the isolation of non-polar and semi-polar organic pollutants from different types of solid matrices, including biota samples.

Although the size of the system can vary, typical procedures use 50 to 200 mL of organic solvent to extract analytes from 1 to 100 g of biological tissue. In extraction, it is essential to match the solvent polarity to the solute solubility, and to thoroughly wet the sample matrix with the solvent. Typical solvents for extracting POPs from animal and plant tissues are *n*-hexane, dichloromethane, and mixtures of toluene–methanol, *n*-hexane–acetone and dichloromethane–acetone.

In addition to solvents, drying chemicals are used for fresh animal and plant tissues in order to reduce their water content; grinding also helps to open up the tissue structure, enabling the solvent to better penetrate the sample matrix. Freeze-drying (water evaporation below 0 °C under vacuum conditions) or liophylization of samples can be performed before extraction as an alternative to chemical drying with sodium sulfate. Unfortunately, this process is extremely energy-intensive.

The advantages of Soxhlet extraction include the following: large amounts of sample (*e.g.* 1–100 g) can be used; no filtration is required after extraction; the technique is not matrix dependent; and many Soxhlet extractors can be set up to perform in unattended operation. Soxhlet extraction is still the preferred option for routine analysis, because of its robustness and relatively low cost. Moreover, Soxhlet extraction is widely used as a standard technique and reference for evaluating the performance of proposed extraction methods. Attempts to automate the technique have had some success, and a few commercial systems are available in which several samples can be extracted in parallel, providing shorter extraction times and using less organic solvent than conventional Soxhlet. The main disadvantages of Soxhlet extraction are that it requires large amounts of solvent; the solvent must be evaporated to concentrate analytes before determination; the process takes several hours or days to complete the extraction; and it generates dirty extracts that require extensive clean-up.

Several alternative extraction techniques that consume less organic solvent and are conducive to automation have been developed to replace conventional Soxhlet methodology. Moreover, they have the potential for efficient and fast on-line clean-up, using selective trapping. However, the comparatively high investment cost of these new techniques is a major reason why conventional Soxhlet extraction, in combination with adsorption columns and/or GPC for purification and fractionation of extracts, is still widely used as the sample preparation reference method in numerous applications.

A comparison of various extraction methods is presented in Table 3.1. Some of the characteristics important in laboratory use are listed, in addition to environmental parameters. The table illustrates the complexity and multi-faceted nature of selecting an appropriate method for sample treatment.

3.2 REDUCED AND SOLVENT-FREE SAMPLE PREPARATION METHODOLOGIES

The growing demand for faster, more cost-effective and environmentally friendlier analytical methods provides a major incentive for improving the conventional sample treatment procedures used in environmental analysis. In most conventional procedures, rapid and powerful instrumental techniques for the separation and detection of analytes are combined with time-consuming, and usually manual, methods for sample preparation that can considerably slow down the total analytical process. Sampling and sample pre-treatment typically account for over 60% of the total analysis time, and the quality of these steps largely determines the success of an analysis from a complex matrix.

According to the principles of Green Chemistry, new methods of sample preparation must meet the conditions mentioned earlier:

- reduce organic solvent consumption;
- reduce emissions and waste generation;
- eliminate toxic reagents;
- save time and energy.

Sample treatment is the part of the analysis that consumes the most solvent and chemicals; this explains the efforts that are being directed toward improving sample processing methods to reduce the amount of solvents and chemicals needed for derivatization. Even small economies in solvent usage and the optimization of the detector can substantially reduce the waste produced in sample preparation.

Table 3.1 Comparison of extraction methods.[10] (Reproduced with kind permission of Elsevier)

	LLE	*SPE*	*SPME*	*LPME*	*Soxhlet*	*SFE*	*MAE*
Recovery	Quantitative		Non-quantitative		Quantitative	Quantitative	Quantitative
Selectivity	Low	Moderate	High	High	Low	High	Low
Amount of sample	10–1000 mL	0.1–100 mL	0.5–50 mL	0.1–1000 mL	10–100 g	1–10 g	2–50 g
Extraction time	20–60 min	5–15 min	20–90 min	20–90 min	6–24 h	10–60 min	5–60 min
Extraction method	Mixing/agitation		Agitation		Heat	Heat + pressure	Heat + pressure
Solvent consumption	20–100 mL	2–5 mL	–	10 μL–2 mL	150–300 mL	10–20 mL	15–50 mL
Operator skills	Low	Moderate	Moderate	High	Low	High	Moderate
Method development time	Low	Moderate	Low	Moderate	Low	High	High
Equipment cost	Low	Low	Moderate	High	Low	High	Moderate
Level of automation	Low	Moderate to high	Moderate	Moderate	Low	Moderate	Low
Possibility of on-line operation	Minimal	Good	Low	Moderate	Minimal	Good	Moderate
Remarks	Formation of emulsions can be problematic		Careful calibration needed	Highly selective clean-up possible	Tedious	Matrix dependent extraction	Matrix dependent

The selection of an extraction technique is based on several criteria. Naturally, sample preparation must be tailored to the analysis. The sample matrix and the type and amount of analytes in the sample are of primary importance. The speed of extraction, the complexity of the instrumentation, the simplicity and flexibility of the method development, and the ruggedness of the method are also crucial. Moreover, a method that is appropriate for target-compound analysis may not be suitable for the comprehensive chemical profiling of a sample. The selectivity of the sample preparation is often a key factor in target-compound analysis, whilst exhaustive extraction is the preferred choice for profiling.

Several approaches provide solutions to this conundrum: solventless methods, microextraction methods and new solvent systems. At-line or on-line coupling of extraction and clean-up steps with the separation system is an important consideration in the development of sample preparation methods; miniaturization has been a key factor in designing integrated analytical systems to provide higher sample throughput and/or automated operation, as well as more economical use of solvents and chemicals.

3.2.1 Solventless Methods

Solventless methods of sample preparation can be divided into three categories: gas extraction, membrane-based techniques, and solid-phase extraction.

Extraction with a stream of gas, which is very often combined with solid-phase extraction,[5] includes:

- direct determination of analytes in a gas phase above the liquid or solid sample – head space analysis;
- purge-and-trap;
- trapping on the sorbent in different forms;
- cryotrapping;
- trapping in the head of a chromatographic column.

Due to the gaseous nature of the analyte-carrying media, the main analysis and identification is performed on a gas chromatograph, and is applicable to volatile analytes. Extractions with supercritical gases like CO_2 are discussed later in Section 3.3 on alternative solvents.

Membrane extraction can be considered a solventless method when the analytes are volatile and the accepting media is a gas phase.

Otherwise, membrane extraction is a form of liquid–liquid extraction that features:

- direct determination of analytes from a stream of gas or liquid flushing the external side of a membrane, with dosage of the analytes into the inlet of the GC, or
- collection of analytes from a stream of gas or liquid flushing the external side of a membrane, on the internal trap side of the membrane.

The solid-phase extraction (SPE) technique is derived from liquid chromatography technology, and in recent years SPE has gradually replaced most liquid–liquid extraction (LLE) methods as the preferred technique, especially for biological samples prior to quantitative analysis.[6] In SPE:

- the sample passes through the sorbent bed (gravitationally or forced), and the analyte is trapped on sorption tubes (inside capillaries, coated columns or syringes), or
- a sorbent is added to a (liquid) sample and the analyte is trapped directly from a medium (gas, liquid), on rods, beads, stir-bars, separating capillary columns, *etc*.

The analyte is later desorbed from the sorbent, either by the appropriate solvent or thermally (the true solventless approach) into the analyzer inlet. With SPE, solvent consumption and costs are reduced and the process can be automated.

Possible sorbents for this type of extraction are:

- synthetic porous polymers, including molecularly imprinted polymers;
- modified silica gels, sometimes in the form of monoliths;
- carbon sorbents, carbon black, activated carbon, soot, *etc*.;
- low-volatility liquids.

Polymeric supports have been used for more than 50 years to simplify and accelerate the work-up of organic reactions. Initially, ion-exchange resins dominated the field, but in the last decade, more specific polymeric supports and linker systems have been used. A number of different polymers, cross-linked (insoluble) and non-cross-linked (soluble), have been used for a large variety of applications.[7] In addition to industrially produced polymers,[8] different kinds of chromatographic material, including those used as the stationary phases in gas chromatographs, and ionic liquids are also widely used. However, even after years of intensive research, universal support for application in sample preparation is still lacking. Every sorbent has its drawbacks, such as chemical stability, polarity or loading capacity.

First, the backpressure must not be too high. Therefore, if particles are used as sorbents, they have to be relatively large, and if a monolith is used, it should have relatively large pores.

Second, the adsorbent plug should be physically stable in the pipette. With silica particles, filters can be used to stabilize the bed. With the use of monoliths as sorbents, it is possible to chemically bind the sorbent to the surface; this approach results in very physically stable sorbent plugs, and no frits are needed.

Third, in order to eliminate carry-over, the pipettes should be disposable. The cost should therefore be relatively low. Silica particles of high quality are quite expensive, and it is unrealistic to use them in disposable tips. Methacrylate sorbents, on the other hand, are inexpensive to prepare. However, they are not acceptable from the standpoint of Green Chemistry, because solid waste is generated. This is also true for all disposable micronized systems, and the reduction of solid waste must be considered during system design.

Fourth, the sorbent should be effective in the clean-up process. Polymethacrylate monoliths can be prepared with a number of different selectivities. Mixed selectivities, *e.g.* reversed-phase and strong cation-exchange functionality, can also be easily prepared if desired.

Fifth, the sorbent needs to have sufficient chemical stability. Organic monoliths are stable over a wide pH range.

Sixth, the amount of sorbent should be related to the task. With smaller sorbent plugs, loadability decreases, but the volume of solvent for elution can, on the other hand, be quite small when the amount of sorbent is small. The need for evaporation and reconstitution can thus be avoided. When higher loadability is needed, the amount of sorbent should be increased. The upper limit is restricted by backpressure.

In the most popular microextraction system, a fused silica fibre is coated with sorbent and the analytes are partitioned between the sorbent rod and the matrix phase.[9] The method is simple and readily automated. It has provided the starting point for the development of a series of new sample preparation methods.

In addition to the effectiveness of an extraction technique, consideration needs to be given to numerous other factors that could affect the analytical scheme.[10] These include equipment and operating costs, the complexity of method development, the amount of organic solvent required and the level of automation.

SPE (in all its modes and formats) is arguably the preparation method of choice for liquid samples, especially where more polar analytes are involved. On-line (and automated) solid-phase extraction–liquid chromatography (SPE-HPLC) is a fully mature technique that can easily be miniaturized. SPE (in all of its modes and formats) is arguably the preparation method of choice for liquid samples, especially when more polar analytes are involved. On-line (and automated) solid phase extraction–liquid chromatography (SPE-HPLC) is a fully mature technique that can easily be miniaturized. Although the technique can be used with a wider range of analytes, less expensive automation and ease of handling seem to be its major advantages.

Soluble polymeric supports have also had a similar, but less pronounced, revival over the past decade. Aliphatic polyethers and non-cross-linked polystyrene are among the most promising candidates in terms of stability. Dendritic and linear polyfunctional soluble polymers have by far the highest loading capacities and show great potential as supports for reagents and catalysts in combinatorial synthesis. Although many separation techniques for soluble polymeric supports have been developed for solution-phase organic synthesis (including some in parallel format), further progress is required to automate solution-phase separation techniques.

There are various methods for separating macromolecules by size: preparative size-exclusion chromatography (SEC), dialysis, membrane filtration and centrifugation.

Size-exclusion chromatography (SEC), also known as gel permeation chromatography (GPC), is used for the separation and fractionation of macromolecules both on an analytical and a preparative scale.[11] Separation occurs predominantly due to the difference in hydrodynamic volume of the macromolecules in solution. However, in some cases the polarity of the molecules can also influence retention times.

Dialysis is a well-established technique used for separating biopolymers in aqueous media, since traditional membrane materials are incompatible with organic solvents. However, with the development of improved membrane materials, dialysis can now be performed in practically any organic solvent. A schematic of a membrane system is presented in Figure 3.1. Typical molecular weight cut-offs (MWCO) are 1000, 5000, and 20 000 $g\,mol^{-1}$ (based on linear polymer standards in water). In organic solvents, however, MWCO can vary slightly because of the diverse swelling ability of the membrane materials in different solvents.

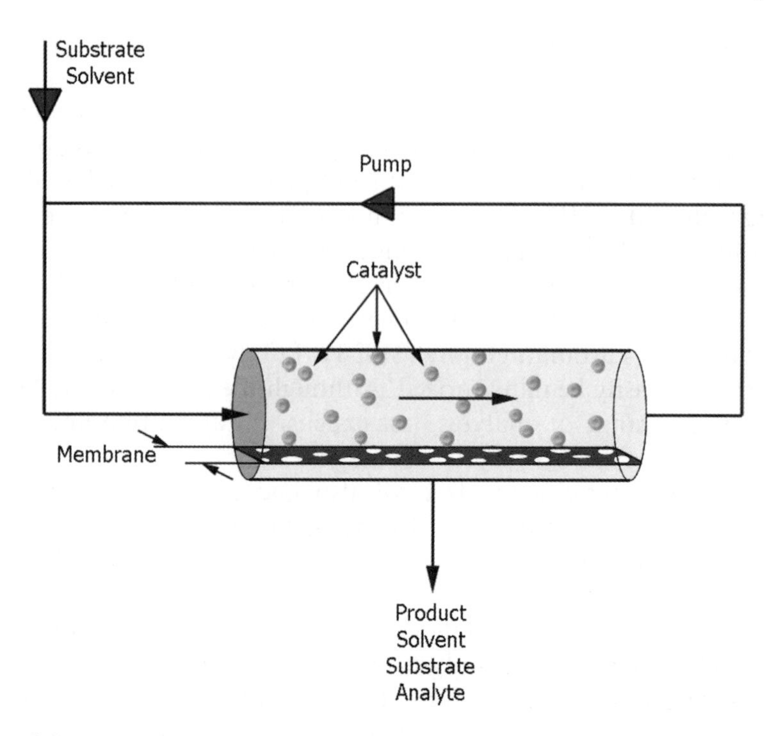

Figure 3.1 Membrane separation system.

Ultrafiltration (UF) is a very efficient membrane-separation technique for soluble macromolecules. Like dialysis, it can be employed to separate low-molecular-weight compounds from soluble polymeric compounds. Membrane materials (organic and inorganic) with high chemical stability and compatible with most organic solvents are now commercially available. In contrast with dialysis, much shorter separation times (*ca.* 1–3 h) can be achieved with UF because of the application of pressure (3–30 bar). A simple membrane system is presented in Figure 3.1. It is necessary to use stirred UF cells or continuous flow systems for efficient separations to avoid clogging of the membrane. For higher pressures and continuous flow membrane reactors, however, HPLC type systems must be used to avoid gas bubbles in the system.

Centrifugation can also generate the required "pressure" for ultrafiltration, and is frequently used in biochemistry for the purification of biopolymers (for example, proteins, DNA). A major drawback of this technique is the limited availability of membrane vials that are stable with a broad range of organic solvents. For analytical purposes, ultracentrifugation can also separate large from small molecules, similarly to

SEC. In both cases, separation times depend on the rotation speed of the centrifuge.

All of the previously mentioned processes can be automated. However, these methods are not considered solvent reducing techniques and the systems must therefore be equipped with solvent recycling systems.

Extraction efficiencies that reduce the volume of solvent can be achieved using new alternative solvents (fluorous solvents, ionic liquids, polymeric solvents, *etc.*) in conventional liquid–liquid extraction systems.

Extraction techniques for solid and semi-solid samples should be exhaustive to guarantee efficient recoveries in different types of samples, and because of the low levels at which analytes (usually micro contaminants) are generally present in the environment. This requirement for quantitative extraction explains the general preference for Soxhlet extraction rather than for more selective techniques, which are usually highly analyte- and/or matrix-dependent. The alternative techniques usually extract analytes more efficiently from the matrix by improving the contact of the target compound(s) with the extraction solvent. Most processes developed to decrease the consumption of solvent but maintain efficiency, use additional energy to accelerate the extraction process. Extraction time and organic solvent consumption are both reduced, and sample throughput is increased. Enhanced extraction efficiency can be achieved by using microwave energy (microwave-assisted extraction; MAE) or ultrasound energy (ultrasound assisted extraction; USE), or applying high pressure and temperature to solvents (*e.g.* pressurized liquid extraction; PLE). These methods can also operate in both static and dynamic mode; additional clean-up is usually required to remove the sample matrix from the analyte-containing solvent. These methods will be discussed in more detail under energy-saving approaches in Section 3.4.

3.2.2 Solvent Microextraction

Several attempts to miniaturize liquid–liquid extraction (LLE) have been made. Three of them will be discussed: single-drop microextraction (SDME), liquid-phase microextractions (LPME), and electro-membrane extraction (EME).

Single-Drop Microextraction (SDME). There is often no need to perform large-scale liquid–liquid extraction for analytical purposes, and

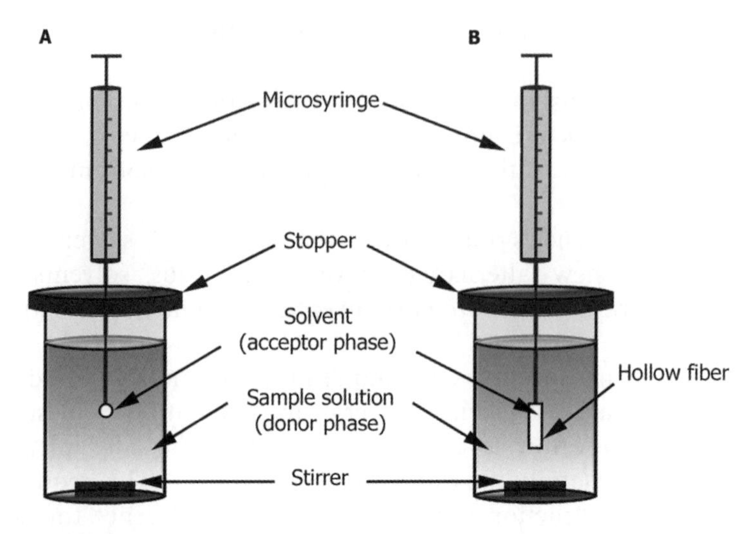

Figure 3.2 Principal schemas of (A) single-drop microextraction and (B) liquid-phase microextraction.

one drop of extracting solvent is enough to "catch" the target analyte and transport it to the analyzer. In this technique, a microdrop of solvent (*ca.* 1–3 μL) is suspended from the tip of a conventional microsyringe and is then either exposed to the headspace of the sample or immersed in a sample solution in which it is immiscible.[12,13] The analytes diffuse into this droplet. Once the extraction is complete, the drop is retracted into the syringe and injected into the chromatographic system for analysis (see Figure 3.2).

Headspace SDME is similar to traditional headspace sampling in that volatiles are sampled from the vapours above the sample, so that interference from the sample matrix is avoided. A variety of methods and specialized equipment are available for this purpose.

The main limitation of SDME – a straightforward technique – is the instability of the drop. Microliter-sized drops are relatively stable, but increasing the temperature to improve extraction efficiency can lead to drop instability and thereby lower reproducibility. The droplet may be lost from the needle tip of the syringe during the extraction, particularly when samples are stirred vigorously, or an emulsion forms, as in the extraction of biological samples.

The main advantage of SDME is the high enrichment factor. This is a consequence of the very small volume of the acceptor phase. The other advantages of the single-drop technique include: its suitability to a wide range of solvents; low solvent consumption and low cost. In addition, no pre-conditioning is required and memory effects are avoided.

Parameter considerations for SDME include: the type of solvent; size of drop; shape of needle tip; temperature of sampling; equilibration and sampling (extraction) time; effect of stirring; and ratio of headspace to sample volume. Of course, pH and salt addition and other parameters that affect the liquid–liquid extraction process must be considered as well.

The choice of solvent is critical. In head-space analysis, the solvent should not be too volatile, or the drop will evaporate during extraction. On the other hand, if it is too non-volatile, it will elute with the analytes. It should also be insoluble in the sample solvent and satisfactorily dissolve the analytes. Typically, *n*-octyl acetate, isoamyl alcohol, undecane, octane, nonane and ethylene glycol are used as solvents.

The poor reproducibility associated with the volatilization of these solvents during the extraction procedure has led to the use of ionic liquids (ILs) to form the microdrop. ILs have low vapour pressure and high viscosity, which allows larger and more reproducible extracting volumes in SDME. However, their non-volatility makes them incompatible with gas chromatography and special interfaces for direct coupling are required.

For the extraction procedure, the sample and a 1cm PTFE magnetic bar are placed in a 10 mL vial, which is subsequently tightly sealed with a silicone septum. The bevel-tip microsyringe is filled with 2 μL of the imidazolium salt, $[C_8C_1im][PF_6]$, and inserted into the vial through the septum. The syringe is clamped in order to place the needle tip in a constant position in the headspace of the vial. Then, the plunger is depressed to form the microdrop, which is exposed at 30 °C for 30 min. After the extraction, the IL is retracted and the syringe is moved to the injection unit where the analytes are swept into the analytical system.

The combination of ionic liquid-based headspace, single-drop microextraction and ion mobility spectrometry (IMS) allows halo-compounds to be determined in a rapid and easy way that takes advantage of their characteristic IMS spectra.[14] An injection unit has been designed to permit the efficient volatilization of the analytes at room temperature and to prevent IL from entering the system. The advantages of this method are the small amount of organic solvent required, the simple and inexpensive experimental and sampling equipment, and the integration of extraction, concentration and sample introduction into a single step. SDME methods are still in the initial stages of development, and are focused on the more volatile compounds, such as alcohols, chlorobenzenes, phenols, amines, organotins, trihalomethanes and BTEX (benzene, toluene, ethylbenzene and xylenes). More development is needed before the method can be routinely applicable to on-line pre-concentration.

Microscale liquid–liquid extraction is another mode of dispersive liquid–liquid microextraction (DLLME). In this method, an appropriate mixture of extraction solvent and disperser solvent is injected rapidly into an aqueous sample, resulting in a cloudy solution. This solution consists of small droplets of disperser solvent and extraction solvent in aqueous solution. The technique can be regarded as multiple-drop microextraction. In DLLME, the instantaneous mixing of the three components ensures equilibration within a few seconds, due to the infinitely large interface between the fine extractor droplets and the aqueous solution. The transfer time of analytes from the aqueous phase to the organic phase is therefore very short compared with typical equilibrium extraction times for solid-phase extraction. Centrifugation is required to separate the phases. The miscibility of the disperser solvent in the organic (extraction solvent) and aqueous phases is the main reason for the selection of the disperser solvent. Typical disperser solvents are acetone, acetonitrile and methanol.

Membrane-Assisted Extraction. One way of minimizing the problem of drop instability in SDME is to add a polymeric membrane to serve as a support for the extracting solvent. This not only enables the use of larger volumes, but also acts as a physical barrier between the phases. The membranes can be microporous, in which the pores are filled with an organic solvent, or nonporous. The extraction mechanism is different with nonporous membranes, as analytes must diffuse through the membrane before extraction to the acceptor phase. The microporous membrane is usually made of polypropylene or some other porous hydrophobic material. The membrane material must be highly compatible with a broad range of organic solvents, and must effectively immobilize the organic solvent in the pores. A further advantage of using a membrane is that because of the pore structure, the concentration of high-molecular-mass compounds in the sample extract is reduced. Membrane-assisted liquid extraction can be performed in a hollow fibre, a flat-sheet membrane module or a membrane bag.

Figure 3.3 illustrates different systems utilized in membrane-assisted liquid extraction. The extraction can be static or dynamic, depending on the membrane module. Two- and three-phase systems are possible using hollow membranes (fibres). The two-phase mode is typically utilized in microporous-membrane LLE (MMLLE) and the three-phase mode in supported-liquid-membrane LLE (SLM). The MMLLE version is mainly used in combination with GC, while SLM, in which the final extract is typically aqueous, is used with LC and electro-driven separations. In the two-phase mode, the analytes are diffused from the

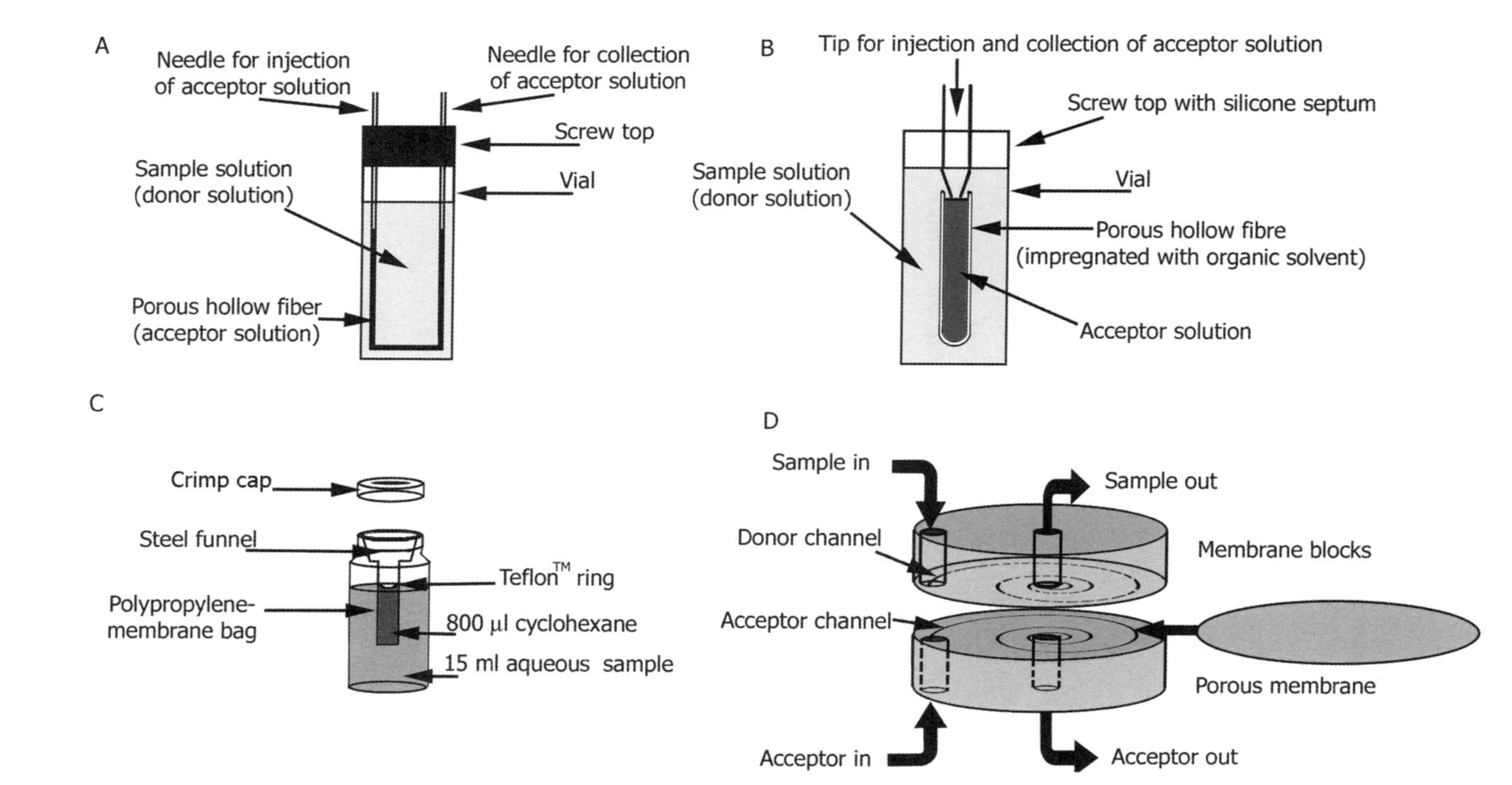

Figure 3.3 Different solutions for membrane-assisted liquid extraction include: (A) a U-shaped hollow fiber, (B) a rod-like hollow fiber, (C) a membrane bag and (D) dynamic MMLLE.

aqueous sample (donor) to the organic acceptor solution, which is also in the pores of the membrane. The extraction process depends on the partition coefficients of the analytes. In the three-phase mode, the solvent in the pores of the fibre differs from that inside the fibre. The three-phase mode is used for extracting polar analytes, and the two-phase mode is used for extracting non-polar and semi-polar analytes.

Systems based on hollow fibres operate similarly to solvent drops, but the solvent is immobilized in the pores of the hollow fibre (see Figure 3.3B). The hollow fibre can also function as a supported-liquid membrane, and an acceptor can be placed inside the lumen.[15] This system is called liquid-phase microextraction (LPME). The advantage of LPME over SDME is the presence of a fibre that supports organic solvent, thereby hindering the process of dissolution or evaporation (in SDME mode) of the solvent. High enrichment factors can also be achieved with this technique because of the relatively small volume of the acceptor phase. The acceptor volume, *i.e.* the volume of organic solvent, is between 5 to 25 μL, and the sample volumes range from *ca.* 0.5 to 4 mL. Extraction times of 15 to 45 min. have been reported.

In dynamic mode, the membrane unit is either a planar membrane or a hollow-fibre membrane. In the planar configuration, the membrane is clamped between two blocks and separates the donor and acceptor flow channels. In the hollow-fibre membrane module, the acceptor phase flows inside and the donor phase outside the membrane (see Figure 3.3A and D).

Membrane-assisted solvent extraction provides a means to combine extraction on-line with a chromatograph. The interface between the extraction unit and the GC is typically an on-column interface or a programmed temperature vaporiser (PTV) and connection is straightforward. The same approach works for HPLC. Interfacing membrane extraction and a chromatograph can be done directly or *via* a sample loop located in a multi-port valve.

The main disadvantage of all membrane-extraction techniques is that the extraction tends to be non-exhaustive. The recoveries are usually at a level similar to those in SPME, and calibration must be performed carefully. Quantitative recoveries are obtained for highly hydrophobic analytes when the flow rate of the donor is kept sufficiently low, or the sample is circulated across the extraction system. If quantitative recoveries are not required, the extraction time can be shortened by increasing the flow rate of the sample, or the sensitivity can be enhanced by using a large amount of sample. Selective extraction can be achieved by a careful choice of the material and of the pore size of the membrane, as well as of the extraction solvent. Pore size is important because, in addition to the extraction process, exclusion by size takes place in the extraction increasing selectivity. Membrane-extraction systems based on hollow fibres and membrane bags can be used for on-site sampling.

Electrically Enhanced Extraction. It is well known from chemical engineering literature that LLE processes can be manipulated and enhanced by the application of electrical fields across the liquid–liquid interface.[16]

There are three different approaches to liquid–liquid extraction (LLE) driven by an electrical field:

- a simple organic–aqueous, liquid–liquid system;
- an aqueous–organic, liquid–gel system;
- an aqueous–organic–aqueous, liquid–liquid membrane–liquid system.

A step towards more rapid automated sample preparation was recently taken with the application of low voltage (50–300 V) over the microextraction cell (hollow fibre).[17] This method is called electro membrane extraction (EME).

The electrical field aids the movement of charged compounds, meaning that the target analytes are charged in the extraction system. In extraction systems, differently charged analytes have different mobilities that can be controlled by electrical potentials – the applied potential controls the selectivity of the extraction. In the case of membranes, it is important that the solvent remains immobilized in the pores of the hollow fibre during extraction, so that it serves as a stable supported-liquid membrane (SLM).

A simple system for analytical extraction by an electrical potential across a SLM is presented in Figure 3.4. In this system target analytes were extracted from an aqueous sample, through an organic solvent immobilized in the pores of a porous hollow fibre, into an aqueous acceptor solution placed inside the lumen of the hollow fibre by applying an electrical potential across the membrane. In order to ensure efficient electrokinetic mobility in the EME system, the pH was adjusted to provide total ionization of the analytes in the two aqueous solutions.[18]

The EME concept appears to be an attractive alternative to more conventional LPME, and supports earlier findings from larger-scale LLE that liquid–liquid distributions can be changed and extractions can be enhanced by applying an electrical field.

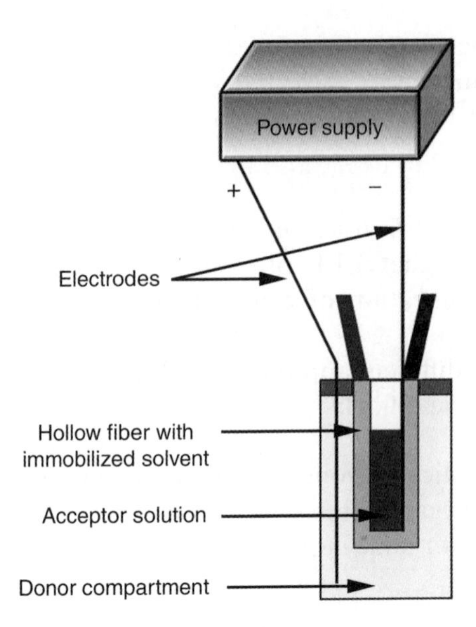

Figure 3.4 Electro-membrane extraction.[18] (Reproduced with kind permission of Elsevier).

First, the electrical potential is the driving force for the extraction, and electrical potentials are easily varied and controlled from the power supply. The extraction can be adjusted by a simple change of applied voltage, and the voltage can be used to optimize the extraction rate.

Second, electrical potentials are easily reversed, opening up the possibility of reversing the extraction system for special back-extraction applications. Furthermore, the electrical potential also determines the selectivity of the extraction system, which can be controlled simply by varying the applied voltage.

Third, the use of electrical potential can accelerate extractions.

The EME system has already been proven feasible for complex samples and has been optimized for different type of analytes (bases, acids, zwitterionic compounds, non-polar compounds, and polar compounds). It is faster than LPME on similar equipment, but more research needs to be conducted to check and validate quantitative procedures with respect to reliability.[19]

3.2.3 Sorption Microextraction

Solid-phase extraction (SPE) has largely replaced liquid–liquid extraction (LLE) in many laboratories, mainly to reduce solvent consumption. Typically, SPE is carried out either in disposable syringes of various sizes or in a 96-well plate format. Both are automatable, but they only accommodate flow in a single direction and are relatively slow. Conventional SPE is scalable, but 96-well plate extraction requires 96 samples, or the unused wells are wasted. To say that SPE is a completely solventless technique is not true, because it usually requires some solvent to desorb the analyte from the sorbent for further detection and identification, generally by some chromatographic method. Sometimes, however, thermal energy is used to desorb the analyte, and in this case, SPE is truly a solventless technique.

Solid-phase extraction involves liquid–solid partitioning, and the analytes are bound to active sites on the surface of a solid sorbent. A wide range of sorbents are available, including C8 and C18 bonded phases on silica, polymeric resins (polystyrene–divinyl benzene copolymer), polar sorbents such as alumina and silica, and ion-exchange sorbents. Mixed-mode sorbents, utilizing both primary and secondary mechanisms for selective retention of analytes, and some highly specific selective sorbents are also available. These phases enable interactions based on adsorption, H-bonding, polarity and non-polarity, cation and anion exchange, or size exclusion to be utilized in extraction.

Solid-phase extraction as a replacement for conventional extraction methods is also changing, and is used in its microextraction form for analytical purposes. Analytical instruments are sufficiently sensitive and require such small amounts of sample that there is no need to process large amounts of starting material, and the size of the probe can be reduced.

The introduction of solid-phase microextraction (SPME) was a significant step towards miniaturization.[20–23] Its main advantages are minimal consumption of harmful solvents and, typically, a high enrichment factor. The improved sensitivity makes it possible to minimize the amount of sample required for the analysis. All these techniques can be readily combined with chromatography – off-line, at-line or sometimes even on-line.

Solid-phase microextraction (SPME) is a technique in which the amount of extraction sorbent is very small compared with the sample volume. It eliminates the use of organic solvents, is significantly quicker and simpler than both LLE and SPE, and also integrates extraction, preconcentration and sample introduction into a single step. Solvent-free SPME is fast, economical and versatile; it is mainly suitable for liquid matrices, and is based on the partition of target analytes between a polymeric stationary phase coated on a support, and the sample extract. SPME in its classical form is a fibre coated with a liquid (polymer), a solid (sorbent) or a combination of both. The fibre coating takes up the compounds from the sample by absorption in the case of liquid coatings or adsorption in the case of solid coatings. The SPME fibre is then transferred *via* a syringe-like device into the analytical instrument for desorption and analysis of the target analytes.

An exhaustive removal of the analyte to the extracting phase does not occur, but equilibrium is reached between the sample matrix and extracting phase. The design of the SPME extraction phase must take into consideration the stability, polarity and thickness of the coating. The thickness of the coating determines the volume and surface area of the stationary phase and, consequently, the amount and rate of absorption. The partitioning of the analyte between sample matrix and extraction phase controls the sensitivity. For practical reasons, the extracting phase, which is a polymeric organic phase, can be cross-linked and permanently attached to an optical fibre made of fused silica. Therefore, the solid-phase technology for SPME fibres presents a more complex problem than it does for conventional extraction techniques.

Solid-phase microextraction has several important advantages compared with conventional sample preparation techniques:

- it provides a rapid, simple, solvent-free and sensitive method for the extraction of analytes;
- it is a simple and effective adsorption/desorption technique;
- it is compatible with analyte separation and detection by HPLC-UV;
- it provides linear results for a wide concentration of analytes;
- it is compact, which is convenient for designing portable devices for field sampling;
- it yields highly consistent, quantifiable results from very low analyte concentrations.

Solid-phase microextraction is easy to use, and has been applied to a range of applications, including environmental, industrial hygiene, process monitoring, clinical, forensic, and drug and food analysis, especially for the extraction of volatile and semi-volatile organic compounds. It has been used routinely in combination with GC and GC-MS, in which the extracted analytes are thermally desorbed in the GC injection port. It has also been directly coupled with HPLC and LC-MS to analyze weakly volatile or thermally labile compounds not amenable to GC or GC-MS. Off-line and on-line SPME coupled to CE have also been developed in recent years. The popularity of method is supported by the commercial availability of SPME fibres with a wide selection of sorbents.[24]

Although the use of SPME fibres is gaining in popularity, it has important drawbacks such as:

- relatively low recommended operating temperature (generally 240–280 °C);
- instability and swelling of the fibres in organic solvents (which greatly limits their use with HPLC);
- stripping of coatings;
- bending of the needle and breakage of the fibre;
- high cost.

The lack of effective chemical bonding of the stationary-phase coating with the fibre surface, and the relative thickness of conventional fibres are responsible for some of the drawbacks, such as the low operating temperature, solvent instability and stripping of coatings. One solution could be to use a hollow-fibre membrane to protect the SPME fibre.

Among the different approaches to the stationary phase development of SPME fibres, the sol–gel approach represents a promising research direction in the preparation of surface coatings for SPME fibres. The sol–gel process provides a versatile method to fabricate size-, shape- and charge-selective materials of high purity and homogeneity for coating SPME supports. Sol–gel-derived organic and organic–inorganic hybrid materials can also be cast on porous silica supports. Ionic and molecular recognition in these materials can be achieved by controlling the pore size and morphology of the silicate host structure, by introducing specific functionalities, such as crown ethers, calix[4]arene, β-cyclodextrin derivatives and polyethylene glycol into the dense framework, or by utilizing molecular imprinting or templating strategies. Sol–gel technology also allows the preparation of support coatings that are stable in strong organic solvents (xylene and methylene chloride), as well as in acidic and basic solutions (pH 0.3 and 13). The correct selection of SPME materials facilitates the clean-up and pre-concentration of strong polar and thermally stable compounds without derivatization.

Common SPME sorbents are polymeric materials that have a gum-like or even liquid-like state, with properties similar to those of organic solvents. The analytes do not usually undergo real (temporary) bonding with the material in SPME, but are retained by dissolution, whereas in SPE, the analyte is often bonded to the sorbent. Retention is based on the distribution equilibria between the sample matrix and a non-miscible liquid or gas phase. In SPME, extraction is rarely quantitative, and thus careful calibration is needed. Several calibration methods are employed in SPME, including classical calibration that relies on equilibrium extraction, or the more novel kinetic calibration. The latter is especially important for the calibration of on-site, *in situ* or *in vivo* analysis, because control of the agitation of the matrix is sometimes difficult, and direct spiking of standards into the matrix is usually not possible in these cases.

The sample capacity of SPME is low because of the small amount of sorbent on the fibre. Novel developments in SPME include miniaturized cold-fibre SPME with CO_2 or Peltier element cooling. In the cold-fibre technique, the sample is heated to a high temperature while the fibre coating is kept at a relatively low temperature. Mass transfer is thereby accelerated and a temperature gap is created between the cold-fibre coating and the hot headspace, increasing the distribution coefficients significantly.

The extraction efficiency of SPME for highly polar analytes is usually poor, therefore direct derivatization of analytes in the sample matrix,[25a] or derivatization in the injection port or on the coating,[25b] are used to improve the extraction. In the first approach, the derivatization reagents are added directly to the aqueous sample. This improves both the affinity of the derivatized analyte for the sorbent and the quality of the subsequent GC separation. In the second approach – in-port derivatization – polar analytes with acid–base properties are extracted into the SPME sorbent as ion pairs, which at the high temperatures of the GC injection port are further decomposed to produce volatile by-products and alkyl derivatives of the target compounds. On-coating derivatization is performed either by preloading the sorbent with the derivatization agent so that the reaction occurs as soon as analytes are incorporated into the sorbent material or, alternatively, by first concentrating the analytes in the sorbent and then exposing the sorbent to the vapour of the derivatization reagent. The most suitable approach depends on the properties of the analytes and the derivatization reaction to be performed.

Most of the different modes and formulations of SPME are shown in Figure 3.5.

In-tube SPME. In-tube SPME uses an internally coated capillary, through which the sample flows, or is drawn repeatedly, instead of a coated fibre. This configuration of SPME is more conducive to automation and on-line use. The tube can be attached to a multiport injection valve. An automated in-tube sorptive extraction device, known as a solid-phase dynamic extraction (SPDE) device, is commercially available.

Covering the sample vessel walls with an appropriate sorbent makes it possible to increase the amount of extracting sorbent and better collect

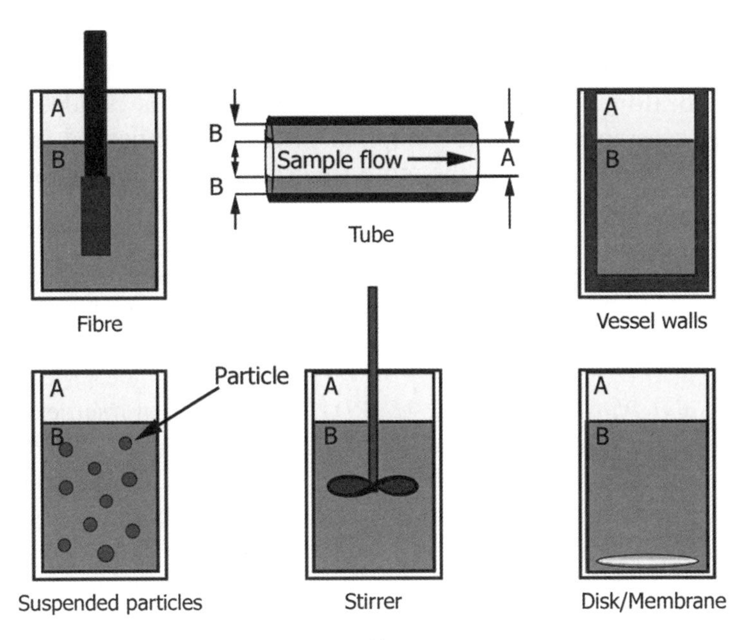

Figure 3.5 Different forms of SPME.[21] (Reproduced with kind permission of Elsevier).

the trace analytes. Special equipment is required to transport the sorbed analytes to the analytical instrument.

Stir-bar Sorptive Extraction (SBSE). Stir-bar Sorptive Extraction is another form of sorptive extraction based on the same principles as SPME. It was developed to support the use of a sorbent stir-bar. The main difference between the two techniques is the design of the extraction system. The sorbent materials are similar and the sorbent is coated on a magnetic stirring bar, which results in a considerably larger amount of sorbent material than on fibre. The analytes are desorbed either thermally (for GC analysis) or with liquid (mainly for LC analysis). In contrast with SPME, a special interface is required for thermal desorption. Because of the large amount of coating, the desorption process is relatively slow. Typical desorption times are *ca.* 10 min, which means that the desorbed analytes must be re-concentrated before GC separation. Contrary to SPME, quantitative recoveries are often achievable with SBSE due to the greater sample capacity. For highly polar compounds, approaches similar to SPME can be applied (*i.e.* derivatization).

Solid-phase microextraction and SBSE devices are easily stored and transported, and they are well suited to on-site sampling. SPME has also been used for *in vivo* sampling. Field sampling is effective because only the fibre or stir-bar with the absorbed analytes needs to be brought back to the laboratory. Transporting large sample volumes is avoided, and no sampling accessories such as pumps or filters (as are needed in on-site SPE) are required. SBSE is the more rugged of the two systems for on-site sampling because SPME fibres are quite fragile.

Using sorbent-coated membranes or disks – they are not actively stirred – is very similar to the stir-bar approach. When membranes are used, they are impregnated with sorbent, which increases the amount of sorbent compared to that of disks.

Matrix Solid-Phase Dispersion (MSPD). This form of microextraction can be used for both liquids and solids. For liquid samples, the sorbent is coated on small polymer or silica particles, such as GC packed-column materials, and dispersed in liquid sample to adsorb the analyte. The sorbent is then separated from the liquid and transferred to a column, and the analyte is desorbed thermally for further analysis, similarly to SPME.

In the case of solid or semi-solid samples, a suitable solid-phase material (*e.g.* derivatized silica, silica gel, sand or Florisil) is blended manually in a mortar. The solid support and sample are then transferred to a column and eluted with an appropriate solvent. The most important factors affecting this kind of dispersion extraction are the characteristics of the solid-phase material (*e.g.* particle size); the ratio of sample to support material; the use of chemical modifiers (*e.g.* acids, bases or chelating agents); the elution solvents; and the elution volume.

Matrix solid-phase dispersion differs from the other techniques in several respects. In MSPD, sample disruption and dispersal onto the particles of the support material takes place in a single step. MSPD is usually performed manually. Elution is also usually accomplished manually, although instrumental techniques (*e.g.* pressurized-liquid extraction (PLE)) can be used for elution as well. The MSPD technique is particularly well suited to biological and plant-derived samples. Its main advantages are the ability to use very small amounts of sample and its low solvent consumption.

Packed Syringe or Pipette Tip Extraction. On-line extraction can be performed with minicolumns that are placed in a syringe or pipette tip packed with suitable solid-phase material. The main difference from conventional SPE cartridges is that an amount of sorbent (*ca.* 1 mg) is inserted into a syringe (100–250 μL) or pipette tip as a plug, which is secured by frits at either end. The sample is withdrawn through the solid-phase plug, and the analytes adsorb onto the SPE material. As the amount of sorbent is small, the quantity of elution solvent required is substantially reduced.

The function of sorbents in pipette tips is largely the same as that of SPE sorbents. The selection of SPE material is the most critical parameter of the extraction, because it must be compatible with elution of the analyte with a suitable organic solvent. The procedure can be performed automatically by an autosampler and even connected on-line to a GC injector if large-volume injection techniques are used. In the case of off-line procedures, more than one washing step can be carried out, and drying can be done by placing the pipette tip into a vacuum or by using a drying agent to remove water from the final extract.

The elution solvent, type of injector, and injection conditions are critical if a packed syringe is coupled on-line. In an automated system, in which drying is usually not possible, the eluent should be miscible with water. It should also be sufficiently volatile. Methanol is used in most applications, although it cannot be considered a very good choice, especially for large-volume injection to GC; ethyl acetate is better for some applications. The injection rate, injection temperature and flow rate are particularly critical for volatile analytes.

Pipette tip extraction is simple in concept and well accepted in bioanalysis, in which the number of probes is usually very large and high throughput is needed. For extraction, a standard pipette tip, either 1 mL or 5 mL, is "loaded" with SPE sorbent. Only a small bed volume is required, which settles into the end of the tip, allowing the amount of solvent per extraction to be reduced. Because a new pipette tip can be used for each extraction, there are no memory effects. The sorbent can be packed between frits or left free to disperse upon the introduction of solvent. The flow into and out of the tip can therefore be bidirectional. This free system with bidirectional flow makes it possible for solvent to be introduced and removed from the tip more rapidly, speeding up the extraction process. The tips can easily be accommodated on a hand-held syringe for small batches of samples, or semi-automated for 20 at a time using a lever-actuated platform. This has the advantage of delivering a uniform, controlled positive flow. Full on-line automation is also available for high-throughput laboratories.

A wide range of SPE materials is available; non-polar as well as polar analytes can be extracted efficiently. These SPE techniques tend to be less selective, and matrix components are often extracted as well. However, relatively low selectivity is an advantage for a profiling type of analysis, *i.e.* when all or most of the sample components are of interest. Micro-SPE-GC can be performed in a fully automated and closed system, in which the whole extract is transferred for GC analysis. This improves the sensitivity and reliability of the analysis, as problems with sample loss and contamination are minimized. Although using the instrumentation is fairly straightforward, the system is complex, particularly in method development. However, because micro-SPE extraction is typically quantitative, calibration is less demanding. This is in contrast to the other techniques in which various calibration methods are employed, including classical calibration relying on equilibrium extraction or the more novel kinetic calibration.

The use and disposal of larger and smaller cartridges and pipette tips must be carefully considered. They are solid wastes and the rules of handling must be applied. This is a factor that diminishes the greenness of these kinds of method.

Passive In Situ Samplers in Sample Pretreatment. Passive samplers are widely used for environmental monitoring. Nearly all passive-sampler techniques share common design characteristics, the most significant of which is the presence of a barrier between the sample and the receiving phase. Samplers can be characterized as diffusion-based or permeation-based. Permeation-based samplers utilize a membrane as the barrier; therefore the sampling mechanism closely resembles that of membrane extraction. SPME can also be used as a passive sampler.

Passive *in situ* samplers combine sampling and (preliminary) sample pre-treatment in a single step, although, in many cases, further sample pre-treatment (*e.g.* with LLE) is required before introducing the sample into an analytical instrument. It should be noted that these samplers collect only the completely dissolved fraction of chemical compounds, because the barrier does not permit the diffusion of colloids and particles into the receiving phase.

Sample Clean-Up. In most cases, particularly with solid samples, extracts require clean-up or fractionation, or both, before final analysis. The complexity of sediments from lakes or seas and biota matrices often requires the use of multi-step purification methods. Many co-extractable and potentially interfering compounds in the raw extract must be removed to ensure a maximum sensitivity of the analysis. The final determination may be affected by:

- lipids, which can cause deterioration of the column and contamination of the injector and detector;
- co-elution with other contaminants;
- other compounds (*e.g.* sulfur and oil) which cause interfering peaks or an erratic response to target analytes.

Therefore, the removal of co-extracted matrix components is critical. For this reason, various clean-up procedures have been developed to minimize their negative effects. When interfering compounds are present, adequate separation schemes or fractionation processes to allow for the isolation of sub-groups of compounds (fractionation of the extract into different classes of compound) are necessary. Although novel methods have been developed for extraction, clean-up still relies heavily on column chromatographic fractionation using classical adsorbents, including silica, alumina, Florisil and carbon. This conventional technique is used in "off-line" mode and involves passing extracts through several adsorbent columns prepared in the laboratory, or through solid-phase extraction cartridges. Alumina, silica gel and Florisil columns of different sizes, mesh sizes, and levels of activity (either separately or in combination) are widely used. An alkaline (saponification) or sulfuric acid treatment is sometimes necessary prior to, or in conjunction with, adsorption columns to remove the bulk of co-extracted lipids. Official EPA Methods 3630C, 3610B and 3620B (using silica gel, alumina and Florisil clean-up, respectively) have been approved for purifying organic extracts from solid environmental samples.

Methods using HPLC and gel permeation chromatography (GPC) have been reported. These chromatographic techniques separate different classes of compound more effectively, and the separation and its optimization can be monitored with a detector. Both HPLC and GPC are easily automated, because the same column can be used for hundreds of samples. An EPA method (Method 3640A GPC Cleanup) has been approved for purifying organic extracts from solid environmental samples.

The various types of solid-phase extraction can successfully replace liquid extraction and substantially reduce the use of organic solvents. Most of the micro methods are completely solvent-free and can provide greener analytical solutions for a wide range of compounds, not only in laboratory studies, but also in the field.

3.2.4 Thermal Desorption

There are two basic methods of releasing analytes from SPE traps: thermal desorption and liquid extraction. The first approach is a solventless technique, where thermal energy is used for liberating the retained analytes.

The following are different SPE options:[26]

(1) Thermal desorption of SPE traps:
- microextraction traps (SPME),
- needle and stir-bar sorptive elements (SBSE).

(2) Thermal desorption from solids:
- direct thermal desorption of the sample (DTD),
- matrix of solid-phase dispersion (MSPD),
- pyrolysis (PY) of solid samples.

The choice of correct desorption parameters is as important as the selection of extraction parameters. Desorption must occur as quickly as possible; therefore, the reactor (usually a GC injector) temperature should be sufficiently high that the least volatile sample components are efficiently desorbed. The stability of the temperature of the fibre or sorbent must also be considered.

The adsorbent used for adsorptive enrichment in combination with thermal desorption should meet the following criteria to guarantee an accurate determination of volatile compounds:[27]

- complete adsorption of the analytes of interest. The specific surface area and the porous structure give a rough indication of the adsorptive strength of a material. The specific breakthrough volume of model compounds gives a more accurate determination of the adsorptive strength;
- complete and fast desorption of the analytes;
- homogenous and inert surface to prevent artefact formation, irreversible adsorption, and catalytic effects during sampling, storage of the loaded adsorbent tubes, and desorption;
- low affinity with water to avoid displacement and hydrolysis reactions and to minimize disturbances of the gas chromatographic analysis. The hydrophobicity of an adsorbent can be determined by the specific retention volume of water;
- high inertness against reactive species, such as nitrogen oxide, sulfur dioxide or ozone;
- high mechanical and thermal stability;
- multiple usability.

Adsorptive enrichment in combination with thermal desorption and capillary gas chromatography is a commonly accepted technique in volatile organic compound analysis. However, the adsorbent has to be chosen carefully according to the compounds to be sampled. Despite the great variety of commercially available adsorbents, a universal adsorbent unfortunately does not exist. As mentioned above, it is advisable to use more than one adsorbent if the analytes have a broad range of volatility. In that case, the materials are arranged in order of increasing adsorbent strength. They could be placed in one adsorbent tube or in separate tubes that are combined for sampling.

In this context, it is necessary to ascertain the adsorption and desorption behaviour of the analytes. Whereas the analysis of non-polar compounds, such as hydrocarbons, is well established, the analysis of more polar compounds by adsorptive enrichment and thermal desorption is still a challenge.

Micronization, such as SPME, has great advantages in this regard because of the higher flexibility of treating materials and small masses thermally. Furthermore, absorption (using polydimethylsiloxane, for example, in which enrichment is based on an absorptive process in the material) is more advantageous than adsorption and can also be used for more polar compounds. The use of SPME fibres does not require special equipment, and an ordinary injector can be used for sampling into a gas chromatograph. Other forms of SPME, however, need some modification of the injector or a special system for injection. This is also important with large-volume injections.

Several large-volume sample introduction techniques are available:

- loop-type injection;
- vapour-overflow technique;
- large-volume on-column injection;
- large-volume, programmed-temperature vaporiser (PTV) injection.

However, the careful selection of an appropriate technique is necessary, depending on the nature of the analytes (volatility, polarity, thermolability) and the sample matrix.[28]

In the majority of cases, PTV-based large-volume injections are carried out in solvent split mode; non-volatile components remain in the insert and solvent does not enter the column. The analytical column and detector will not be attacked by the large volume of solvent vapour or by aggressive volatile components, since they leave the system *via* a split outlet. The separation of solvent and analytes occurs in the PTV insert, assisted by adsorbents. This mechanism is not suitable for analytes of higher volatility. In such cases, better results are obtained by using stationary phase focusing or solvent-trapping with a retaining pre-column, (*e.g.* an on-column interface). PTV injection produces intermediate results compared with other injection techniques for testing thermolabile compounds.

The PTV consists of the same elements as a classical split/splitless inlet (see Figure 3.6), but is equipped with an efficient heating and cooling system, in which air, liquid nitrogen, CO_2 or electrical systems are used

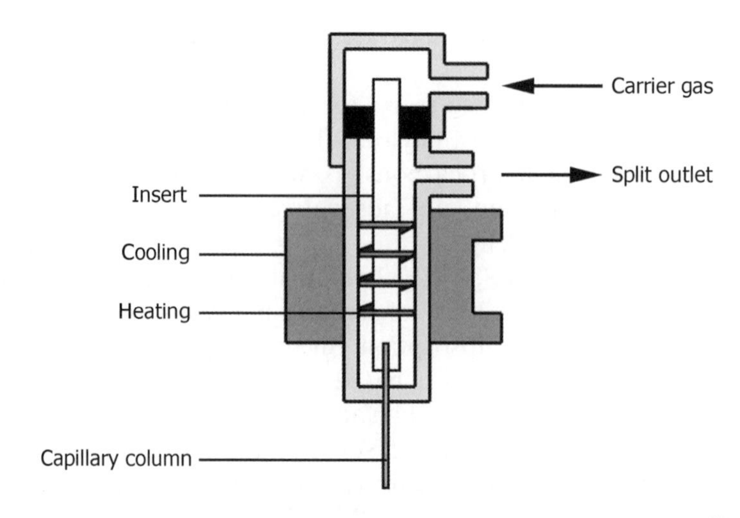

Figure 3.6 Schematic diagram of the programmed-temperature vaporizer.[28] (Reproduced with kind permission of Elsevier).

for cooling. During injection of the sample, the temperature control in the PTV is time-programmed; sample components with different volatilities are evaporated in succession. Whereas the thermal mass needs to be large in conventional split/splitless inlets in order to prevent a decrease in temperature during the evaporation process, it should be as small as possible with PTV in order to enable changes in heating capacity to be quickly transmitted to the sample.

Large-volume PTV injection improves the detectability of the analyte by increasing the injection volumes. The application of microextraction techniques or combination with derivatization procedures leads to simplified sample preparation. Water samples and aqueous eluents can be injected directly. Furthermore, PTVs can be used as an interface for on-line coupling with sample preparation methods or other separation techniques. Occasionally, the injectors serve as a cryotrap or thermoreactor.

The sample preparation technique called direct thermal desorption (DTD) is similar to the thermal desorption of solid-phase extraction sorbent for introducing analytes into a gas chromatograph. It has already been applied to a variety of difficult matrices and can be fully automated, although at considerable cost. Its main limitations are the determination of thermolabile and very high-boiling compounds. DTD is a valuable alternative to headspace techniques for the isolation of volatile compounds from non-volatile solid, semi-solid and, occasionally, liquid matrices, and a wide variety of applications have been reported.

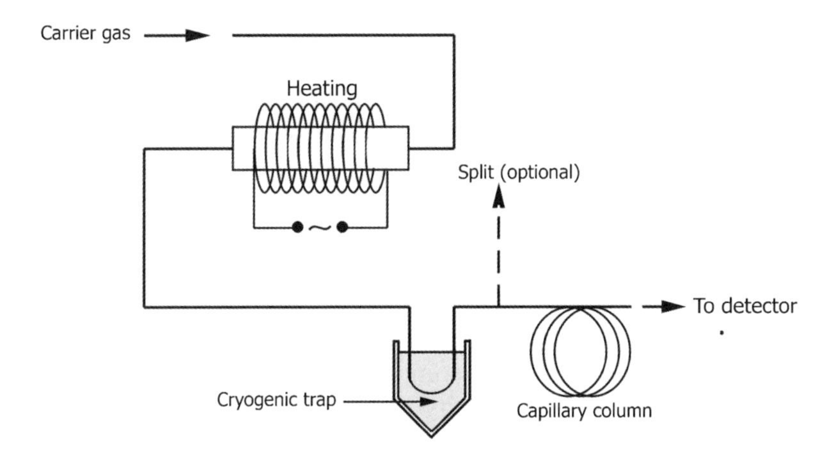

Figure 3.7 Desorption from a trap and introduction into a gas chromatograph.

Direct thermal desorption is usually connected to large-volume injection systems like on-column trapping or cold trapping. A typical simple system is presented in Figure 3.7, in which a 1–40 mg sample is placed in a desorption cartridge between two glass-wool plugs. By heating the cartridge for a pre-set time, the volatiles are desorbed and then adsorbed on a cold trap. Heating of the trap affects a rapid transfer of the analytes to the GC for further analysis. DTD-GC-MS can also be used as a screening method, e.g., for chlorinated hydrocarbon contamination in soil, without any special treatment of the soil. In some cases, a dual-tube system can be used to concentrate the analytes on a combined trap, prior to their release and transfer to the GC system. The total analysis including sample preparation requires less than 1 hour. DTD can produce a similarly fast characterization of (oil-containing) rocks in which the low-molecular-mass components provide a fingerprint of the rock. Solvent residues in polymers can be directly analyzed by thermal desorption-GC.

As indicated above, the basic instrumentation needed for (D)TD studies is rather simple. However, because of the (semi-) solid nature of most samples, automating sample introduction is difficult. Eliminating the clean-up step accelerates sample processing and significantly reduces solvent usage.

When using thermal desorption from traps, special care must be taken to avoid the thermal degradation of the sample matrix or sorbent, as well as the analyte. There are special techniques for using controlled thermal degradation of the sample and analyzing volatile products. Pyrolysis (PY) is an analytical technique that has been used extensively in the last 20–30 years, in which large molecules are degraded into smaller volatiles species using only thermal energy. The ultimate objective of analytical pyrolysis is to use chromatographic information on pyrolysis products to determine the composition or structure of the original sample. Pyrolysis techniques are used to study very-high-molecular-mass structures that are not amenable directly to GC. The complexity of polymeric materials can vary greatly and they can be very challenging to analyze. Pyrolysis combined with analytical methods, such as gas chromatography and/or mass spectrometry (PY-GC/MS), can offer a quick, convenient and powerful tool for characterising polymers or other non-volatile, complex heterogeneous samples.[78] As a consequence of the high sensitivity of pyrolysis, the amount of sample needed for analysis is very small (micrograms).

Stable and well-controlled heating of the sample is required to obtain reproducible sample degradation, and major instrumental developments in laser pyrolysis systems are being made in this regard. Laser energy used as a fragmentation source has facilitated controlled pyrolysis of specific regions of a sample, and has provided useful data on the molecular compositional units of macromolecules *in situ*.

A major disadvantage of all conventional pyrolyzers is that because they are external to the GC system, there is a significant possibility for the deposition of higher-boiling-point pyrolyzates and condensation of reaction products in the transfer line. This often results in sample loss and discrimination of high molecular-weight components.

Pyrolysis is a simple and undemanding resource method for macromolecular compounds. Current developments in PY-GC technology are directed toward constructing pyrolysis reactors with fast and reproducible heating profiles; non-discriminating connections to chromatographs; and the use of comprehensive two-dimensional GC (GC×GC) for the enhanced separation and detection of pyrolyzates. These are combined with extensive use of chemometrics for data processing.

3.3 ALTERNATIVE SOLVENTS

There are several factors motivating research on the use of solvents in chemical processes, including chemical analysis:

- more regulation of common industrial solvents has increased the attractiveness of alternative, non-toxic, environmentally acceptable industrial solvents;
- increasingly stringent pollution-control legislation has caused the industry to look for alternative means of waste treatment;

- rising energy costs have made energy-intensive separation techniques, such as distillation, more expensive;
- increased performance demands are being made on materials, which conventional processing techniques cannot meet.

This illustrates the paramount importance of the selection of an appropriate solvent or solvent mixture for chemical and physical processes. Solvent mixtures are often more appropriate media than pure solvents. Solvent mixtures, compared with their pure counterparts, can have improved physical properties, such as solvation power, density, viscosity, vapour pressure, relative permittivity, refractive index, and freezing or boiling point. However, a particular property of solvent mixtures must always be taken into account: the composition of the solute solvation shell can differ from that of the bulk solvent mixture. Preferential solvation leading to molecular-microscopic, solute-induced local inhomogeneities in multi-component solvent systems can occur in solute molecules or ions.[29]

Solvent effects are usually explained in terms of so-called solvent polarity. Simple electrostatic solvation models often fail because solute–solvent interactions take place on a molecular-microscopic level with the individual solvent molecules surrounding the ions or molecules of the solute in a noncontinuous, usually highly-structured, medium.

For this reason, solvent polarity is simply defined as the "overall solvation capability for:

- adducts and products which influence chemical equilibrium;
- reactants and activated complexes (transition states) which determine reaction rates;
- ions or molecules in their ground and first excited state, which are responsible for light absorptions in the various wavelength regions.

This overall solvation capability depends on the action of all, nonspecific and specific, intermolecular solute–solvent interactions, excluding such interactions leading to definite chemical alterations of the ions or molecules of the solute".[30]

The number of solvents available to chemists working in academia and industry is between 250 and 300 (plus an infinite number of solvent mixtures), and this number is increasing. The replacement of present technologies by new ones requires a deep knowledge of the molecular basis of all the processes and products, as the definition of polarity indicates. Hence, the characterization of a new alternative solvent requires understanding its liquid phase behaviour from a molecular viewpoint, *i.e.* understanding how intermolecular forces determine its properties.

As a consequence of the issues mentioned above, several large companies have directed considerable effort toward the development of solvent selection tools. The best-known system is by Pfizer Global Research and Development.[31] According to this tool, solvents are thoroughly and systematically assessed in three general areas:

- Worker safety – including carcinogencity, mutagenicity, reprotoxicity, skin absorption/sensitisation, and toxicity.

The highest priority for solvent evaluation was given to worker safety issues, as the handling of large quantities of solvents in manufacturing facilities carries the highest potential health and safety risk to the workforce. Within this group of solvents, the Carcinogenicity, Mutagenicity, Reprotoxicity (CMR) classifications are rated as the highest concerns, due to the potential for chronic effects on human health. Sensitisation and toxicity were also accorded high priority in the evaluation. Skin absorption properties increase the likelihood for sensitisation due to the potential carrier effects of these solvents. Toxicity (mainly assessed through published LD50 figures) has a potential acute and direct impact on the health and safety of the workforce.

- Process safety – including inflammability, potential for high emissions through high vapour pressure, static charge, potential for peroxide formation, and odour issues.

Environmental and regulatory considerations were considered next. Because regulatory considerations vary globally, this work incorporated both major EU and US classifications, such as the EU risk phrases and the US hazardous air pollutant and toxic chemical lists. Solvents with ecotoxic properties, such as those designated by the EU R50, R51 and R53 risk phrases, are difficult to treat in wastewater facilities or very expensive to dispose of. There is increased public attention to the potential environmental impact of industrial operations. This is supported by publicly available polluter registers in some countries, such as the Toxic Release Inventory in the United States. Some solvents with ozone-depleting and photoreactive potential are getting more public and government attention, as they are regularly discussed at professional forums and their use regulated by various countries. Solvents classified as very toxic, and/or classified with CMR properties, or as environmentally problematic materials, (*e.g.* with the potential for persistence and bioaccumulation), are increasingly subject to regulatory attention.

This may include restrictions or prohibitions and/or increased requirements to control and report use. Certain regulated compounds such as US HAPs, and chemicals subject to EU Integrated Pollution Prevention and Control (IPPC), can trigger expensive and technically challenging control requirements. In summary, the use and handling of such substances is monitored very closely by environmental protection agencies worldwide.

- Environmental and regulatory considerations – including ecotoxicity and ground water contamination, potential EHS regulatory restrictions, ozone depletion and photoreactive potential.

Based on this type of analysis, a table has been proposed (see Table 3.2) which divides solvents into three categories – preferred, usable and undesirable.

The properties of undesirable solvents that place them in the red category, are described in Table 3.3.

Solvents in the red category must be accompanied by a list of proposed replacements. The replacements are either chemically similar, (*e.g.* heptane as a replacement for the highly inflammable pentane) or functionally equivalent, (*e.g.* ethyl acetate, methyl tert-butyl ether (MTBE) or 2-methyltetrahydrofuran (2-MeTHF) as alternative extraction solvents to dichloromethane). One list of possible replacements is presented in Table 3.4.

Table 3.2 Division of solvents.[31] (Reproduced with kind permission of The Royal Society of Chemistry)

Preferred	*Usable*	*Undesirable*
Water	Cyclohexane	Pentane
Acetone	Heptane	Hexane(s)
Ethanol	Toluene	Diisopropyl ether
2-propanol	Methylcyclohexane	Diethyl ether
1-propanol	Methyl t-butyl ether	Dichloromethane
Ethyl acetate	Isooctane	Dichloroethane
Isopropyl acetate	Acetonitrile	Chloroform
Methanol	2-methyl THF	Dimethyl formamide
Methyl ethyl ketone	Tetrahydrofuran	N-methylpyrrolidinone
1-butanol	Xylenes	Pyridine
t-butanol	Dimethyl sulfoxide	Dimethyl acetate
	Acetic acid	Dioxane
	Ethylene glycol	Dimethoxyethane
		Benzene
		Carbon tetrachloride

Table 3.3 Red category solvents.[31] (Reproduced with kind permission of The Royal Society of Chemistry)

Red solvent	*Flash point*	*Reason*
Pentane	–49 °C	Very low flash point, good alternative is available.
Hexane(s)	–23 °C	More toxic than the alternative heptane, classified as a hazardous airborne pollutant (HAP) in the US
Diisopropyl ether	–12 °C	Very powerful peroxide former, good alternative ethers are available
Diethyl ether	–40 °C	Very low flash point, good alternative ethers are available
Chloroform	N/A	Carcinogen, classified as a HAP in the US
Dichloroethane	15 °C	Carcinogen, classified as a HAP in the US
Dimethyl formamide	57 °C	Toxicity, closely regulated by EU Solvent Directive, classified as a HAP in the US
Dimethyl acetamide	70 °C	Toxicity, closely regulated by EU Solvent Directive
N-methyl-pyrrolidinone	86 °C	Toxicity, closely regulated by EU Solvent Directive
Pyridine	20 °C	Carinogenic/mutagenic/reprotoxic (CMR) category 3 carcinogen, toxicity, very low threshold limit value (TLV) for worker exposure
Dioxane	12 °C	CMR category 3 carcinogen, classified as a HAP in the US
Dichloro-methane	N/A	High volume use, regulated by EU solvent directive, classified as a HAP in the US
Dimethoxy-ethane	0 °C	CMR category 2 carcinogen, toxicity
Benzene	–11 °C	Avoid use: CMR category 1 carcinogen, toxic to humans and environment, very low TLV (0.5 ppm), closely regulated in the EU and the US (HAP)
Carbon tetrachloride	N/A	Avoid use: CMR category 3 carcinogen, toxic, ozone depleter, banned by the Montreal protocol, not available for large-scale use, closely regulated in the EU and US (HAP)

The company claimed that they obtained good results by using this relatively simple guide and replacement scheme, including a 50% reduction in chlorinated solvent use across the whole research division (comprised of more than 1600 laboratory-based synthetic organic chemists, and four scale-up facilities) during the time period 2004–2006.

The chemical industry bases its selection of solvents for chemical processes, and subsequent waste-solvent management, mainly on economic, safety and logistical considerations, but it is also necessary to consider the health and environmental impact of solvents. More complex approaches have been proposed that cover the major aspects of the environmental impact of solvents in chemical production, as well as

Table 3.4 Solvent replacement table.[31] (Reproduced with kind permission of The Royal Society of Chemistry)

Undesirable solvents	*Alternative*
Pentane	Heptane
Hexane(s)	Heptane
Diisopropyl ether or diethyl ether	2-MeTHF or *tert*-butyl methyl ether
Dioxane or dimethoxyethane	2-MeTHF or *tert*-butyl methyl ether
Chloroform, dichloroethane or carbon tetrachloride	Dichloromethane
Dimethyl formamide, dimethyl acetamide or *N*-methylpyrrolidinone	Acetonitrile
Pyridine	Et3N (if pyridine used as base)
Dichloromethane (extractions)	EtOAc, MTBE, toluene, 2-MeTHF
Dichloromethane (chromatography)	EtOAc/heptane
Benzene	Toluene

important health and safey issues.[32] The environmental, health and safety (EHS) hazards assessment[33] is a screening method that aims to identify the potential hazards of chemicals, which are assessed in nine categories of effects: release potential, fire/explosion and reaction/decomposition (representing safety hazards), acute toxicity, irritation and chronic toxicity (representing health hazards), persistency, air and water hazards (representing environmental hazards). This is combined with life-cycle assessment (LCA),[34] a detailed method which is used to assess environmental emissions and to track resource use over the full life-cycle of a solvent, including production, use, potential recycling, and disposal. Both assessment methods for the selection of the most environmentally benign solvents or solvent mixtures are areas of intense development, and there are more extensive databases and software for estimating respective parameters.[35]

Developing solvent-free processes is obviously the best solution from an environmental viewpoint. However, although remarkable advances are being made in this area, organic solvents are still being used in chemistry, and it is clear that we need to find new groups of solvents to replace current ones, most of which are volatile and often hazardous both to humans and the environment. In the quest for more sustainable technologies, the replacement of organic solvents alone is not sufficient. The use of alternative reaction media, where alternative does not necessarily mean new or newly discovered solvents, must also be explored.

Lactate Ester Family of Solvents. There are other new and insufficiently researched organic solvents that have suitable properties and could be included in this replacement list. Alternatives could be considered from the lactate ester group of solvents. Ethyl lactate (EL) is the most important member of the lactate esters group; it has a very favourable toxicological and environmental profile because it is readily biodegradable, but also has excellent solvent properties. It can easily be obtained from carbohydrate feedstock, and recently developed purification procedures have produced pure fluids at very competitive prices. Millions of litres of toxic industrial solvents could therefore be replaced by this environmentally friendly substance. EL has the ability to develop intramolecular hydrogen bonding among neighbouring hydroxyl and carbonyl groups; this interaction competes with intermolecular hydrogen bonding with close EL molecules and with water molecules when mixed.[36]

The widespread processing of renewable resources has led to the use of new liquids that can also be considered as alternative solvents. One of these is glycerol – the by-product of the transesterification of a triglyceride in the production of natural fatty acid derivatives. Glycerol has promising physical and chemical properties: a very high boiling point and negligible vapour pressure; it is compatible with most organic and inorganic compounds; and does not require special handling or storage. Glycerol and other polar organic solvents such as DMSO and DMF permit the dissolution of inorganic salts, acids and bases, as well as enzymes and transition metal complexes, and also dissolve organic compounds that are poorly miscible in water. Glycerol is a non-toxic, biodegradable and recyclable liquid. Different hydrophobic solvents, such as ethers and hydrocarbons, which are immiscible in glycerol, allow products to be removed by simple extraction.

Polyethylene Glycols. There are some other classes of solvents that have environmentally benign properties and must also be taken into consideration. One of these classes is the linear polymers, formed from the polymerization of ethylene oxide, polyethylene glycols (PEG). Polyethers usually have a molecular weight of less than 20 000 and are inexpensive, thermally stable, recoverable, biologically compatible and non-toxic.[37] Complete toxicity profiles are available for a range of polyethylene glycol (PEG) molecular weights. Furthermore, PEG and its monomethylethers have low vapour pressure, are not inflammable, require simple work-up procedures, and are recyclable. For these reasons, PEG is considered to be an environmentally benign alternative to volatile chemical solvents and an eminently practical medium for organic reactions. PEG is most commonly employed as a support for various transformations (PEG 2000–20 000), and it is a biologically acceptable polymer used extensively in drug delivery and bio-conjugates as a diagnostic tool. It can also be used as an efficient medium for phase-transfer catalysts. Its application as a solvent is relatively recent and it is usually used at low molecular weights (<2000), because it is either liquid at room temperature or has a low melting point. Mixtures of PEG and water have been widely used in many different kinds of reaction and separation systems.[38] Their low toxicity, low volatility and biodegradability are important environmentally benign characteristics, which combined with their relatively low cost, make them particularly attractive as bulk commodity chemicals.

Fluorous Solvents. Amongst these alternatives to common organic solvents is a group of so-called neoteric solvents,[39] which includes perfluorinated (fluorous) solvents, supercritical fluids and room-temperature ionic liquids. Fluorous (perfluorinated) solvents, such as perfluoroalkenes, perfluoroalkyl ethers and perfluoro-alkylamines, are generally chemically inert, non-toxic, nonflammable, thermally stable and possess a superior ability to dissolve oxygen gas, which is an advantage in medical technology (1-bromoperfluorooctane is used as a component of artificial blood). Fluorous solvents have a higher density than non-fluorinated solvents, high stability (mainly due to the stability of the C–F bond) and low solvent strength.

Since the intermolecular reactions in fluorous solvents are very weak, the *n*-perfluoroalkanes have lower boiling points than *n*-alkanes, and extremely low polarities. The solubility of water in fluorous solvents is very low, due to the lack of suitable H-bonding interactions. Gases, such as carbon dioxide, oxygen and hydrogen, are much more soluble in fluorous solvents than in water and have much greater solubility than in hydrophobic organic solvents.

The poor solubility of fluorinated solvents is due to their low surface tension, low intermolecular interaction, high density and low dielectric constant. They usually have a limited, temperature-dependent miscibility with conventional organic solvents, forming biphasic solvent systems with such solvents at ambient temperatures. Because of the different solubilities for reactants, catalysts and products, such biphasic organic–fluorous solvent combinations can facilitate separation of the product from the reaction mixture. Some organic–fluorous biphasic systems can become single phase at elevated temperatures, which can then serve as a homogeneous reaction medium. This thermomorphic effect can be used to switch a reaction from heterogeneous to homogeneous, with concomitant mass transfer advantages. After completion of the reaction, cooling the reaction mixture again leads to the formation of two separate phases. Ideally, the product is dissolved in one phase and the remaining reaction components in the other, making isolation of the product very easy. This new experimental technique, using fluorous biphasic systems (FBS) with fluorous bi-phase catalysis (FBC), has already found many applications in synthetic organic chemistry.[40]

Fluorous solvents have been produced for commercial use in a range of roles: low boiling point compounds as alternatives to chlorofluorohydrocarbons (CFCs) as refrigerants (R-134, R-227ea), and greases and lubricants based on polyfluoropolyethers that operate at >300 °C. However, interest in fluorous solvents on an industrial scale is currently limited due to their high cost.

Although fluorous phase systems are solvents, they are not straightforward solvent replacements. Due to their extremely non-polar characteristics, they are not suitable for most organic reactions and tend to be used in conjunction with a traditional organic solvent (or some sort of immiscible solvent), forming a bi-phase system. For these reasons, fluorous bi-phase systems are effective in synthesis to separate the catalyst from reactants and products.

Despite this advantage, a shadow of doubt exists with regard to the "greenness" of fluorous solvents due to their persistence in the environment, and this is still a matter of debate. Furthermore, the manufacture of fluorous solvents is not simple, and generally requires the use of huge amounts of highly volatile organic solvents and toxic reagents such as fluorine gas and hydrogen fluoride.

Two other classes of solvents – supercritical fluids and ionic liquids – could have the properties required to make processes environmentally safer. Some of them are environmentally friendly and can replace organic solvents in liquid–liquid extractions, catalysis and separations. Their use would reduce or eliminate the related costs, disposal requirements and hazards associated with volatile organic compounds (VOCs). Their main novelty, compared with common solvents, is their ability to fine-tune the properties of solvent media, which allows them to replace specific solvents in a variety of different processes.

Furthermore, the unique properties of both liquids have opened new fields of research. Supercritical CO_2 and ionic liquids operate together because ionic liquids do not dissolve in supercritical fluid; however, the fluid is miscible with ionic liquid, and that allows supercritical fluid to be used for the extraction of compounds from ionic liquid.[41]

3.3.1 Supercritical Fluids

The main source of organic waste is the use of solvents in sample treatment and the separation of mixtures into components. The solution is to use alternative solvents that are more benign to the environment and make an analysis "greener".

An important direction is to seek replacements for existing organic solvents, as discussed above. A second approach is to use common gases as well as solvents in different conditions – pressures and temperatures. The use of solvents above their normal temperature and pressure conditions will continue to expand the range of analytical methods and should be seen as part of an overall spectrum of solubility, polarity and volatility properties of solvents and mobile phases. Every compound has a critical point on the temperature and pressure scale where differences between its gaseous and liquid states disappear and the compound is in a supercritical state. Supercritical fluids link gases and liquids, providing a continuum of mobile phase properties to the analyst.

The history of supercritical fluids dates back to the beginning of the nineteenth century, and carbon dioxide attracted much attention in the latter half of the century. Dr. Thomas Andrews from Queen's College, Belfast, described his observations of the critical properties of CO_2, and the values he reported in 1875 for the critical point of carbon dioxide (30.92 °C and 73.0 atm.) are in close agreement with those presently accepted.[42]

It was soon clear that supercritical fluids could dissolve compounds. Researchers carrying out experiments in the late 1800s and early 1900 met and overcame many obstacles, especially the generation of high pressure. They were able to publish reviews about supercritical fluid solubility at the end of the nineteenth century. The first patents on extraction with supercritical fluids were issued in 1940, and this method has received wider attention since then.[43] Several informative textbooks on this subject are available.[44]

The most popular substances used in supercritical fluid processing are water (T_c = 374.1 °C; P_c = 218.3 atm) and carbon dioxide (T_c = 31.1 °C; P_c = 72.8 atm). Water is a safe solvent, but it has high critical temperature and pressure values; therefore, special instrumentation is required to work with water in its supercritical state. CO_2 has much lower critical parameters. Other advantages of CO_2 as a solvent are that it is non-toxic, easy to purify and relatively inert. CO_2 is produced in large quantities as a by-product of fermentation, combustion and ammonia synthesis, which makes it very inexpensive. Its release into the atmosphere could be postponed by using it as a supercritical fluid. Because the CO_2 molecule is non-polar, it is considered an intermediate solvent, between truly non-polar solvents, such as hexane, and weakly polar solvents. However, it has a limited affinity with polar solutes because of

Table 3.5 The critical parameters of some common compounds.

Solvent	*Critical temperature, °C*	*Critical pressure, atm*	*Critical density, g cm^{-3}*
toluene	318.9	40.6	0.292
ethane	32.4	48.3	0.203
n-pentane	196.6	41.7	0.554
methanol	240	79.9	0.272
ethanol	243	63.0	0.276
water	374.1	218.3	0.315
CO_2	31.1	72.8	0.468
N_2O	36.4	71.5	0.452
NH_3	132.3	111.3	0.235
SF_6	45.5	37.0	0.738
Xe	16.6	57.6	1.113
CHF_3	25.9	46.9	0.52

its large quadrupole moment. To improve its polar properties, CO_2 is modified with polar additives, such as methanol.

The critical parameters of some commonly used compounds are presented in Table 3.5.

Tunability is another dimension that makes alternative solvents like supercritical fluids even more attractive for researchers. Around the critical point of supercritical fluids, small changes in temperature and/or pressure cause significant changes in density and other physical parameters that make it possible to tune the solubility and other parameters of the solvent. Therefore, using solvents in their supercritical condition expands the overall spectrum of solubility, polarity and volatility properties of solvents and mobile phases. The ability to fine-tune the properties of the solvent medium allows it to replace specific solvents in a variety of different processes, or to create new methods for processing (analyzing) samples. The same solvent can be used in different applications and procedures.

Supercritical solvents exhibit rather small viscosities and high self-diffusion coefficients, which resemble those of gases more than those of liquids. The mean values of important parameters for gases, fluids and liquids are presented in Table 3.6.

Even though a supercritical fluid solvent possesses the density of a liquid over much of the range of industrial interest, it also exhibits the diffusivity and viscosity of a gas. Supercritical fluids can have gas or liquid properties according to the temperature and pressure, but not both simultaneously. Furthermore, the zero surface tension of supercritical fluids allows facile penetration into microporous materials – a property that is advantageous in extracting solid samples and processing

Table 3.6 Physical properties of fluids compared with gases and liquids.

	Density, ρ (g cm^{-3})	*Viscosity, η eP*	*Self-diffusion coefficient, D12 (cm^2 s^{-1})*
Gas, 30 °C, 1 atm	$(0.6–2)\times10^{-3}$	$(1–3)\times10^{-4}$	0.1–0.4
Supercritical fluid			
Near T_c, P_c	0.2–0.5	$(1–3)\times10^{-4}$	0.7×10^{-3}
Near T_c, $4P_c$	0.4–0.9	$(3–9)\times10^{-4}$	0.2×10^{-3}
Liquid, 30 °C, 1 atm	0.6–1.6	$(0.2–3)\times10^{-2}$	$(0.2–2)\times10^{-5}$

porous materials. These properties, combined with the pressure-dependent solvent power of a supercritical fluid, have provided the impetus for applying supercritical fluid technology to numerous separation problems experienced in many areas of industry.

Using supercritical fluids instead of organic solvents for extraction is becoming more popular for liquid–solid extractions, especially when supercritical CO_2 is used as a solvent. With supercritical fluids, a greater range of solvent properties can be achieved with a single solvent, through careful manipulation of temperature and pressure. However, there are two competing factors, fluid density and solid sublimation pressure, which affect the solubility of solids in supercritical fluids and, consequently, the efficiency of extraction. Supercritical solvent is also highly soluble in the coexisting liquid phase, which makes viscosities lower and diffusion coefficients higher. This greatly improves mass transfer and the equilibrium rate in extractions.

The multitude of parameters that can affect the extraction process in supercritical fluid extraction (SFE) lends flexibility to the method, but the lack of fundamental knowledge about how these parameters affect extraction makes straightforward method development difficult and essentially empirical.

However, supercritical CO_2 is being used to replace organic solvents in sample preparation to extract target compounds in some US EPA methods. This replacement saves analysis time and minimizes the consumption of organic solvent. Another green aspect of using gases in supercritical condition is that they save energy because of the absence of distillation and solvent evaporation.

Supercritical fluid extraction has been adopted by the EPA as a reference method for extracting PAHs (EPA Method 3561) and PCBs (EPA Method 3562) from solid environmental matrices, and also for extracting hydrocarbons (EPA Method 3560).[45]

The usual limiting factor for the treatment of solids is how analytes are incorporated into the sample matrix. Extractable compounds are often trapped within the organic matrix of the environmental solid. The extraction of compounds from one matrix to surrounding media involves different transport processes, the slowest of which determines the rate of the extraction process. Transport processes are dependent on a variety of physical properties, such as diffusion, viscosity, partition, solubility and surface tension. These properties generally become more favourable at higher temperatures.

Extraction is controlled by either a kinetic/desorption step or a solubility/elution step:

- analytes have to diffuse from the core of the particle to the surface;
- analytes are transferred from the particle surface into the extraction fluid (distribution between matrix and fluid);
- analytes have to diffuse through the surface film to the flow of extractant;
- analytes are eluted out of the extraction cell by the flow of supercritical extractant.

The SFE extraction rate is determined by the slowest of these four steps. Identification of the rate-determining step is therefore an important aspect of method development in SFE.

The efficiency of SFE is affected by a wide range of parameters, such as the nature of the supercritical fluid, the temperature and pressure, the extraction time, the shape of the extraction cell, the sample particle size, the matrix type, the moisture content of the matrix and the analyte collection system. SFE has particular advantages in terms of extraction selectivity, low temperatures and minimization of post-extraction sample clean-up procedures.

Due to the numerous parameters that affect extraction efficiencies, SFE affords a high degree of selectivity and the extracts are relatively clean (thus, they require only moderate additional clean-up). In fact, combined with solid adsorbent traps, SFE could provide extraction and clean-up in a single step. However, the need to control so many operating parameters makes SFE optimization tedious and complicated in practice.

Supercritical fluids have significant advantages over methods that are often slow and largely manual, in ease of handling and the ability to eliminate organic solvents, despite the limited sample size (usually 10–20 mL).

Solubility is very sensitive to changes of pressure and temperature in the supercritical fluid. Collection of the analyte after extraction is a critical step in performing an analysis, because of possible losses of analyte. Various systems have been developed, such as liquid trap, solid-phase trap, cold trap, *etc.* Despite its attractiveness, it is necessary to evaluate supercritical fluid technology on a case-by-case basis. The process of extraction using supercritical fluids has not been fully studied, and much work remains to be done to establish reliable analytical methods based on SFE. One possible direction of research is the on-line coupling of the extraction and separation units with suitable detectors, in which the potential analyte loss during transfer from one instrument to another is minimized.

Supercritical fluids can be used as mobile phases in chromatography because of their advantageous properties. They can act as substance carriers similar to the mobile phases in gas chromatography (GC), and also dissolve these substances like solvents in liquid chromatography (HPLC).[46] This chromatographic variant has been known as supercritical fluid chromatography (SFC) and will be discussed in Chapter 5.2.1.4.

During the development of packed-column SFC, it was realized that the characteristics of chromatographic separation are present irrespective of whether the fluid is defined as a liquid, a dense gas or a supercritical fluid. In some instances, the initial pressure used in SFC is actually below the critical pressure. The differences between SFC, subcritical fluid chromatography, enhanced fluidity chromatography and high-performance liquid chromatography (HPLC) have been overstated in the past. When the outlet pressure is elevated and pressure and temperature are controlled for the mobile phase, the resulting techniques are similar and the behaviours of conventional GC and HPLC are completely and seamlessly bridged.[47] Each type of chromatography represents part of a continuum of increasing mobile phase solvating power coupled with increasing mobile phase viscosity and decreasing mobile phase diffusivity. In principle, with a supercritical carbon dioxide carrier it is possible to perform GC, SFC, and HPLC in the same chromatograph (see Figure 3.8).

Safety considerations must influence any technical choice and operation, and a detailed analysis of potential hazards must be conducted for each case. Supercritical fluids and liquefied gases present important hazards that must be taken into account, both for equipment design and construction, and for operation and maintenance:

- mechanical hazards: the high pressures employed increase hazards due to tubing connection rupture, metal fatigue and fragilization, as well as corrosion;
- thermodynamical hazards: drastic decreases in temperature can result in tubing becoming plugged by ice; vapour is susceptible to explosion due to the expansion of boiling liquid;

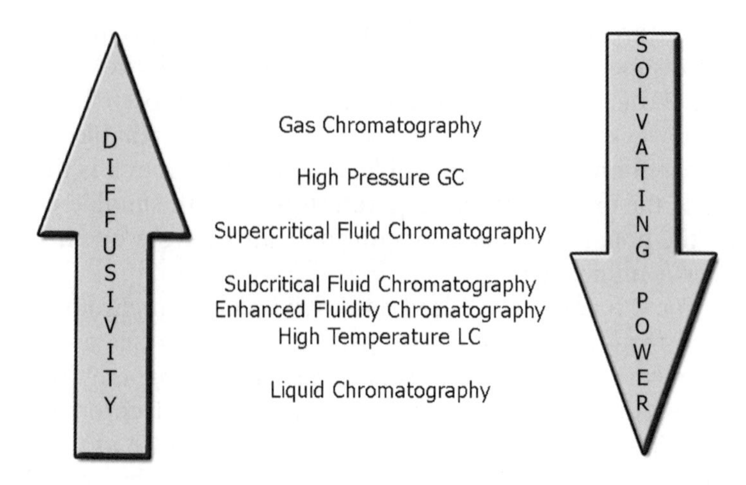

Figure 3.8 Continuous scale for chromatography.

- chemical hazards: oxidation of organic compounds by nitrous oxide, inflammable fluids (co-solvents, products), corrosion;
- biological hazards: potential asphyxia requires ventilation for CO_2, possible leaks when working with toxic materials.

Despite the positive features of methods using supercritical fluids, they are not as widespread as they could be, and there is a lack of standard procedures in control laboratories. This is because the methods are technically complicated and the process (extraction or chromatography) parameters must be carefully and precisely controlled, which also makes instrumentation costly. However, the number of research publications is growing and new instruments are appearing on the market. Supercritical fluids are the most suitable solvents for a growing number of applications in both chromatography and extraction.

3.3.2 Ionic Liquids

The history of ionic liquids (ILs) began almost a hundred years ago, from an interest in some low melting-temperature organic salts as new electrolytes. They have been known since 1914, when Paul Walden in St. Petersburg, Russia, prepared ethylammonium nitrate ($[C_2H_5NH_3]^+$ $[NO_3]^-$) for electrical conductivity measurements. This compound has a melting point of 12 °C but has not found a use because of its high reactivity.[48] Ionic liquids are defined as salts with a melting point below 100 °C, a convenient, arbitrary temperature limit. They form liquids composed of ions. This gives these materials the potential to behave very differently from conventional molecular liquids when used as solvents.

Two more discoveries had to be made before ILs were broadly recognized as interesting media possessing great possibilities for wide application: first, salts which remained liquid at room temperature and could be easily synthesized and, second, ILs which were stable in both water and air. These were true breakthroughs and after they were made, ILs were no longer treated as a curiosity of interest to only a small group of scientists.

Here is a chronology of the discoveries concerning IL:

- Ethyl ammonium nitrate: the description of a low melting point salt, 1914.
- N-ethylpyridinium bromide–aluminium chloride melt: the first room-temperature ionic liquid, 1951.
- 1,3-dialkylimidazolium salts: the most stable and conductive salts, 1982.
- 1-ethyl-3-methylimidazolium tetrafluoroborate: the hydrophobic ionic liquids, 1992.

The most popular ionic liquids are 1-alkyl-3-methyl imidazolium salts and 1-alkyl pyridinium salts with a multiple selection of anions. The number of publications on ionic liquids is increasing almost exponentially, as salts with different ion constituents can easily be prepared.

The most common classes of ionic liquids are presented in Table 3.7.

Ionic liquids have two main advantages: non-volatility and thermal robustness. This gives most of them a wide thermal operating range (typically −40 °C to 200 °C) with the possibility of no measurable vapour pressure, enabling a wide range of kinetic control for reactions occurring in ionic liquids without solvent vapours. The main property that supports the claim of ILs as "green solvents" is negligible vapour pressure.

Their specific properties (fluidity over a broad temperature range, non-inflammability, non-explosiveness, non-coordination, slight corrosiveness, extremely low vapour pressure, miscibility or non-miscibility with water and other solvents, high dissolving ability for many organic and inorganic materials, electrical conductivity, high electrochemical stability) can be tuned by an appropriate choice of anion and cation. This ability might best be described as the "chemical tunability" of ionic liquids – a class of solvents possessing similar physical properties, but different chemical behaviour. Finally, ILs can contain specific functional groups, which has given rise to task-specific ionic liquids (TSILs) which

Table 3.7 The most common ionic liquids.

Cations: differently substituted		*Anions*	
Ammomium	$[NR_xH_{(4-x)}]^+$ 1	$[AlCl_4]$-tunable Lewis Acid/Base and miscibility properties, reactive with water	
Sulfonium	$[SR_xH_{(3-x)}]^+$ 2	$[PF_6]$-moisture stable, water immiscible IL	
Phosphonium	$[PR_xH_{(4-x)}]^+$ 3	$[BF_4]$-moisture stable, but water miscible IL depending on the ratio of ionic liquid: water, system temperature, and length of the alkyl chain in the cation.	
Lithium	Li^+ 4	Triflate [TfO]	$CF_3SO_2^-$
Imidazolium	 5	Nonaflate [NfO]	$CF_3(CF_2)_3SO_2^-$
Pyridinium	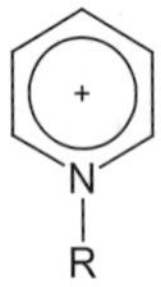 6	Bis(triflyl)imide [Tf_2N]	$(CF_3SO_2)_2N^-$
Pyrrolidinium	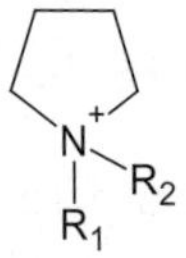 7	Trifluoroacetate [FAcO]	$CF_3CF_2CO_2^-$
Thiazolium	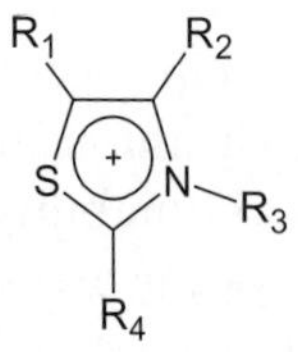 8	Heptafluorobutanoate [HB]	$CF_3(CF_2)_3CO_2^-$

Table 3.7 (*Continued*).

Triazolium	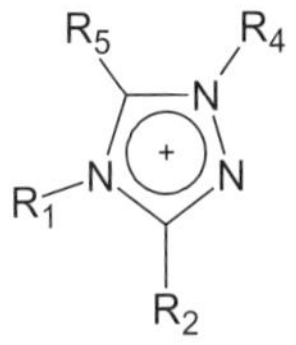 9		
Oxazolium	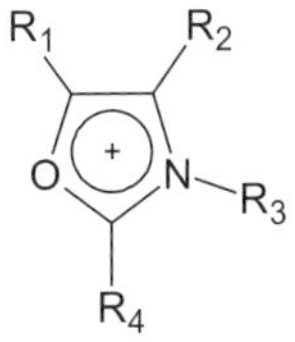 10		
Pyrazolium	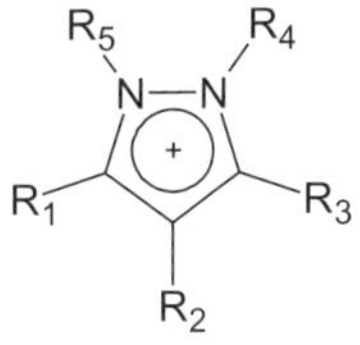 11	Dicyanoamid [dca]	$N(CN)_2^-$

have gained wide attention. Three different generations of ionic liquids can be identified:[49]

1. ILs with unique tunable physical properties;
2. task-specific ILs with targeted chemical properties and selected physical properties;
3. task-specific ILs with targeted biological properties and selected physical and chemical properties.

Ionic liquids can be used not only to improve the sensitivity and selectivity of analysis in existing methods; their different behaviour and properties can provide original solutions in chemical analysis as well. The properties of ILs vary widely as shown in Table 3.8.

Ionic liquids have good solvating properties that, combined with a broad range of spectral transparency (UV cut-off wavelengths of alkylimidazolium-based ILs are in the range of 230 to 250 nm), make them

Table 3.8 Properties of ionic liquids.

Density	1.1–1.6 g cm^{-3}
Refractive index	1.5–2.2
Viscosity	52–450 cP (water = 0.89 cP)
Surface tension	40 dyn cm^{-1} (water = 73 dyn cm^{-1})
Conductivity	in order of 0.1 S m^{-1} Depending on ion size, structure, degree of dissociation
Polarity	Common solvatochromic probes show that ionic liquids behave similarly to low alcohols
Stability towards water	Depends on anion:
Water is always present in hydrophobic liquids	$[C_4C_1ImCl][AlCl_3]$-hygroscopic $[C_4C_1Im]^+[PF_6]^-$-hydrophobic $[C_4C_1Im]^+[CF_3COO]^-$-water-soluble Solubility depends on the length of the alkyl chain in the cation, decreasing as the length increases
Solubility of gases in IL	Henry's Law Constants for gas solubilities in various ILs range from 1 bar to 6000 bar depending on the gas

suitable solvents for spectroscopic measurements especially in the visible region.[50] A variety of transition metal complexes, which are unstable in other media, can be studied in room-temperature ionic liquids without the problems of solvation and solvolysis. This permits reliable solution spectra to be recorded for these species.

The main advantage of ionic liquids over organic solvents for other applications in analytical chemistry is their low volatility, which makes them useful as solvents at high temperatures (GC stationary phases). The wetting ability and viscosity of ILs allow them to be coated onto fused-silica capillaries. Task-specific ionic liquids have high thermal stability up to 260 °C, symmetrical peak shapes, and, because of their different ranges of solvation-type interactions for anions and cations, they exhibit dual-nature selectivity behaviour.[51] The difference in selectivity between ionic liquids and the conventional GC stationary phase, methylphenyl polysiloxane, is due to the unique solvation characteristics of ILs which allow them to serve as useful dual-nature stationary phases. Mixed ionic liquid stationary phases also provide more flexibility of control and optimization of selectivity for complicated analyte mixtures. ILs can also be used for chiral separations, which can be accomplished in two ways: (1) a chiral selector can be dissolved in a non-chiral ionic liquid, or (2) the ionic liquid itself can be chiral.[52] Using ionic liquid as the stationary phase in supercritical fluid chromatography could have advantages because of its non-solubility in supercritical carbon dioxide.

Due to their high viscosity, pure ionic liquids are not good chromatographic media for separation; they are used as specific additives in liquid chromatography and capillary electrophoresis. In LC and CE, when added to mobile phases at a concentration of 0.5–1.5% (v/v), ILs block silanol groups on the surface very effectively and provide excellent separations of strongly basic drugs that are otherwise not eluted.[53,54] Both gas and liquid chromatography are well-established techniques with many standardized methods. This makes them good candidates for the use of ionic liquids.

Imidazolium-based ionic liquids have good solubility in organic solvents. They are used in non-aqueous capillary zone electrophoresis as ionic additives for the adjustment of analyte mobility and separation. The separation of different analytes in organic solvents occurs as they become charged in the presence of ILs in the separation media.[55]

The liquid state of ILs makes them potential solvents for use in counter-current chromatography (CCC), which employs solvents as both the mobile and stationary phases. The CCC technique is mainly used to produce significant amounts of purified chemicals, using as little solvent as possible. Since ionic liquids are able to form biphasic systems with a number of solvents, they have significant potential for use in CCC. Unfortunately, the viscosity of most popular ILs is too high for direct use as a liquid phase in CCC, but they can be combined with organic solvents. The situation is likely to change in CCC with the availability of low-viscosity ILs.

The low volatility of ionic liquids makes them useful as solvents operating in a high vacuum and, together with the more amorphous solid varieties, they merit further study as MALDI matrices. These ionic matrix systems allow homogenous sample preparation with a thin ionic liquid layer that has negligible vapour pressure. The vacuum-stable, liquid consistency of ionic liquid matrix sample preparations considerably enhances MALDI-MS analysis in terms of shot-to-shot reproducibility. This enables a facilitated qualitative and quantitative measurement of analytes that is not possible with classical solid matrices.[56,57]

Suitable electrolytes should have high conductivity, large electrochemical windows, excellent thermal and chemical stability, and negligible evaporation; most of these properties are characteristic of ionic liquids. By using ILs as an electrolyte medium, it is possible to achieve a wider range of operational temperatures and conditions, relative to other conventional electrolytic media, and makes them promising materials for various electrochemical devices, such as batteries, fuel cells, sensors and electrochromic windows.[58] The production of microchips for analysis and combining the electrowetting phenomena of ionic liquids are also topics for more intensive research. Novel and exciting applications may emerge in areas that were not even considered in the original concept. Furthermore, nanotechnology will affect the processes and separations with respect to scale and materials.

In addition to the advantages provided by their physical properties, ILs usually afford higher reaction rates, higher yields and better selectivities than conventional organic solvents when used in organic synthesis. They have been applied in a large number of reactions, especially in catalysis, and particularly in biocatalysis and biochemical technology, because many enzymes retain their activity in these solvents. ILs have been explored for many applications in the chemical, petroleum and allied fields, for reaction media, electroplating, acid-gas scrubbing and desulfurization of transportation fuels, often motivated by the perception that they are "green solvents". Therefore, the replacement of volatile solvents with ILs is expected to prevent the emission of VOCs. However, the environmental impact of the life cycle phases of ILs and comparisons with alternative methods have not been studied. A life cycle assessment (LCA) is essential before any legitimate claims of "greenness" can be made. Despite the popularity of ILs, there is significant uncertainty regarding the potential environmental impact of ILs including, but not limited to, toxicity; thermal decomposition and hydrolysis during usage; difficulty of recovery; and resistance to environmental degradation. Certain ILs are as toxic as phenol and more toxic than benzene.[59] The thermal and hydrolysis stability of certain ionic liquids have also raised questions. For example, 1-alkyl-3-methylimidazolium halides, such as $[C_4C_1Im][AlCl_4]$ can decompose to halomethane and alkylimidazole. ILs containing $[PF_6]^-$ can be hydrolyzed to HF, POF_3, *etc.* when in contact with moisture. Thus, it is reasonable to assume that ILs are at least as hazardous as their decomposed or hydrolyzed products.

Previous research has focused mainly on obtaining a fundamental understanding of ILs and developing their application and disposal phases. Many upstream processes in the life cycle of ILs involve volatile and hazardous organic chemicals. Life cycle analysis takes into account VOC emissions, the use of organic solvents, and fossil fuel consumption per kilogram of the selected solvent. Apart from VOC emissions from intermediate processes, the life cycle may generate more than twice the amount of VOCs compared with VOCs from the life cycles of benzene or acetone.[60] Since most ILs are synthesized with a long supply chain involving organic compounds and require purification steps, it can be expected that ILs may have an adverse environmental impact. However, the high environmental impact of the synthesis phase of ILs may be counter balanced during their use, because of their stability and potential to increase yields of desired reaction products and improve separation efficiency, as well as their ease of recycling.

It can be said that ionic liquids are in their infancy, despite enormous academic interest. The use of ionic liquids is opening new possibilities in different areas of analysis, and completely new approaches to chemical and biological analysis are appearing.[61] Even using ionic liquids as additives in liquid chromatography and electrophoresis has great advantages compared with conventional materials.

The variability of ionic liquids presents a challenge not only for instrumental separation methods like chromatography, but also for industrial methods currently in use. With regard to extraction, this is associated with the possible differences in mechanisms of solute transfer, mutual solubilities in biphasic systems, solvation abilities and probably even bulk phase structures. The success of IL-based extraction systems, especially for metal ions, lies in their ionic nature, ion exchange ability and non-volatility. Task-specific ionic liquids used as separation media in liquid–liquid extraction processes achieve higher efficiency and selectivity of separation than common solvents. The combination of ionic liquids with supercritical carbon dioxide (sc-CO_2) as an extractant has the potential to combine chemical reaction and downstream separation in one system. Ionic liquid maintains solvent strength even at fairly high loadings of CO_2, whereas the viscosity in the vicinity of the solute is dramatically reduced, leading to enhanced mass transport and facilitated separation. They can be used with organic co-solvents that "solvate" the constituent ions of the ionic liquid, resulting in decreased aggregation of these ions (lower viscosity and higher conductivity).

A better understanding of ionic liquid-based extraction systems that take full advantage of the design of ILs is required. This will undoubtedly be beneficial for emerging technological and analytical applications.

When using ILs, one must be mindful that the properties of these compounds can be dramatically altered by the presence of impurities. The influence of water and chloride anions on the viscosity and density of ILs has already been extensively discussed by many authors. Furthermore, the water content of ILs can affect the rate, direction and selectivity of reactions, and can act as co-solvents in the extraction process.

At present, the number of synthesized ionic liquids is large, and among them many different cations, and even more anions, are used. There is an almost limitless variety of liquids with a wide range of possible applications still to be discovered. There are several proofs of principle for the advantageous use of some ILs, but the road to design and optimization of task-specific ILs is neither short nor smooth. A thorough study of the physico-chemical properties of ILs, such as acidity, basicity, viscosity, solubility, *etc*., and their thermodynamic behaviour when mixed with other solvents, or even other ionic liquids, is required to fill gaps in our knowledge. Tailoring ILs for specific applications must be accompanied by theoretical work in order to develop acceptable thermodynamic models, like COSMO or modified UNIFAC. A QSPR approach must be used to its full extent to provide a sound approach for the design of ILs. High-quality data on reference systems and the creation of a comprehensive database have to be included in strategies to make progress in the field of ionic liquids. At the moment there are more opportunities than results. Ionic liquids are commercially available and intensive studies are underway in every area of chemistry to find their proper niche.

3.4 ENERGY SAVING PROCEDURES

Most chemical processes use thermal sources of energy originating from fossil (or nuclear) fuels. This energy input is non-specific to the process, *i.e.* it is not targeted at the chemical bond. Much of the energy is "wasted" in heating up reactors, solvent and even the surrounding environment, instead of targeting only the molecules undergoing reaction. This makes energy saving one of the most important and, at the same time, most difficult laboratory issues to solve.

Thermal energy conservation measures are closely connected with Good Laboratory Practice regarding the economical use of equipment time and optimized laboratory procedures. Costs and environmental impact can be reduced by:

- Control and maintenance. The equipment must operate at design specifications, and temperature controls have to be optimized. This can easily be achieved by modern computerized digital monitoring and control systems. Regular maintenance and lubrication of equipment can improve energy efficiency and extend equipment life.
- Good housekeeping. Leaking valves and pipelines should be repaired to save energy.
- Using heat produced in one process for other processes through exchanges, or even for keeping the workplace temperature optimal.
- Matching energy sources to process requirements and having a variety of energy sources to meet the specific demands of processes used in the laboratory.

Whereas this approach is part of the Laboratory Quality Assurance Program and has little to do with Green Analytical Chemistry, challenging and measurable energy savings can be obtained by introducing new, less energy-consuming analytical methods and procedures.

The first step in analysis – sample preparation – depends on the physical state of the sample and the matrix. Typical sample preparation processes may include sample homogenisation and filtration, centrifugation, distillation, extraction, fractionation and concentration. The successful execution of these processes will ensure that the analyte is present in a form compatible with the analytical system. Sample preparation is still the most time-consuming step in the whole analytical procedure, accounting for approximately two thirds of the total analysis time. Therefore, selecting and optimizing a sample preparation scheme is a key factor in the success of an analysis and greatly affects reliability, accuracy and cost.

All of the processes mentioned above require additional energy for analysis, such as using additional equipment or processes, evaporating solvent, drying the sample, *etc.* Traditional sample preparation techniques rely on extraction with solvents, such as liquid–liquid extraction (LLE) and Soxhlet extraction (SOX). From an environmental standpoint, however, these procedures use large amounts of organic and often undesirable chlorinated solvents, and have long extraction times (usually 6–24 h). For these reasons, pressurized-liquid extraction (PLE), in its static (SPLE) and dynamic (DPLE) forms, has been developed to replace these traditional sample pre-treatment techniques. It is evident that if applying high temperature is to be selected as an energy-saving procedure, the time must be short and result in very high process efficiency. In PLE – also known as pressurized-fluid extraction (PFE), pressurized solvent extraction (PSE), accelerated solvent extraction (ASE) and enhanced solvent extraction – extractions are fast because of higher diffusivity, improved solubilization capability and more efficient analyte interactions in a liquid solvent at temperatures above its boiling point. PLE provides quantitative extractions with reduced consumption of solvents and substantially shorter extraction times than the Soxhlet method. For a number of organic trace pollutants in various solid and semi-solid matrices, recoveries with PLE are comparable to, or even better than, those with Soxhlet extraction. Solvents similar to those used for Soxhlet extraction usually give good results, so it is relatively straightforward to replace old methods with PLE. Mixtures of low-polar and high-polar solvents generally provide more efficient extractions of analytes than single solvents. The temperatures are usually in the range of 60–200 °C. With the exception of labile analytes or samples, higher extraction temperatures will increase the efficiency of PLE as a result of enhanced sample wetting and better penetration of the extraction solvent, as well as higher diffusion and desorption of the analytes from the matrix to the solvent. It is possible to use purely aqueous solvents in PLE, *i.e.* to use pressurized hot water extraction (PHWE) – also called sub-critical water extraction (SWE), hot-water extraction, high-temperature water extraction, superheated water extraction and hot liquid water extraction. This technique uses water as an extraction solvent at temperatures of 100–350 °C and at a pressure high enough to keep it in a liquid state. Typical extraction times for PLE are 10–30 min, which substantially reduces the energy requirement per sample. PLE instruments can seldom be used for PHWE because the sealing materials cannot withstand the high temperatures.

When pressures and temperatures are above the critical parameters of the solvent, pressurized extraction becomes supercritical fluid extraction (SFE). If CO_2 is used as an extraction solvent, the process becomes even more environmentally friendly. Due to the relatively low critical parameters of CO_2, the additional amount of thermal energy needed for this process is not substantial. However, there are several parameters that affect extraction recovery, including temperature and pressure, extraction time, flow rate, choice of modifier, and collection mode (solvent, SPE trap or empty vessel). Furthermore, optimal conditions are strongly dependent on the matrix, (*i.e.* fresh optimization may be required for different types of sample matrices). Both static and dynamic modes can be used, and the two modes are often combined for efficient extraction. Typical extraction times vary from 5 min to 60 min and temperatures are in the range of 40–60 °C. SFE can therefore be considered more energy efficient than Soxhlet extraction. A significant advantage of SFE with CO_2 is that various (*e.g.* sulfur and lipid) clean-up steps can easily be added.[62] Commercial systems are available for both processes, but it is also possible to use conventional supercritical fluid extractors for pressurized liquid extraction.

Various solid-phase extraction (SPE) techniques have been developed to replace conventional sample pre-treatment techniques. In addition, membrane-based techniques (dialysis, ultrafiltration, supported-liquid membrane extraction) and other miniaturized extraction techniques have been developed to overcome the problems of traditional methods. These methods can be considered energy saving treatment methods, because high temperatures and pressures are not required. Using other forms of energy such as ultrasound or microwaves for accelerating sample preparation steps will be discussed later.

All the other steps of an analysis contribute to the laboratory energy requirements and must be taken into account to produce an energy score for the method. A ranking has been proposed by Douglas Raynie[63] with three levels of energy consumption per sample (see Table 3.9):

According to the energy assessments in this table, wet chemistry methods such as titration and biochemical assay-based methods are in the green area. Simple instrumental methods such as GC and HPLC are in the yellow area. More complicated and combined instrumental methods that are energy intensive are in the red zone. With instrumentation, high sample throughput is of obvious importance since this reduces the energy requirement per sample. The use of solvents usually creates additional energy demands, since energy is required for solvent evaporation.

The major factors for the efficient functioning of a laboratory are: the cost of equipment; operating costs, including the cost of energy; the complexity of method development and the amount of organic solvent required; and the level of automation. In addition, the number of samples to be analyzed is also important – the question is whether the planned procedure will be unique or whether it will be used to perform routine analysis. In the latter case, techniques facilitating automation and low cost per analysis are to be preferred. Miniaturization and automation of

Table 3.9 Ranking instruments according to their energy consumption.

Green rating = 1; green ≤0.1 kWh per sample	*Green Rating = 2; yellow ≤ 1.5 kWh per sample*	*Green Rating = 3, red > 1.5 kWh per sample*
SPE (vacuum assistance)	Hot plate solvent evaporation less than 2.5 h with green instrument	Hot plate solvent evaporation more than 2.5 h with green instrument
Rotavap	ASE	
Needle evaporator	SFE	Hot plate solvent evaporation more than 1 h with yellow instrument
Sonicator	Ultrasound probe	
FTIR	Microwave	Soxhlet
UV-Vis spectrometer	AA spectrometer (flame or furnace)	
Fluorescence spectrometer	ICP-MS	NMR
UPLC	GC	GC-MS
Titration	LC	LC-MS
Immunoassay		X-Ray diffractometer
Microbiological assay		
Hot plate solvent evaporation less than 10 min	Hot plate solvent evaporation less than 1 h with yellow instrument	

laboratory equipment are developments that also result in energy savings. Smaller analytical instruments consume much less energy and produce faster analyses, which deliver higher sample throughput.

3.4.1 Microwaves

In order to replace wasteful thermal energy with a more economical type that can be targeted specifically at the molecules being analyzed, other forms of energy, *e.g.* photochemical and microwave, need to be applied.

Microwave ovens are in almost every kitchen, and they are now found in laboratories as well.[64] Microwave-assisted sample preparation techniques have been widely used in analytical laboratories all over the world.[65] The latest advances in the application of microwave techniques to various fields of analytical chemistry include: sample digestion for elemental analysis, solvent extraction, sample drying, measurement of moisture, analyte desorption and adsorption, sample clean-up, chromogenic reaction, speciation, and nebulization of sample solutions.[66]

Microwaves are between 1 m and 1 mm (0.3–300 GHz) long, and hence have similar frequencies to radar and telecommunication devices. The systems for household and industrial appliances are regulated and can operate on fixed frequencies (commonly 2.45 GHz). Microwave energy is a non-ionising type of radiation that causes molecular motion through the migration of ions and the rotation of dipoles without altering the molecular structure. The mechanism whereby microwave energy is absorbed by substances is complex and varies for different compounds. The absorbed energy is transferred to molecular kinetic energy and the sample heats up almost instantaneously. Substances that do not have a dipole moment (or in which one cannot be induced) are not directly heated by microwaves. Because of the large distances between molecules, gases are also not heated by microwave irradiation. Microwaves may be considered a more efficient source of heating than conventional heating since energy is directly imparted to the medium where the process takes place, rather than through the walls of a vessel.

Analytical equipment for microwave-assisted sample preparation can be classified according to: the type of microwave energy applied to the sample (multi-mode in cavity or focused by waveguide); the subjection or lack of subjection to overpressure (from open atmospheric pressure vessels to closed pressure vessels of more than 80 atm); or whether a dynamic (a flowing stream of mixture passing through a microwave field) or static (the stream of mixture is stopped when it is contained within a microwave field) approach is used. Generally, closed-vessel systems (under controlled pressure and temperature) are of the multi-mode type (the microwave radiation is dispersed into a cavity where the sample is located, so that it is randomly irradiated), and open-vessel systems (under atmospheric pressure, open or closed to the atmosphere) are of the focused type (the microwave radiation is focused on a restricted zone where the sample is placed).

Consequently, the effect of microwave energy is greatly dependent on the nature of both the solvent and the matrix. Usually, a solvent is chosen that strongly absorbs microwave energy. However, in some cases (for thermolabile compounds), the microwaves may be absorbed only by the matrix, resulting in heating of the sample and release of the solutes into the cold solvent.

Microwave-Accelerated Extraction (MAE). Microwave-accelerated extraction is based on heating an organic solvent by applying microwaves to the sample and extraction solvent. The main variables affecting the efficiency of MAE are the nature of the solvent, the temperature of the extraction and the extraction time. The solvent should selectively and efficiently solubilize the analytes in the sample but, at the same time, it should absorb the microwaves without over-heating (in order to avoid the eventual degradation of the analyte compounds). Thus, it is common practice to use a binary mixture, *e.g.* hexane–acetone (1 : 1), in which only one of the solvents absorbs microwaves. Moreover, to obtain reproducible results, the water content of the sample needs to be carefully controlled to avoid excessive heating, because water strongly absorbs microwaves. Typically, MAE is performed in static, batch mode, but it is also possible to use dynamic extraction. The simplest way to perform MAE is using a consumer microwave oven and glassware to hold the SPME fibre as shown in Figure 3.9.

High-efficiency extraction is obtained because of elevated temperature, as a result of the increased diffusivity of the solvent into the internal part of the matrix, and the enhanced desorption of the components from the active sites of the matrix. As a result, typical extraction times are 5–20 min. MAE uses organic solvents or solvent mixtures for extraction, the most common being methanol, isopropanol and hexane–acetone mixtures. The use of aqueous solutions and non-ionic surfactants as extractants is also possible. A major advantage of MAE is the ability to extract components selectively – either by selective heating of a phase or individual component of a system (*e.g.* free water molecules in vegetable cells), or by using an organic solvent which selectively extracts a particular component. In closed systems, pressure is also an important variable; however, this is directly dependent on the temperature. It must be remembered that under the high temperatures

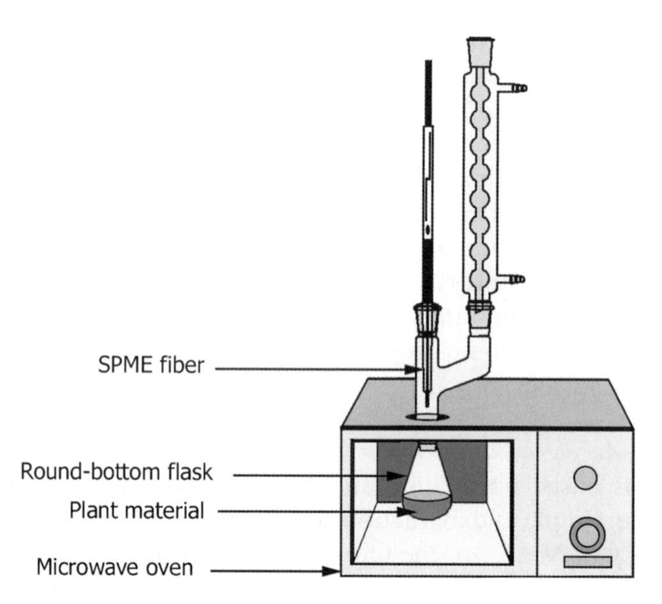

Figure 3.9 Simple system of microwave-accelerated extraction combined with SPME.

reached in closed systems with microwave heating, the properties of many solvents change significantly and they can even reach supercritical conditions.

Microwave-accelerated extraction (mainly in its focused energy form) is currently considered a good alternative to traditional Soxhlet extraction for different analytes in solid samples, because it reduces extraction time (*e.g.* 20–30 min per batch for as many as 12 samples); uses small amounts of solvents (30 mL in MAE *vs.* 300 mL in Soxhlet extraction); and improves extraction yields. The EPA has approved an MAE method (EPA Method 3546) for the extraction of organic compounds from solid environmental samples.[45]

The benefits of microwave-accelerated extraction include:

- speed – extraction can be completed within minutes;
- lower solvent usage – the costs of solvent purchase and disposal are reduced;
- higher analyte recoveries than older methods;
- a simple system, with no need to modify the existing chemistry;
- precision-computer control of all process parameters ensures method reproducibility.

The use of laboratory microwave ovens for organic extraction elevates sample preparation to the same level of sophistication as sample analysis. However, MAE does have several drawbacks: the extract must be filtered after extraction; polar solvents are needed; extract clean-up is almost always required (since MAE is very efficient); and the equipment is moderately expensive. Since the application of microwave energy to inflammable organic compounds can pose serious hazards, it is strongly recommended that only equipment approved for laboratory applications be used.

Microwave-Assisted Digestion (MAD). Conventional wet-sample preparation methods for the decomposition of solid samples for the analysis of metal content are usually carried out in vessels containing the sample with a large volume of digestion reagent(s) (typically 15 to 100 mL of nitric acid, alone or in combination with hydrochloric or hydrofluoric acid). These are then heated for long periods using a hot plate, heating mantle or oven. This type of open-vessel digestion has many drawbacks, including: the use of large volumes (and multiple additions) of reagents; a significant potential for contamination of the sample by materials and the laboratory environment; and the exposure of the analyst and the laboratory to corrosive fumes.

Closed-vessel microwave decomposition uses an entirely different technology to accomplish sample decomposition.[67] Decomposition of most solid samples can be achieved using near stoichiometric quantities of reagents, typically 10 mL, and can usually be completed in 10 to 15 minutes. This decrease in sample preparation time can be attributed to the higher pressure in closed vessels and the rapid heating of the sample mixture. The higher temperatures achieved almost instantaneously give microwave digestion a kinetic advantage over hot plate digestion. In a closed-vessel system, it is possible to retain even volatile elements. Good dissolution procedures depend on the specific temperature profile and the reagents that are used. Digestion is more complete because many acids exhibit improved oxidation potential at elevated temperatures. The temperature profile and reagents establish the mechanisms and kinetics of the reaction. This is not only important for complete dissolution, but also for reproducible extraction (leaching), analyte solubility, analyte volatility, species stability and, most importantly, safety. The most commonly used digestion reagents are nitric acid, hydrochloric acid, hydrofluoric acid, sulfuric acid, perchloric acid and hydrogen peroxide. The analyst will have to evaluate the reactivity of the acids with each specific matrix, because every matrix will have unique chemical interactions.

The application of microwave digestion has been described for a wide variety of sample types, including geological, biological, botanical, food, environmental, sludge, coal and ash, metallic and polymeric, as well as liquid samples. The EPA has approved several methods using microwave digestion: EPA Method 3052 – Microwave-Assisted Acid Digestion of Siliceous and Organically Based Matrices; EPA Method 3051a – Microwave-Assisted Acid Digestion of Sediments, Sludges, Soils, and Oils; and EPA Method 3015a – Microwave-Assisted Acid Leach of Aqueous Samples and Extracts.[45]

Microwave heating with high-pressure vessel technology allows reactions to be conducted at pressures and temperatures up to 200 atm and 280 °C. Research is also being undertaken on the automation of microwave dissolution procedures that should maintain the accuracy and precision of sample preparation and increase sample throughput. New equipment such as the focused microwave system, low-volume-sample microwave digestion, and microwave digestion for speciation especially in continuous-flow, is at the forefront of this field.

Continuous-flow microwave digestion generally involves the introduction of the sample in the form of slurry *via* an autosampler. Slurries are usually made by the pre-treatment of samples with an appropriate acid or mixture while being stirred in an open vessel. The slurry is then guided into the closed system in the microwave oven by a peristaltic pump. The slurry is digested as it passes through a coil made of a microwave–transparent material located inside the oven's cavity. After the dissolution stage, some systems have incorporated filters, cooling steps and gas vapour traps generated from the dissolution step. Finally, the digestate may be injected into a carrier stream *via* an injection valve where it is carried to the appropriate instrument for analysis, or collected in an analyzer autosampler tray. This digestion system is easy to use and provides a cost-effective alternative to the direct introduction of untreated samples, without decreasing the sample throughput.

High-pressure, microwave-assisted acid leaching methods provide rapid, reliable and very convenient sample preparation techniques for the screening of hazardous, bioavailable trace metals (*e.g.* cadmium, cobalt, chromium, copper, mercury, nickel, lead and zinc) in environmental samples (*e.g.* soils, sediments and sludges). However, the dependence of element recovery on applied leaching parameters (such as temperature, time and medium) is rather strong, but at optimized conditions, extraction recoveries for different standard reference materials are almost quantitative, with an excellent precision of analytical results.

Microwave heating accelerates reactions in digestion and it can substantially improve the dissolution of samples in less reactive conditions. When microwave energy penetrates a sample, the energy is absorbed by the sample (not by the solvent) and migration of the analytes out of the matrix (destruction of the macrostructure of the matrix) takes place. This process is quite different from conventional extraction, in which the solvent diffuses into the matrix and the analytes are removed from the matrix by solubilisation.

Microwave Drying, Microwave-Assisted Desorption and Adsorption. Microwave drying is a rapid and simple method for removing volatile compounds from a sample and it is more efficient than hot air drying. Usually, water is the major volatile compound that needs to be removed from a sample. During drying, owing to the direct absorption of radiation energy by water, heating and evaporation of the water occurs much more quickly than under standard conditions. As the moisture is removed, sample heating decreases and the sample temperature drops; therefore, the sample cannot be overheated. Microwave-assisted drying can also proceed at reduced pressure – water (or other microwave-absorbing solvents) is removed at lower temperatures, so the decomposition or loss of volatile organic compounds and the formation of localized hot spots are avoided to some extent. Microwave drying has also been applied to the drying of precipitates in gravimetric analysis.

The advantages of using microwave drying under vacuum conditions include: a milder thermal mode of drying; rapid removal of vapours during the shorter analysis time; and the minimization of degradation or oxidation of samples. Traditional techniques like oven- or air-drying are time consuming (8 h to a few days) and microwave vacuum drying of sediment provides a faster turn-around time for a higher number of samples. The same kind of procedure can be applied after digestion, to evaporate the solvent and concentrate the sample for further analysis. In recent years, microwave drying has gained popularity as an alternative drying method for marine sediments and a variety of food products, such as fruits, vegetables, snack foods and dairy products. There are some precautions to be taken in sample drying when using microwave procedures because of possible interactions between the microwave energy and the sample.

Nebulization with Microwaves. In atomic spectrometry, the sample introduction step has long been an important area for research. The thermospray (TSP) ionization technique has recently attracted increased attention because it generates a hot and very fine primary aerosol, thus making it possible for a high percentage of analyte to reach the atomization cell. Therefore, TSP performs much better than conventional pneumatic nebulizers in terms of sensitivity and limits of detection for most elements. Up until now, most studies on TSP adopted thermal conduction to heat a quartz or stainless-steel capillary to generate aerosols. With this heating method, there is a heat gradient from the capillary wall to the centre of the liquid stream. If the liquid contains a relatively large amount of analyte, or the temperature is not controlled accurately, the analyte will be deposited on the capillary wall. As a result, it is difficult to use a thermospray technique to analyze slurries or sample solutions with a high salt content. If the capillary is heated with microwave radiation, and since the quartz capillary wall is almost transparent to the microwave, the whole liquid vein is heated at the same rate. There is no heat gradient from the wall to the centre of the liquid stream, and analyte deposition on the wall will thus be minimized. Moreover, the solvent vapour will mix well with the remaining liquid, and, theoretically, the droplet diameter distribution of the aerosols produced will be much narrower. In most studies on TSP, to avoid localized overheating, the capillary diameter is very small (25–200 mm), whereas with microwaves, the capillary diameter can be up to 1 mm. Thus, microwave thermospray nebulization (MWTN) can work at a lower liquid pressure, and can be used to analyze sample solutions with high salt content or even slurries.[68]

Ionic liquids are attractive media for microwave (MW) radiation processes; because of their ionic nature, their ability to absorb MW radiation energy is extremely high.

Microwave irradiation has already been used in sample preparation in bioanalysis. Until the mechanism is better understood whereby microwave irradiation mediates an alternative catalytic mechanism to traditional heating methods, it is hard to predict which biochemical and analytical processes will benefit from microwave assistance. Nevertheless, it is known that microwave assistance increases the denaturation of proteins to allow easier accessibility of either proteolytic cleavage sites, for removal, post-translational modifications or amino acid residues for derivatization.[69]

3.4.2 Ultrasound

Ultrasound (US) is mainly used in analytical laboratories to improve or accelerate the preliminary steps of the analytical process. US aids in conditioning the sample (*e.g.* mixing or solubilizing) and transportating it to detectors (*e.g.* with US nebulization, levitation or slurry formation).[70] Ultrasonic equipment operating in the range of 20–100 kHz is relatively inexpensive and readily available (some high frequency equipment (500 kHz) has also been used for more fundamental mechanistic investigations).

Ultrasound energy, when applied to solutions, causes acoustic cavitation, *i.e.* bubble formation and implosion. The collapse of bubbles formed by ultrasound results in the generation of extremely high temperatures and pressures at the interface of the collapsing bubble and the other phase. Moreover, the influence of extremely high effective temperatures and pressures at the interface between an aqueous solution and a solid matrix, combined with the oxidative energy of hydroxyl radicals and hydrogen peroxide created during sonolysis of water, leads to enhanced chemical reactivity. Thus, acoustic cavitation provides a unique interaction of energy and matter, and a higher extractive power is obtained from the solvent.

Ultrasound-Accelerated Extraction (SAE). Ultrasound-accelerated extraction uses acoustic vibrations that cause cavitation in the liquid. Cavitation enhances the removal of analytes from the matrix surface and creates microenvironments with high temperatures and pressures. Therefore, sonication-assisted extraction is faster (5–30 min per sample) than the Soxhlet mode and allows large amounts of sample to be extracted at a relatively low cost. Solvents used in Soxhlet extraction are suitable for SAE.

Ultrasound-accelerated extraction is usually done statically in a discontinuous batch mode. However, dynamic SAE (DSAE) is also possible. DSAE has several advantages over conventional batch mode, one of which is that the sample is continuously exposed to fresh solvent, which improves the extraction kinetics. Furthermore, filtration and rinsing steps after extraction are avoided, and solvent consumption and the danger of loss and/or contamination of the extracted species during manipulation are minimized. The greater activity of the system (including solvents, analytes and matrix) increases efficiency, which may be better or similar to that obtained by Soxhlet.

Unfortunately, SAE uses as much solvent as Soxhlet extraction, and filtration is required after extraction. Moreover, it is labour intensive, since, apart from the polarity of the solvent, the efficiency of the extraction is dependent upon the nature and homogeneity of the sample matrix, the ultrasound frequency and the sonication time. Sonication-assisted extraction has been approved by the EPA in EPA Method 3550B (ref. 45).

Ultrasound Assisted Leaching. Ultrasonic radiation is another solid sample pre-treatment alternative, since it accelerates some steps, such as dissolution, fusion and leaching. The collapse of bubbles created by the sonication of solutions results in the generation of extremely high local temperature and pressure gradients, which may be regarded as localized "hot spots". It has been reported that the ultrasonic leaching method produces high recoveries of organics from sediment and soils, plants and biological materials, and elements from atmospheric particulate and street dust in a much shorter time than is required for other extraction procedures (1.5 h instead of 8 h). Ultrasonic leaching of metals from sediments, although not yet sufficiently exploited, could be an attractive alternative to conventional, acid bomb and microwave digestion, since, apart from the time required for digestion, unloading reactors is faster when neither high temperature nor pressure are present. The type and concentration of acid in the liquid extractant was found to be the most critical parameter affecting ultrasound leaching.

Ultrasound-assisted leaching offers fast, easy, reliable and efficient sample preparation for the direct determination of lead, copper, nickel, manganese and zinc in sediment samples by flame AAS.[71] All the parameters studied (sonication time, sample amount, particle size and extractant) influence leaching efficiency. Under optimum conditions, quantitative recoveries for all metals are reached in less time with a smaller amount of reagents, and the results obtained are comparable to those obtained by conventional sample pre-treatment based on acid digestion.

Ultrasonic-Assisted Matrix Solid-Phase Dispersion (UA-MSPD). An ultrasonic-assisted matrix solid-phase dispersion (UA-MSPD) method has been developed for extracting and cleaning up pesticides (OPPs) and triazines in a variety of fruits. No additional clean-up step is required before GC-MS analysis,[3] which increases the speed and efficiency of the MSPD process. For pesticides, ultrasonic energy has been reported to accelerate and improve extraction efficiency during solid–liquid extraction (SPE), and also when the cartridge is placed inside an ultrasonic bath. Compared with conventional MSPD, the proposed method improves extraction efficiency, decreases dispersion of results and allows complete sample treatment within a few minutes. These results, combined with low levels of detection (LODs), prove that the proposed procedure is suitable for the accurate determination of target analytes at levels set in current legislation, using an amount of sample as small as 100 mg.[72]

Sono-electroanalysis. Ultrasound has been applied to a range of modern analytical electrochemistry problems, and has allowed the sensitive determination of a wide number of analytes from a variety of otherwise hostile matrices. This combination of technologies has led to the improvement of techniques that have hitherto been relatively limited, mainly because of electrode passivation, reliability and sensitivity problems. Moreover, sono-electroanalysis offers a valid alternative to AAS and ICP-MS – standard methods currently used in the industry that are costly and time consuming. An additional advantage is that sono-voltammetric techniques are well suited to miniaturization, and portable versions of sono-voltammetric equipment are currently undergoing field trials.

Electrochemistry in biphasic systems is an area that had been exploited only minimally. This is because of the difficulties associated with creating and maintaining emulsions in the absence of surfactants that, if present, can detrimentally affect the analysis. The use of US to form emulsions ensures that, regardless of the relative densities of the two liquids, both kinds of droplet are in contact with the electrode surface during voltammetric analysis.

Ultrasound has been used to overcome the effects of electrode passivation, allowing sensitive electroanalysis to be carried out in hostile media such as blood, effluents, foodstuffs and fuels.[73] The introduction of ultrasound into voltammetric cells has a marked effect on the mass transport and surface activation characteristics of an electrochemical system. Mass transport is greatly increased *via* acoustic streaming and micro-jetting, resulting from cavitational collapse local to the electrode surface. In addition, surface activation *via* cavitational erosion can be used to activate otherwise passivated electrodes. The coupling of power-ultrasound with electroanalytical techniques dramatically improves reaction rates, analysis times, detection limits and electrode life-span, compared with those achieved under silent conditions.

The advantages of coupling power-ultrasound and electroanalytical techniques are greatly enhanced mass transport characteristics and electrode activation. The methods have been developed using readily available and relatively inexpensive equipment. Sono-electroanalysis combines a number of techniques that can be tailored to specific analytical situations, and have a wide range of real-world applications. Recently developed sono-electroanalysis protocols include the determination of copper in beer and blood; lead in wine, petrol and river-bed sediment; vanadium in aqueous media; nitrite in eggs; and manganese in instant tea granules.

Ultrasound-Assisted Levitation and Sample Transport. Ultrasound can expedite the sample reaching the detector through levitation, nebulization and slurry formation.[74] A crucial aspect of sampling and sample storage relates to the contact between the sample and the walls of the container. This situation is particularly problematic in microanalysis, where such effects can be decisive. An important aspect in this case is maintenance of the unaltered composition of the sample. Changes can be caused by adsorption of the analyte at the walls or desorption of either the analyte or interfering substances from the walls. The best way to minimize these alterations is to perform at least some of the steps of an analytical method, especially sample pre-treatment, without any contact between the sample and container walls. One way to treat small samples without such contact is to apply acoustical levitation. This sampling technique simplifies the analytical process and allows direct determination when used with suitable detectors. Since it maintains the levitated object in a fixed position, the process is not influenced by any surface contact other than the surrounding medium

Acoustical levitation requires no specific physical properties of the sample, in contrast to most levitation techniques, such as those based on electrostatic or magnetic fields. Consequently, almost all samples – solid or liquid, conductor or insulator, magnetic or nonmagnetic – are acoustically levitable. The maximum diameter of a levitated sample is a function of the US wavelength and is approximately half the wavelength under ambient atmospheric conditions. Usually, levitators operate with US frequencies of 15–100 kHz, resulting in wavelengths of 2.2–0.34 cm. However, the maximum volume of a drop is related to its liquid properties: specifically, surface tension and specific density.

The use of levitation techniques in analytical chemistry should comply with the following requirements:

- a stable sample position with effortless adjustment and measuring, *i.e.* a vertical and horizontal gradient of the levitating force is desirable;
- easy access to the sample;
- no special sample properties;
- a wide range of sample volumes;
- low cost of supply and operation.

Only acoustical levitation meets all the above requirements, at least to a certain degree. The use of this model of levitation in analytical chemistry is still in the early stages, so other techniques are currently more popular.

The free levitation of microsamples is a good alternative for trace and ultra-trace analysis because, after the necessary pre-concentration and clean-up steps using solvent evaporation in a levitated drop, it can result in a final sample volume of 1–2 μL (preceding injection into a GC-MS or CE, for example). With this containerless approach, the sample is easily accessed and effects such as contamination by compounds desorbing from, and analyte adsorption on container walls are reduced. It offers important advantages over many approved trace-analysis methods.

Ultrasonic slurry sampling has the following attractive qualities: decreased sample preparation time; avoidance of environmentally unfriendly solvents; decreased risk of contamination and analyte losses; and the benefits of solid–liquid sampling. Moreover, ultrasonic probes for homogenizing slurries provide additional advantages such as: the ability to automate slurry sample preparation and insert samples into an atomic detector; improved reproducibility and analyte extraction; and improved representativeness of the aliquot assayed. Typical shortcomings of slurry sampling, such as the difficulty of homogenizing slurries, are either minimized or removed by ultrasound assistance. Nevertheless, the influence of particle size is significant. Substantial improvements in the precision of the slurry technique have been achieved by using a small US probe to mix the slurry in the autosampler cup just before introducing it into the detection system. This agitation helps in the formation of fine slurry through the rupture of particles because of the cavitation phenomenon.

Direct injection by nebulization of an aqueous or organic phase containing the analytes of interest has been widely used with practically all types of absorption, emission or mass detectors based on atomization (*e.g.* by flame or plasma). In addition to the pneumatic method, ultrasound nebulization (USN) is becoming a common way to produce the aerosols that are used in plasma techniques. US nebulization in combination with a desolvation system has lower limits of detection than those provided by conventional pneumatic systems. USN has also made it possible to improve the introduction of organic solvent by overcoming overloading and pyrolysis effects. This is achieved by a combination of more efficient nebulization and desolvation of the organic aerosol by a heater and cooler system.

The use of microflow US nebulizers enables the coupling of liquid chromatographs (LC) and flow injection (FI) systems (for carrying out an on-line pre-concentration step) with atomic spectrometers. This also applies to mass spectrometers with electrospray ionization. In LC-MS, electrospray use has been limited by restrictions on mobile phase composition (low conductivity and without surfactants) and volumetric flow-rate (not more than $5\,\mathrm{mL\,min^{-1}}$) that make it amenable to electrospraying. The spray produced by US rather than an electric field makes spray formation independent of the physical properties of the LC mobile phase and flow-rate.[75] In this way, LC mobile phases can be used, extending the range to include high flow-rates, high electrical conductivities and high-surface tension solutions.

Coupling on-line, high-intensity focused ultrasound (HIFU) with a sequential injection-flow injection analysis (SIA-FIA) system[76] for metal analysis decreases the volume of sample and the amount of reagent required. High throughput can be achieved with automation and implementation of the sonication process in on-line systems. Note that inorganic mercury and total mercury can be separately determined in the same solution. When HIFU is turned off and the hypochlorite ion is not added to the solution, the signal corresponds only to the presence of inorganic mercury. On the other hand, when the sonication process takes place in the presence of the hypochlorite ion, the signal corresponds to all mercury species in the solution. These successful demonstrations are expected to stimulate new analytical experiments. For instance, columns allowing pre-concentration can easily be added to the system. Furthermore, the proposed methodology can also be used with ICP, ICP-MS, AFS, and, as a general rule, in any FIA or SIA system.

Acoustic forces are used to separate particles based on their size and density. The method has been shown to be suitable for biological and non-biological suspended particles. The microfluidic separation chips were fabricated using conventional micro-fabrication methods. Particle separation was accomplished by combining laminar flow with the axial acoustic primary radiation force in an ultrasonic standing wave field. Dissimilar suspended particles flowing through the 350 μm wide channel were thereby laterally translated to different regions of the laminar flow profile, which was split into multiple outlets for continuous fraction collection.[77]

Safety precautions must be observed with high-energy intensity techniques such as ultrasound: ultrasonication must be performed with care; manufacturers' safety instructions must be followed; experiments must be carried out in a fume cupboard; and hearing protection must be used.

REFERENCES

1. K. Cammann, *Fresenius J. Anal. Chem.*, 1992, **343**, 812.
2. K. Eckschlager and K. Danzer, *Information Theory in Analytical Chemistry*, Wiley, New York, 1994.
3. N. Fidalgo-Used, E. Blanco-Gonÿalez and A. Sanz-Medel, *Anal. Chim. Acta*, 2007, **590**, 1–16.
4. R. M. Smith, *J. Chromatogr. A*, 2003, **1000**, 3.
5. B. V. Ioffe, A. G. Vitenberg and I. A. Manatov, *Head-Space Analysis and Related Methods in Gas Chromatography*, Wiley-Interscience, 1984.

6. E. M. Thurman and M. S. Mills, *Solid-Phase Extraction, Principles and Practice*, Wiley, New York, 1998.
7. J. E Mark, *Physical Properties of Polymers Handbook*, Springer, 2nd edn, 2006.
8. M. Chanda and R. K. Salil, *Industrial Polymers, Specialty Polymers, and their Applications*, CRC, 2008.
9. J. Pawliszyn, *Solid Phase Microextraction: Theory and Practice*, Wiley-VCH, New York, 1997.
10. J. R. Dean and G. Xiong, *TrAC*, 2000, **19**, 553.
11. K. E. Geckeler, *Adv. Polym. Sci.*, 1995, **121**, 31–79.
12. L. Xu, C. Basheer and H. K. Lee, *J. Chromatogr., A*, 2007, **1152**, 84–192.
13. G. Ouyang, W. Zhao and J. Pawliszyn, *Anal. Chem.*, 2005, **77**, 8122–8128.
14. E. Aguilera-Herrador, R. Lucena, S. Cárdenas and M. Valcárcel, *J. Chromatogr., A*, 2009, **1216**, 5580–5587.
15. J. Lee, H. K. Lee, K. E. Rasmussen and S. Pedersen-Bjergaard, *Anal. Chim. Acta*, 2008, **624**, 253–268.
16. J. Stichlmair, J. Schmidt and R. Proplesch, *Chem. Eng. Sci.*, 1992, **47**, 3015–3022.
17. S. Pedersen-Bjergaard and K. E. Rasmussen, *TrAC*, 2008, **27**, 934–941.
18. S. Pedersen-Bjergaard and K. E. Rasmussen, *J. Chromatogr., A*, 2006, **1109**, 183–191.
19. I. J. Østegaard Kjelsen, A. Gjelstad, K. E. Rasmussen and S. Pedersen-Bjergaard, *J. Chromatogr., A*, 2008, **1180**, 1–9.
20. C. L. Arthur and J. Pawliszyn, *Anal. Chem.*, 1990, **62**, 2145–2148.
21. H. Lord and J. Pawliszyn, *J. Chromatogr., A*, 2000, **885**, 153–193.
22. J. Pawliszyn, *Solid Phase Microextraction: Theory and Practice*, Wiley-VCH, New York, 1997.
23. F. M. Musteata and J. Pawliszyn, *TrAC*, 2007, **26**, 36–45.
24. http://www.sigmaaldrich.com/analytical-chromatography/analytical-products.html (last accessed 03/01/2010).
25. (a) J. B. Quintana and I. Rodriguez, *Anal. Bioanal. Chem.*, 2006, **384**, 1447. (b) M. Kawaguchi, R. Ito, K. Saito and H. Nakazawa, *J. Pharm. Biomed. Anal.*, 2006, 40, 500.
26. S. de Koning, H.-G. Janssen and U. A. Th. Brinkman, *Chromatographia*, 2009, **69**, S33–S78.
27. K. Dettmer and W. Engewald, *Anal. Bioanal. Chem.*, 2002, **373**, 490–500.
28. W. Engewald, J. Teske and J. Efer, *J. Chromatogr., A*, 1999, **842**, 143–161.

29. Ch. Reichardt, *Solvents and Solvent Effects in Organic Chemistry*, Wiley–VCH, Weinheim, 3rd edn, 2003.
30. P. Müller, *Pure Appl. Chem.*, 1994, **66**, 1077–1184.
31. K. Alfonsi, J. Colberg, P. J. Dunn, T. Fevig, S. Jennings, T. A. Johnson, H. P. Kleine, C. Knight, M. A. Nagy, D. A. Perry and M. Stefaniak, *Green Chem.*, 2008, **10**, 31–36.
32. C. Capello, U. Fischer and K. Hungerbühler, *Green Chem.*, 2007, **9**, 927–934.
33. G. Koller, U. Fischer and K. Hungerbühler, *Ind. Eng. Chem. Res.*, 2000, **39**, 960–972.
34. *Environmental Management – Life Cycle Assessment – Principles and Framework, EN ISO 14040*, European Committee for Standardisation, Brussels, Belgium, 1997
35. C. Capello, S. Hellweg and K. Hungerbühler, *The Ecosolvent Tool*, ETH Zürich, 2006; http://www.sust-chem.ethz.ch/tools/ecosolvent (last accessed 03/01/2010).
36. S. Aparicio, S. Halajian, R. Alcalde, B. García and J. M. Leal, *Chem. Phys. Lett.*, 2008, **454**, 49–55.
37. J. M. Harris and S. Zalipsky, *Poly(ethylene glycol): Chemistry and Biological Applications*, ACS Books, Washington, DC, 1997.
38. *Aqueous Biphasic Separations: Biomolecules to Metal Ions*, ed. R. D. Rogers and M. A. Eiteman, Plenum Press, New York, 1995
39. meaning: "modern" or "recent in origin" derived from Greek *neoterikos* meaning "younger"
40. *Handbook of Fluorous Chemistry*, ed. J. A. Gladysz, D. P. Curran and I. T. Horváth, Wiley–VCH, Weinheim, 2004
41. L. A. Blanchard, D. Hancu, E. J. Beckman and J. F. Brennecke, *Nature*, 1999, **399**, 28–29.
42. T. Andrews, *Proc. R. Soc. London*, 1875, **76**(24), 455.
43. S. Pilat and M. Godlewicz, *US Pat.* 2188013, 1940
44. (a) M. A. McHugh and V. J. Kurkonis, *Supercritical Fluid Extraction, Principles and Practice*, Butterworths, 1986; (b) G. H. Brunner, *Gas Extraction, An Introduction to Fundamentals of Supercritical Fluids and the Application to Separation Processes*, Darmstadt, Springer, New York, 1994; (c) L. T. Taylor, *Supercritical Fluid Extraction*, Wiley Interscience, 1996; (d) J. M. Desimone and W. Tumas, *Green Chemistry Using Liquid and Supercritical Carbon Dioxide (Green Chemistry Series)*, Oxford Press, 2003; (e) *Supercritical Fluid Technology in Materials Science and Engineering: Syntheses: Properties, and Applications*, ed. Ya-Ping Sun, CRC, 2002; (f) G. H. Brunner, *Supercritical Fluids as Solvents and Reaction Media*, Elsevier Science, 2004.

45. http://www.epa.gov/SW-846/pdfs/ (last accessed 03/01/2010).
46. E. Klesper, A. H. Corwin and D. A. Turner, *J. Org. Chem.*, 1962, **27**, 700–701.
47. T. L. Chester and J. D. Pinkston, *Anal. Chem.*, 2004, **76**, 4606–4613.
48. P. Walden, *Bull. Acad. Imp. Sci. St. Petersburg*, 1914, 405–422.
49. W. L. Hough, M. Smiglak, H. Rodríguez, R. P. Swatloski, S. K. Spear, D. T. Daly, J. Pernak, J. E. Grisel, R. D. Carliss, M. D. Soutullo, J. H. Davis Jr and R. D. Rogers, *New J. Chem.*, 2007, **31**, 1429–1436.
50. D. Appleby, C. L. Hussey, K. R. Seddon and J. E. Turp, *Nature*, 1986, **323**, 614.
51. J. L. Anderson and D. W. Armstrong, *Anal. Chem.*, 2003, **75**, 4851–4858.
52. J. Ding, T. Welton and D. W. Armstrong, *Anal. Chem.*, 2004, **76**, 6819–6822.
53. R. Kaliszan, M. P. Marszall, M. J. Markuszewski, T. Baczek and J. Pernak, *J. Chromatogr., A*, 2004, **1030**, 263–271.
54. W. Qin, H. Wei and S. F. Y. Li, *J. Chromatogr., A*, 2003, **985**, 447–454.
55. M. Vaher, M. Koel and M. Kaljurand, *Electrophoresis*, 2002, **23**, 426–430.
56. D. W. Armstrong, L. -K. Zhang, L. He and M. L. Gross, *Anal. Chem.*, 2001, **73**, 3679–3686.
57. M. Zabet-Moghaddam, E. Heinzle and A. Tholey, *Rapid Commun. Mass Spectrom.*, 2004, **18**, 41–148.
58. M. C. Buzzeo, R. G. Evans and R. G. Compton, *ChemPhysChem*, 2004, **5**, 1106–1120.
59. S. Stolte, S. Abdulkarim, J. Arning, A. -K. Blomeyer-Nienstedt, U. Bottin-Weber, M. Matzke, J. Ranke, B. Jastorff and J. Thöming, *Green Chem.*, 2008, **10**, 214–224.
60. J. Ranke, S. Stolte, R. Störmann, J. Arning and B. Jastorff, *Chem. Rev.*, 2007, **107**, 2183–2206.
61. *Ionic Liquids in Chemical Analysis*, ed. M.Koel, CRC Press, Taylor & Francis Group, 2009.
62. T. Hyötyläinen, *J. Chromatogr., A*, 2007, **1153**, 14–28.
63. D. Raynie and J. L. Driver, presented at the 13th Green Chemistry and Engineering Conference, Maryland, USA, 2009.
64. *Microwaves in Organic Synthesis*, ed. A. Loupy, Wiley VCH, 2nd edn, 2006
65. *Microwave-Enhanced Chemistry: Fundamentals, Sample Preparation, and Applications*, ed., H. M. Kingston and S. J. Haswell, ACS, 1997.

66. K. Srogi, *Anal. Lett.*, 2006, **39**, 1261–1288.
67. R. Richter, *Clean Chemistry: Techniques for the Modern Laboratory*, Milestone Press, 2003.
68. Q. Jin, F. Liang, H. Zhang, L. Zhao, Y. Huan and D. Song, *TrAC*, 1999, **18**, 479.
69. J. R. Lill, E. S. Ingle, P. S. Liu, V. Pham and W. N. Sandoval, *Mass Specrom. Rev.*, 2007, **26**, 657–667.
70. F. Priego-Capote and M. D. Luque de Castro, *TrAC*, 2004, **23**, 829–838.
71. H. Güngör and A. Elik, *Microchem. J.*, 2007, **86**, 65–70.
72. J. J. Ramos, R. Rial-Otero, L. Ramos and J. L. Capelo, *J. Chromatogr., A*, 2008, **1212**, 145–149.
73. A. J. Saterlay and R. G. Compton, *Fresenius J. Anal. Chem.*, 2000, **367**, 308–313.
74. M. D. Luque de Castro and J. L. Luque-García, *Acceleration and Automation of Solid Sample Treatment*, Elsevier, 2002.
75. J. F. Banks, J. P. Quinn and C. M. Whitehouse, *Anal. Chem.*, 1994, **66**, 3688.
76. C. Fernandez, A. C. L. Conceio, R. Rial-Otero, C. Vaz and J. L. Capelo, *Anal. Chem.*, 2006, **78**, 2494–2499.
77. F. Petersson, L. Åberg, A.-M. Swärd-Nilsson and T. Laurell, *Anal. Chem.*, 2007, **79**, 5117–5123.
78. Th. P. Wampler, *Applied Pyrolysis Handbook*, CRC Press, Boca Raton, 2nd ed., 2006.

CHAPTER 4

Green Instrumental Analysis

4.1 ASSESSMENT OF ANALYTICAL METHODS FOR "GREENNESS"

Could the concepts of Green Chemistry or the concepts of a green world view be applied to instrumental analysis methods? As Keith *et al.* state in their conceptual introduction to Green Analytical Chemistry: "...the path toward greening analytical methodologies includes incremental improvements in established methods as well as quantum leaps that completely rethink an analytical approach. Strategies used include changing or modifying the reagents and solvents, reducing chemicals used through automation and advanced flow techniques, miniaturization, and even eliminating sampling by measuring analytes *in situ*, on-line, or in the field".[1]

Various means of assessing chemical methods for greenness have been developed. One such approach, that includes assessment methods proposed by Keith *et al.*[1] and Raynie,[2] has been described in Chapter 2. It allows the comparison of one method with another. Since the use of such proposals is just beginning and scores for most methods are lacking, the authors will take a more qualitative approach in the following paragraphs. We will attempt to illustrate how the ideas that are put forward by Keith *et al.*[1] could be implemented in analytical practice, to demonstrate that many instrumental methods meet the principles of Green Chemistry, and to present existing analytical methods that are inherently green. The opportunities to transform traditional analytical methods to green ones will also be discussed. The general structure of the following paragraphs is as follows: first, the principles of the green

Green Analytical Chemistry
By Mihkel Koel and Mihkel Kaljurand

Published by the Royal Society of Chemistry, www.rsc.org

method are described (if the method itself is not widely known), then its green aspects are discussed, and finally, some recent applications are cited to demonstrate how the method is used.

4.2 GREENING FLOW INJECTION ANALYSIS

Flow injection analysis (FIA), developed by Ruzicka and Hansen, is a simple automated analytical technique intended to replace the manual handling of solutions (known as "beaker chemistry") in instrumental analysis.[3,4] It is computer compatible and allows automated handling of sample and reagent solutions with strict control of reaction conditions. In its simplest form, the sample zone is injected into a flowing carrier stream of reagent. As the injected zone moves downstream, the sample solution disperses into the reagent, causing the product to form. If a flow-through detector is placed downstream, it records the desired physical parameter, such as absorbance or fluorescence. FIA is a widely used analytical method that can significantly reduce sample consumption and is easy to automate. Figure 4.1 illustrates various opportunities to use advanced FIA manifolds to achieve the goals of Green Analytical Chemistry.[5] However, their application in analytical chemistry is often hampered by the complexity of the required manifolds, particularly for multi-component analysis.

A variant of FIA, sequential injection analysis (SIA), can overcome this dilemma to some extent and may be an alternative approach to process analytical chemistry.[6] The apparatus for SIA consists of a bi-directional pump, a holding coil, a multi-position valve, and a flow-through detector (see Figure 4.2). The valve in the SIA manifold is initially positioned in such a way that it enables the whole system to be filled with a carrier stream. After the sample zone has been drawn up into the holding coil, the selection valve is advanced to a port connected to a reagent reservoir and a reagent zone is drawn up into the holding coil. In this way, it is possible to construct a stack of well-defined zones that can be mixed together to produce a detectable species. Flow acceleration and reversal help to blend these zones, due to the parabolic profile induced by differences between the flow velocities of the adjacent streamlines. Eventually, the multi-position valve is switched to the detector position and the flow direction is reversed, forcing the sample/reagent zones to flow through the detector cell. Unlike FIA, which requires re-plumbing each time if a more complex chemical addition scheme is required, all that is required in SIA is a change in the flow program because the manifold remains the same. The versatility of FIA/SIA equipment can be increased even further by one of the recent innovations in flow systems

a) Replacement of toxic reagents
Determination of Fe using Guava leaf extracts by UV

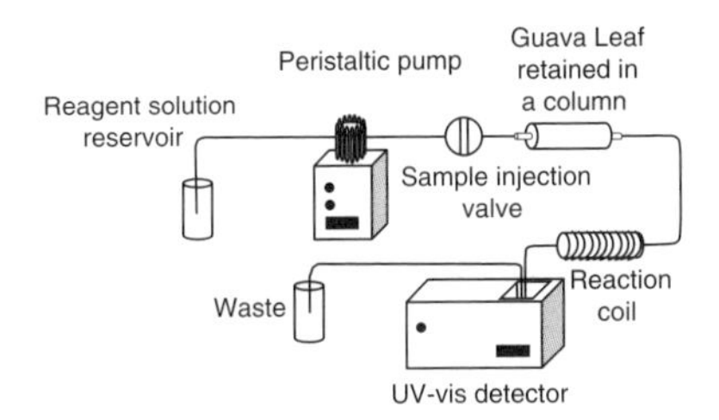

b) Minimization of wastes
Determination of chloride in waters by UV

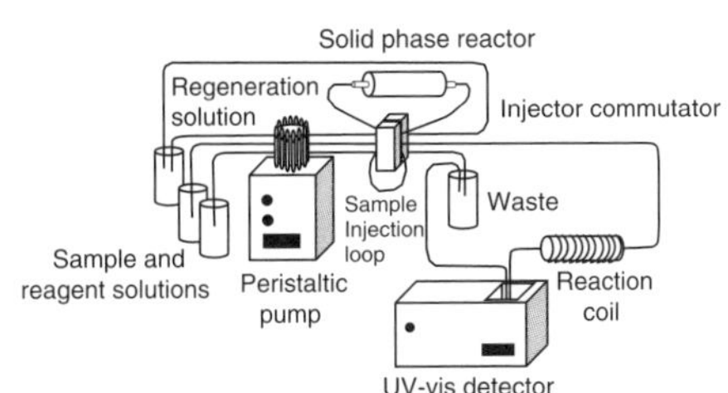

c) Recovery of reagents
Determination of Pb in gasoline by UV

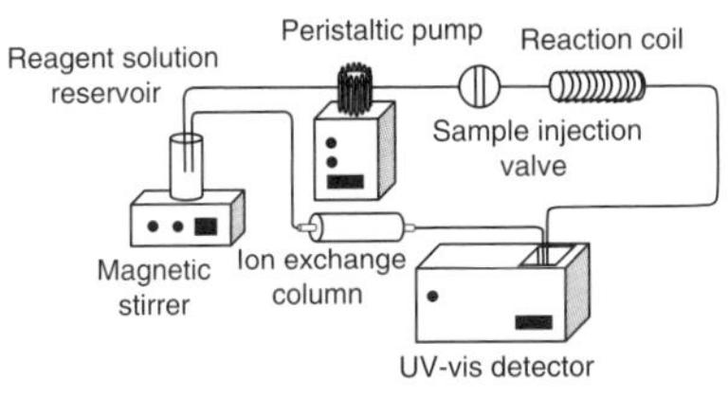

Determination of propyphenazone and caffeine in pharmaceuticals by FTIR

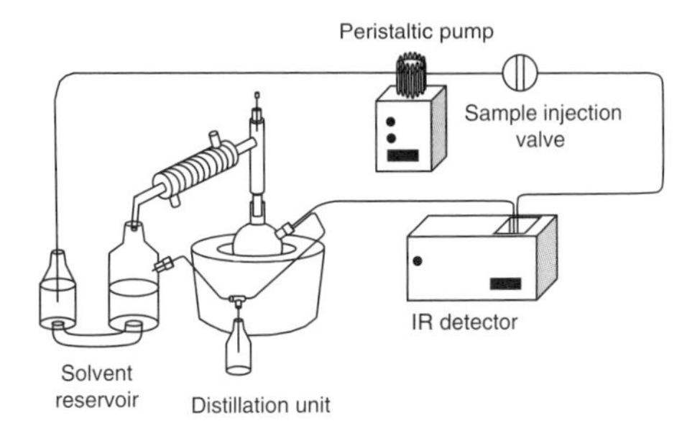

d) On-line decontamination of wastes
Determination of formetanate in waters by UV

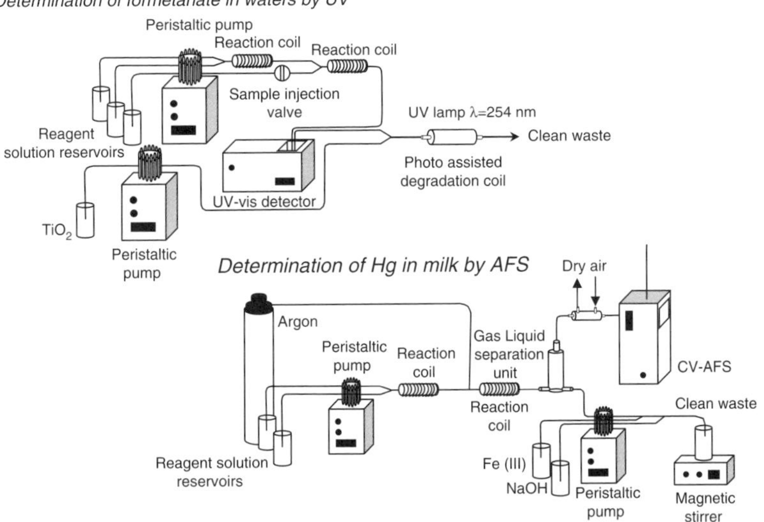

Figure 4.1 Basic components of manifolds used in multi-commutation FIA designed to: (a) replace toxic reagents, (b) minimize wastes, (c) recover reagents and (d) decontaminate wastes on-line.[5] Schematics are self-explanatory. (Reproduced with kind permission of Elsevier).

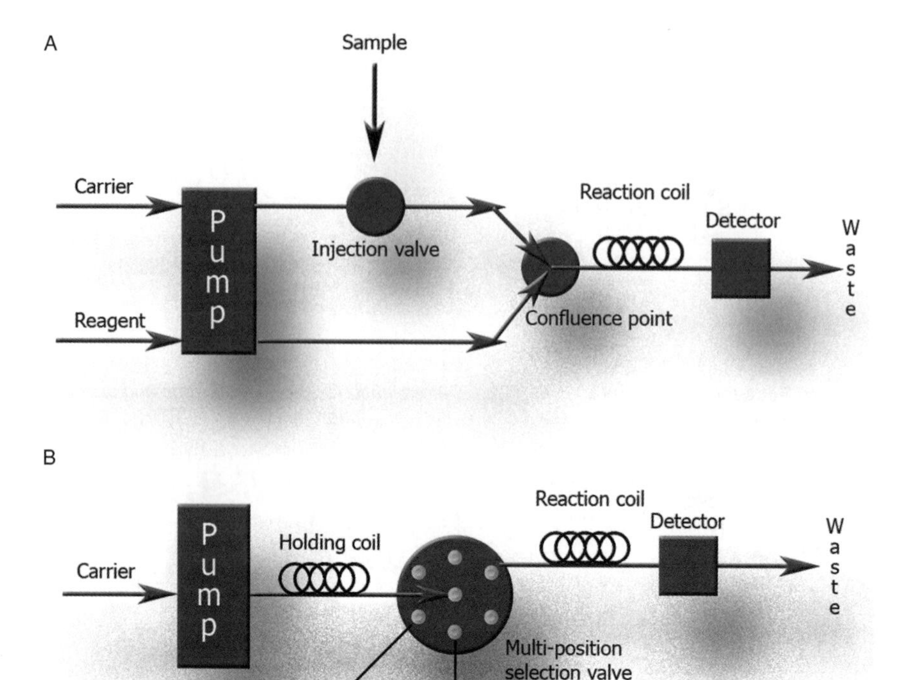

Figure 4.2 FIA (A) *versus* SIA (B) manifolds.

called *multi-commutation.*[7] This technique uses discrete commutation devices (*e.g.* three-way solenoid valves or mini-pumps) to build dynamic manifolds that can be reconfigured by software.

Sequential injection analysis has several advantages over FIA from the standpoint of Green Analytical Chemistry:[8]

- reagent use is drastically reduced. Typical FIA experiments make use of at least 1 mL of reagent per measurement. SIA typically makes use of 50 μL. This means that in a 24 hour period, assuming one measurement per minute, the FIA analyzer would consume 1440 mL of reagent and the SIA analyzer would consume 72 mL. This also means that the production of potentially hazardous wastes is minimized;
- flow manifolds are simple and robust – comprised of a pump, selection valve and detector connected by tubing. The same manifold can be used for widely different chemistries simply by changing the flow program rather than the plumbing;

- analyzer maintenance is simplified;
- components used in SIA manifolds are amenable to laboratory, field, and plant operation. In addition to these, SIA has all of the advantages of FIA.

A drastic reduction in both reagent consumption and waste generation are therefore common features of analytical procedures based on advanced FIA/SIA methods. These methods present increasing potential, given that attention is now being paid to analytical procedures that are friendly to the environment, in keeping with the "green" chemistry paradigm. For example, an SIA system has been coupled on-line to a high-intensity focused ultrasound probe and used for trace analysis of inorganic and total mercury in waters and urine by cold vapour atomic absorption spectrometry.[9] This resulted in a reduction of the amount of chemicals used and handled in the laboratory, thereby minimizing the risk of contamination.

We list below more examples of reduced solvent consumption FIA methods that are characterized by the authors of these publications as green methods. In one of these green methods, flow injection chemiluminescence using controlled-reagent-release technology was investigated for rapid and sensitive monitoring of sub nanogram amounts of chlorpyrifos[10] and tsumacide.[11] The analytical reagents were immobilized on an anion-exchange column. A green FIA method for ultra-sensitive determination of mercury in natural waters by electrothermal AAS was proposed by Gil *et al.*[12] Correia *et al.*[13] designed a flow-injection spectrophotometric method for the determination of creatinine, in which reagent consumption was reduced by 60% compared with the corresponding batch procedure. This experiment serves as a good example for educational purposes, since it illustrates a Green Analytical Chemistry approach to determining creatinine in real or synthetic urine samples. It can be used as an introduction to Green Chemistry for undergraduate students because the amount and type of reagents consumed and wastes generated can easily be calculated. A T-shaped micro fluidic manifold on a PMMA chip has been fabricated by laser ablation.[14,15] The chip served as a basis for a simple, inexpensive and reagent-less, colorimetric micro-flow analysis system that was applied in the determination of iron in water samples. A new, green flow-injection method for the determination of hypochloride in bleaching products was developed by March and Simonet.[16] In order to achieve high selectivity, a mini-column containing cobalt oxide was inserted in the flow system. The column catalyzed the decomposition of hypochlorite to chloride and oxygen. A green, flow-based procedure for the fluorimetric determination of acid-dissociable cyanide in natural waters exploiting multi-commutation was proposed by Infante *et al.*[17]

A greener analytical procedure based on flow-injection solid-phase spectrophotometry has been proposed by Teixeira and Rocha for iron determination.[18] This research focused on selecting a buffer and reducing agent that took into consideration both analytical performance and toxicity. Iron(II) was reversibly retained on an immobilized reagent, 1-(2-thiazolylazo)-2-naphthol on C18-bonded silica, and eluted as iron(II) with a small volume of dilute acid solution without removing the immobilized reagent, which could be used for at least 100 determinations. This procedure was ten times more sensitive than the analogous procedure based on measurements in solution. It had a good level of accuracy and precision for the determination of iron in water samples. The proposed procedure resulted in reduced effluent generation (3.6 mL per determination) and consumed micro amounts of reagents. In a paper by Passos *et al.,* the application of SIA in forensic science and legal medicine was discussed in accordance with the principles of Green Analytical Chemistry, and an SIA methodology was developed for the estimation of time since death (known as the postmortem interval).[19] The method was based on the determination of hypoxanthine and potassium in the same sample of vitreous humor, since the concentrations of both parameters change with the post-mortem interval and vitreous humor is regarded as the ideal extracellular fluid for these kinds of determinations. The method had a low environmental impact as only 2.7 mL of chemical waste were produced during both determinations.

4.3 CHEMICAL SENSORS

In general terms, a sensor can be defined as a device that allows the transduction of physical and/or chemical information at an interface of two phases to an electrical signal. Chemical sensors are devices that respond to changes in sample concentration in the chemical environment. They are small devices comprised of a recognition element, a transduction element and a signal processor capable of continuously and reversibly reporting a chemical concentration (see Figure 4.3).[20] Besides being a device that rapidly transforms the concentration of information about a specific compound into an electric signal, and maintains its activity over a long period of time, the sensor is usually cheap and small, which makes it of interest from the standpoint of Green Analytical Chemistry. From this perspective, the small dimensions are conducive to portability, minimal solvent/ reagent consumption and waste production (to the nanoliter level). A high degree of integration, efficiency, speed and

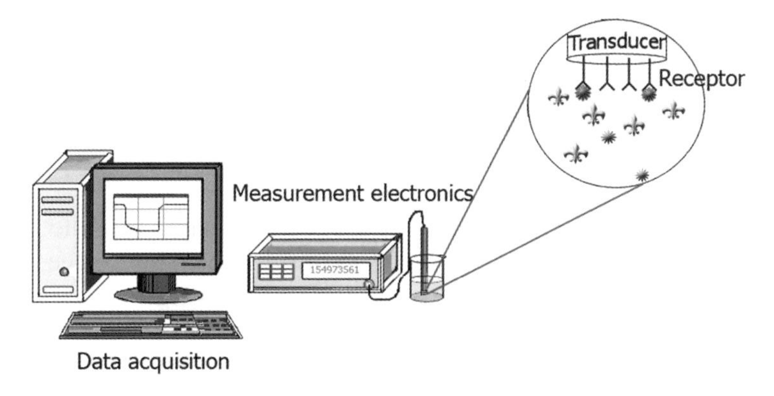

Figure 4.3 Typical arrangement of a chemical sensor.

disposability make such systems attractive for on-site environmental or industrial applications. Sensors hold considerable promise for faster and simpler *in situ* monitoring of priority pollutants. They facilitate the design of real-time, on-site monitoring technologies that successfully address the time constraints associated with classical laboratory analysis. The new generation of sensors offers tremendous potential for obtaining the desired analytical information in a faster, simpler and cheaper manner compared with traditional laboratory-based instruments.

According to Lieberzeit and Dickert,[21] despite the great potential of sensors, sensor research nevertheless remains essentially confined to research laboratories, and often has not yet reached the stage of technical development. The amount of literature published in recent years on real life examples is very small. They point out that a key issue for measurement in real life environments is the complexity of the surrounding media and the possibility of abrupt changes in the background matrix. Therefore, instead of one single sensor, the use of sensor arrays (each with rather low selectivity for each individual analyte) followed by chemometric analysis of the data is becoming increasingly popular. This means that chemometric results rather than concentrations of compounds, *e.g.* principal component analysis of sensor array data (see Chapter 6.1), are correlated with reference data.

The following sensor groups have been developed:

- optical sensors based on absorbance, reflectance, fluorescence, refractive index, optothermal effects and light scattering;
- electrochemical sensors based on voltammetry and potentiometry;
- electrical sensors based on metal oxides, organic semiconductors and electrolytic conductivity;

- mass sensitive detectors based on piezoelectric devices and surface acoustic waves;
- paramagnetic sensors for oxygen;
- thermometric sensors based on the thermal effects of chemical reactions.

Although most optical and semiconductor sensors meet the criteria for green analytical devices, we will not consider them here since there is abundant literature on these sensors. In the following two paragraphs we will restrict our discussion to two lesser known but emerging types of sensors: surface acoustic wave sensors and surface plasmon resonance sensors. They will be described in detail.

4.3.1 Surface Acoustic Wave Sensors

Surface acoustic wave (SAW) sensor-based analytical methods are examples of direct analytical procedures that do not require a sample preparation step. Therefore, SAW sensors may be considered as typical examples of environmentally friendly devices used for the determination of volatile organic compounds (VOCs). These sensors are made *via* a photolithographic process by depositing metal (usually aluminium) electrodes onto a carefully polished and cleaned piezoelectric substrate. The specific pattern of metal remaining on the device is called an interdigital transducer ((IDT); see Figure 4.4).

As the acoustic wave generated by the input electrodes propagates on the surface of the material, any changes to the characteristics of the

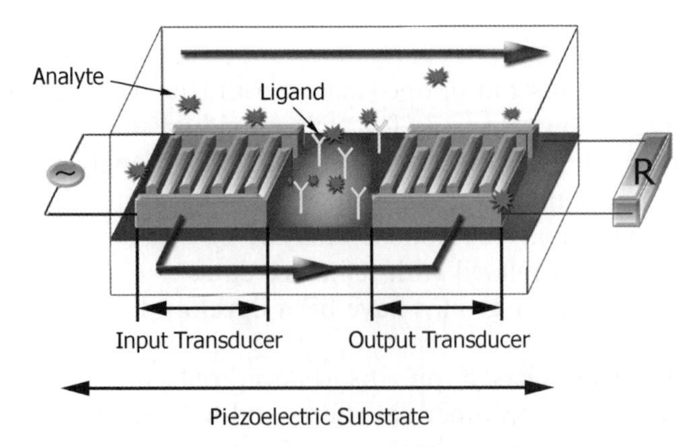

Figure 4.4 A typical acoustic wave device consists of two sets of interdigital transducers. One transducer converts electric field energy into mechanical wave energy; the other converts the mechanical energy back into an electric field.

propagation path affect the velocity and/or amplitude of the wave which can be detected by output electrodes. Changes in velocity can be monitored by measuring the frequency or phase characteristics of the acoustic wave and correlated to the corresponding physical quantity being measured. By changing the length, width, position and thickness of the IDT, the sensitivity of the sensor to a certain analyte can be maximized. SAW sensors become vapour sensors when a coating is applied to the sensor surface which absorbs only specific chemical vapours. SAW devices work by effectively measuring the mass of the absorbed vapour. A detection limit of about $0.5\,mg\,m^{-3}$ (dimethylmethylphosphonate) has been reported.[22] A biosensor can be constructed by coating the SAW surface with a material that absorbs specific biolochemicals in liquids.

There are thousands of publications on SAW sensors. The recent trend in this field involves coupling sensors into arrays capable of detecting different chemicals. Ensuring internal security is an important application for such sensors.[23] In this vein, an "electronic nose" was developed that is based on an array of SAW sensors implementing three different polymer coatings for the classification of warfare agent simulants using chemometrics.[24] Similarly, information was published on various SAW handheld sensor systems for the detection and quantification of explosives and chemical warfare agents.[25] Landmines can be detected by SAW sensor arrays, since fluoropolyol-coated sensors can detect as little as 235 ppt 2.4-DNT.[26] Besides internal security, food odour and the classification of food products is an area for the possible application of SAW sensor arrays, *e.g.* wine analysis using a zinc oxide SAW sensor array,[27] SAW immunosensors could be developed based on immobilized C60-proteins.[28]

4.3.2 Surface Plasmon Resonance Sensors

Surface plasmon resonance phenomena can be observed at an interface between two transparent media of different refractive indices (*e.g.* glass and water). If the light comes from the side of higher refractive index, then a total internal reflection is observed above a certain critical angle of incidence. While incident light is completely reflected, the electromagnetic field component penetrates a short distance (tens of nanometers) into a medium of a lower refractive index, creating an exponentially decreasing evanescent wave. If the interface between the media is coated with a thin layer of metal (typically gold), light is monochromatic and p-polarized. The intensity of the reflected light is reduced at a specific incident angle (producing a sharp shadow), due to the excitation of the metal electron gas (or specifically, surface pseudo-particles – plasmons) by an evanescent wave. This resonance energy transfer between light and surface plasmons is known as surface plasmon resonance (SPR). As is the case with SAW sensors, the resonance conditions are greatly influenced by the material adsorbed onto the thin metal film manifesting as a shift of the resonant angle. The direction of the reflection of light with minimum intensity shifts when an analyte molecule binds to the ligand attached to the surface of the gold layer. This fact makes SPR a prospective method for the study of biochemical interactions if one component of the reaction is bonded to the metal film and the solution of the other component flows over the metal surface (see Figure 4.5). Interactions of relevant molecules, such as proteins, sugars and DNA, can be studied. The SPR signal, which is expressed in resonance units, is therefore a measure of sample concentration at the metal surface. This means that analyte and ligand association and dissociation can be observed and, ultimately, rate constants as well as equilibrium constants can be calculated.

The bulkiness (the size of a gas chromatograph) and cost (several hundred thousand US dollars) of SPR laboratory instruments is an incentive to miniaturize SPR methodology.[29–32] Commercial SPR sensors are based mainly on Texas Instruments' Spreeta 2000 SPR sensor components, first described by Chinowsky.[33] The dimensions of this SPR sensor are $3\,cm \times 0.7\,cm \times 1.5\,cm$, making it an ideal device for Green Analytical Chemistry. The light source, detector and gold surface necessary for SPR sensing are combined in one small, inexpensive moulded package. Spreeta's construction is simple and self-evident from Figure 4.6. The light beam (infrared LED at $\lambda = 830\,nm$) passes a plastic sheet polarizer, strikes the active sensor surface at a range of angles both above and below the critical angle, reflects from the sensor surface, then reflects from the sensor's top mirror back down onto the diode array detector. Each detector pixel will collect light that strikes the sensor surface at a different angle; a reflectivity *versus* angle spectrum may be obtained by reading the detector array.

The cost of Spreeta is low (fifty US dollars) making it attractive for disposable applications and making SPR sensor technology feasible for medical diagnostics, environmental monitoring, and food and water safety. It also facilitates inexpensive, quantitative, point-of-care (POC) biosensing instrumentation for use in the hospital, physician's office, emergency transport, or anywhere POC testing is needed. For environmental monitoring, SPR sensor instrumentation can be sufficiently lightweight and low powered to be battery operated and used in the field to check for soil or water contamination. For food and water safety

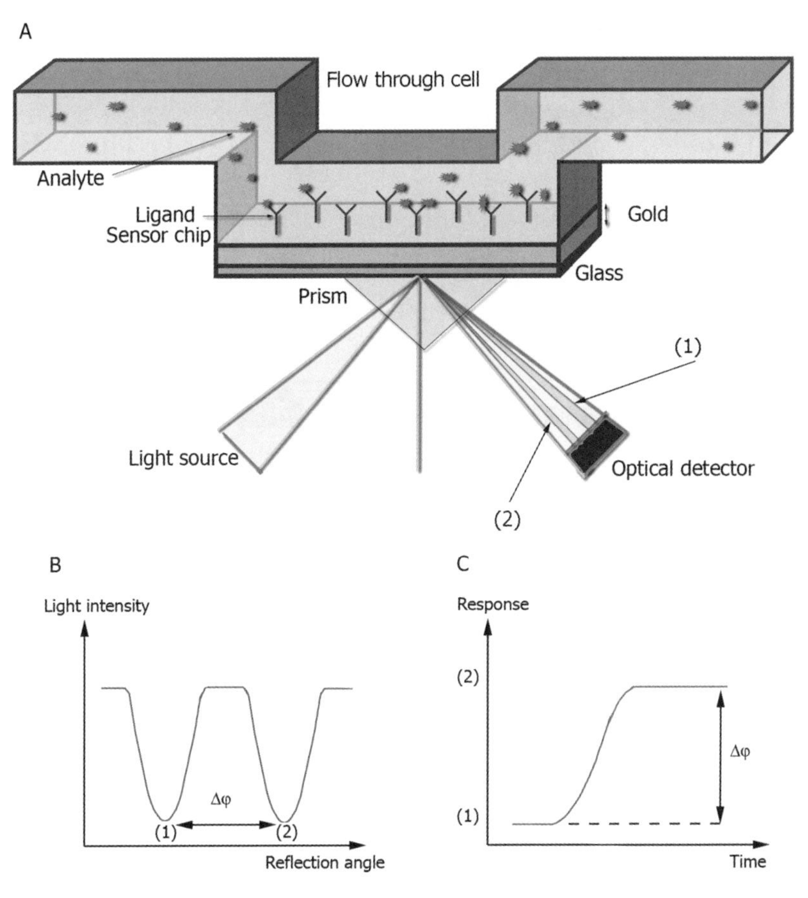

Figure 4.5 Experimental set-up for observing SPR (A), intensity of reflected light as a function of incident angle (B); the sensorgram is a plot of the change in the angle of reflected minimum over time (C).

applications, disposable SPR sensors enable real-time process control, for instance, in monitoring antibiotic content in meat and dairy production. Some examples of recent applications of SPR sensors include: airborne analyte detection using an aircraft-adapted SPR sensor;[34] the detection of the toxin domoic acid from clam extracts;[35] the rapid detection of six distinct analytes ranging from small molecules to whole microbes during the course of a single experiment;[36] the detection of cortisol in saliva;[37] and antibody-linked magnetic nanoparticles.[38]

4.3.3 Microelectrode Sensor Arrays

Electrochemical processes occurring on the surface of an electrode are the basis of many chemical sensors. The combination of

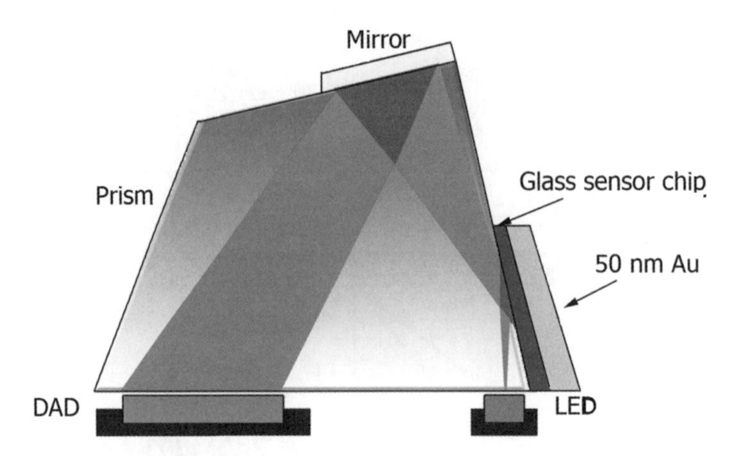

Figure 4.6 SPEETA surface plasmon resonance sensor. The sensor consists of a plastic prism moulded to a printed circuit board.[33]

modern electrochemical methods with advances in microelectronics and in microfluidics (see Chapter 5.2.3 for fabrication details) makes possible the utilization of powerful analytical devices for effective process or pollution control. The fact that miniaturized electrochemical systems (EC) can meet the needs of Green Analytical Chemistry was pointed out by Wang as early as 2002.[39] He wrote that the demand for greener analytical chemistry provides new challenges for electroanalytical chemistry, and for developing greener analyses in general.

A good example of greening electroanalysis is discontinuing the use of mercury. For a century, mercury has been the most common electrode material in electrochemical (EC) analysis due to its very attractive properties (reproducible, renewable and smooth-surfaced), but routine applications of mercury-based procedures can result in the daily generation of a few grams of mercury waste.[39] Due to the toxicity of mercury, greener electrode materials like bismuth-film[40] or environmentally friendly carbon electrodes need to be found. The latter have been widely used for more than three decades.

New electrochemical sensor systems should be capable of monitoring an environment without polluting the system or affecting it in any negative manner. However, common electrochemical working electrodes are large – on the scale of millimeters or centimeters. From the point of view of Green Analytical Chemistry, practical advances in the miniaturization of electrodes offer some fundamental advantages: reduced resistance (ohmic drop), ability to incorporate many electrodes in a small area, and increased ability to facilitate measurements in low ionic strength water samples. Photolithography offers needed opportunities for the microminiaturization of electrodes through the fabrication of electrode systems on planar glass substrates. The downscaling and integration of chemical assays make these analytical microsystems attractive as Green Analytical Chemistry screening tools. Other attractive factors are: high sensitivity, approaching that of fluorescence; low power requirements; low cost; ability to be miniaturized; and adaptability to advanced micromachining and microfabrication technologies. The amount of waste generated could be reduced by about 4–5 orders of magnitude compared with that of conventional liquid chromatographic assays (*e.g.* 10 μL *versus* 1 L per daily use).[39] Such a significant improvement in the rate of waste generation and material consumption has enormous implications for Green Chemistry.

The principal advantage of using miniaturized electrodes is the mass transport enhancement due to the nonlinear diffusion properties, which are further amplified when multiple electrodes are utilized in an array. Due to the larger perimeter-to-surface area, and nonlinear radial diffusion through the diffusion zones, microelectrodes attain larger response changes and improved signal-to-noise ratios. By combining microelectrodes into an array (MEA), one can benefit from the enhanced properties inherent in the smaller microelectrode size. With the use of a multi-channel potentiostat, one can use all the microelectrodes independently, cycling through each electrode, scanning at different potential ranges, or use them simultaneously with all electrodes performing in unison to increase their individual contribution while allowing greater current flow.

An example of such a microelectrode array device was reported by Gardner *et al.*[41] It utilized miniaturized microelectrodes that reside in a planar array that allowed trace levels of various ions to be detected. The sensors were constructed by implementing basic photolithography methodology utilized in integrated circuit fabrication. First, layers of titanium (bonding agent) and gold were deposited by electron beam sputtering to the exterior of a glass microscope slide. Second, gold-coated slides were stripped to exact geometries (specified for the 54 individually isolated electrodes) by spin coating the glass slide with a photoresist and exposing it to a UV lamp through a mask that defined the electrode configuration. Third, the slide was developed with gold and titanium etching solutions. The remaining metal comprised the contact pads, circuitry and electrode pads on the slides. The final step in sensor construction was the adhesion of sterol plastic wells over the microelectrode arrays (see Figure 4.7).

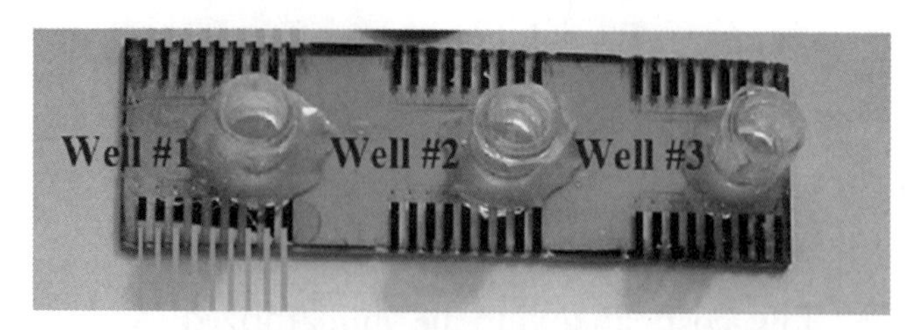

Figure 4.7 A microelectrode array.[41] (Reproduced with kind permission of Elsevier).

The ability of the sensor platform to detect copper and lead heavy metal ions in aqueous solutions was also proven by utilizing the electrochemical method of differential-pulse anodic stripping voltammetry.

In a publication by Guenat *et al.*,[42] the fabrication of an ion-selective microelectrode array platform for *in vitro* intracellular recording is described. The electrodes had inner and outer diameters of 5 and 6 μm, respectively. The Ca^{2+}-selective microelectrodes were made of silicon nitride. The array had a linear range from 10^{-6} to 0.1 M for a Ca^{2+} solution. A paper by Liu *et al.*[43] describes how a gold microelectrode array, manufactured with a microfabrication technique, was applied to the measurement of dissolved oxygen profile in an aerobic granule. In this instance, the MEA contained five gold microelectrodes which had a linear response to dissolved oxygen. The experimental results demonstrate that the MEA was able to measure dissolved oxygen levels in aerobic granules accurately and precisely, and that the MEA could be used to determine constituents, profiles and functions *in situ* in small spaces. In another publication, a microelectrode array for the indirect amperometric measurement of iodate content in very small volumes (typically 50–200 μL) was described.[44] Anjos and Hahn constructed and researched two different gold microelectrode arrays – one with 37 electrodes with a diameter of 20 μm and the other with 256 electrodes with diameters of 5 μm.[45] They describe the use of a practical sensor design, incorporating a porous polytetrafluoroethylene membrane-covered gold microelectrode array, for the simultaneous measurement of oxygen and carbon dioxide. It was demonstrated that gold microelectrode arrays can be successfully used to interrogate the electrochemical reduction of both oxygen and carbon dioxide individually, and measure their concentrations in tertiary gas mixtures. A sensor array based on a 64-channel indium tin oxide (ITO) electrode array modified with an enzyme, horseradish peroxidase, and an electron transfer mediator was developed by Nahoko *et al.*[46] for determining hydrogen peroxide (H_2O_2). H_2O_2 has been found to be associated with neurological disease pathology because it is a by-product of a degenerative reaction of reactive oxygen species, one of the major causes of oxidative stress in mammalian cells. An electrochemical sensor array has also been proposed by Mugweru *et al.*[47] for glucose monitoring.

The use of low temperature co-fired ceramics technology – a technology that bonds together thin ceramic sheets – made possible the fabrication of an integrated microelectrode array on a ceramic support as reported by Ciosec *et al.*[48] Potassium-selective membranes were applied on the surface of PdAg/AgCl electrodes formed on the ceramic substrate. The array of microelectrodes covered with polymeric layers of various selectivities was applied as an "electronic tongue" to differentiate between various diet supplements. In another design, a complete electrochemical "cell-on-a-chip" that uses a micro disk electrode array was proposed. This design was evaluated as a prototype for long-term biosensors for eventual intramuscular implantation for the continuous amperometric monitoring of glucose and lactate.[49] The device was comprised of two discrete electrochemical "cells-on-a-chip", each with a reference, counter and working electrode array. Each MEA is comprised of 37 microdisks (positioned in a hexagonal arrangement).

4.3.4 Bioanalytical Methods (Immunoassay Techniques)

Immunoassay techniques are well-known; in this section therefore we will discuss only their green aspects. Immunoassays implement antibodies for the molecular recognition of analytes. An immunoassay is a direct analytical technique (a preparation step is not necessary) and it may be considered as a typical example of a procedure that is more friendly to the environment than many other methods. The second advantage of biology-based analytical screening techniques is the complete replacement of organic solvents with aqueous media, which reduces toxic waste. In addition to bioanalysis, immunological methods are used in environmental chemistry. Environmental immunoassays have been developed and evaluated for analytes including the major classes of pesticides, organic compounds as polychlorinated biphenyls (PCBs), polyaromatic hydrocarbons (PAHs), pentachlorophenols (PCPs), BTEX (benzene, toluene, ethylbenzene and xylene), dioxins and furans, and some inorganic substances, for example, cadmium, lead, mercury, and microbial toxins.[50]

4.4 SPECTROSCOPIC METHODS

Spectroscopy was originally conceived as a study of the interaction between electromagnetic radiation and matter. The intensity of the interaction is represented as a function (spectrum) of wavelength. The concept has been expanded to include the study of the distribution of the amount (intensity) of a certain feature of matter as a function of the value of that feature (*e.g.* mass). Analytical spectroscopy has rarely been examined from the standpoint of greenness. One example, however, is the paper by He *et al.* who list several features of green spectroscopic methods.[51] The following list is based on the general ideology of Green Analytical Chemistry:

- the analytical method produces no pollutants, *i.e.* has low or even no reagent cost, waste release and elimination of highly toxic reagents from analytical procedures;
- it is time, labour and energy efficient;
- it requires little or no sample preparation;
- it involves little or no sample destruction;
- it allows real-time process analysis/monitoring *in situ*/*in vivo*;
- it can be implemented by simple and portable instrumentation while maintaining high selectivity and sensitivity.

Obviously, one method cannot meet all the requirements. Also, it is probable that not all analytes can be determined by green analytical procedures.

4.4.1 Transforming Spectroscopy to Green Analytical Chemistry through Simplification and Portability

Portability is an important and obvious feature of a green analytical instrument, since the consumption of resources (either power or chemical) and the generation of waste are central issues. Portable instruments can be taken to the analysis site, *i.e.* to the point of care (POC). The POC can be a hospital, home or crime scene. The portability of an instrument is generally understood to be its potential to work in the field. It is believed that portable instruments are more economical than their bulky counterparts.

If portability is not an option and commercially available bulky laboratory instruments are used, a step in the direction of greener methods is usually modification of sample preparation. Several popular instrumental measurements use continuous-flow sample introduction that requires a considerable amount of sample. Flame atomic absorption spectrometry (FAAS) is an example of a widely used spectroscopic technique for metal determination. The sample used in FAAS is usually an acidified solution of metal ions. Flow sample introduction consumes a significant amount of sample solution (50 mL) that is mostly discarded due to the nebulizer construction. By contrast, electrothermal vaporization is another sample introduction technique in AAS that requires only a small amount of sample (50 μL), and can perform direct solid sampling. Thus, electrothermal vaporization AAS could be considered as a green alternative to FAAS that should be used whenever possible. However, the situation is not so straightforward: electrothermal vaporization requires special sample cells made of graphite that are expensive and are not durable. Moreover, not all metals lend themselves to analysis by electrothermal vaporization.

A green alternative is to analyze the sample *in situ*, as advocated by He *et al.*[51] In fact, many field analyses could be accomplished without traditional sampling. Field analytical chemistry (FAC) is a growing trend in analytical chemistry because it requires little or no sample collection and sample preparation, and eliminates transportation of the sample to a laboratory. The requirements of *in situ* analysis have already been addressed above. As we saw in Chapter 2, it was pointed out that in order to perform field analyses, the ideal analytical instruments should meet several requirements; they should:

1. have a fast instrumental response time for the acquisition of necessary information on a real-time or near real-time basis;
2. be capable of *in situ*/at-site rather than just on-site* analysis, as well as little or no requirement for sample preparation;
3. be portable for field use with a minimum requirement for power supply (battery-powered desirable), consumables (gases/solvents) and clean space for handling samples;
4. perform a cost-effective analysis.

However, to be useful at the point of care, portable instruments must meet several demands. A step out of the laboratory into the field requires a giant leap from technology design to capability. Turl and Wood list the following considerations for portable instruments:[52]

- ***Performance***: sensitivity; selectivity; suppression of response by interferents; false alarm rate; capacity for quantification; intra-instrument comparability; dynamic range; alarm threshold setting;

---

* He *et al.*[51] provide a definition of the terms "*in situ*" and "on site". In this book "on-site analysis" is understood to be a common analytical procedure that involves sample collection/preparation using a field-portable instrument. "*In situ*" analysis leaves the sample site (virtually) undisturbed. "*In situ*" analysis would be possible with an X-ray spectrometer, but it is difficult to imagine that it could be done by chromatography.

sampling time; ease of interpretation for non-technical operators; and servicing and cleaning intervals.

- ***Proper Sampling Method***: the instrument must not decline in performance; increase in operational false alarm rate; diminish in sensor lifetime; or increase in downtime due to maintenance.
- ***Power Consumption***: power consumption; availability and suitability of main supply and battery or fuel cell; life and recharging time.
- ***Environmental Issues***: thermal effects on efficacy; effect of contaminants such as dust or dirt on instrument; noise emission; vibration of mobile units causing system to fail; robustness; weather conditions – resistance to water, cold, overheating and emission of and susceptibility to electromagnetic interference.
- ***Portability***: weight and size; manual handling; number of personnel required to operate; possibility of modular design for ease of carriage; vehicle mounting and impact on deployment and response time.
- ***Health and Safety***: third-party health and safety liability; operator health and safety risks - both equipment- and threat-related; compliance with regulations such as Control of Substances Hazardous to Health; and ionizing radiation.
- ***Legal Aspects***: irradiating people with or without consent; taking and storing of images and other personal data with or without consent; legal powers which define the operation – search, surveillance, *etc.*; use of hazardous substances or radiation sources which require notification of regulatory bodies.
- ***Continuity***: effect of total failure or loss; Mean-Time-To-Failure (MTTF) and Mean-Time-Between-Failure (MTBF); cost of consumables; ease of servicing.
- ***Security***: classification of technical equipment in public view; interoperability with existing infrastructures – radio, Global System for Mobile communications (GSM), other agencies' (secure) systems and availability of performance data from manufacturers which could highlight capability gaps.
- ***Operator Training***: training at the correct level for non-technical personnel; initial and follow-up training; accreditation of training if deemed necessary and ongoing quality assurance of operator competence.
- ***Human Factors***: operator fatigue (can be minimised by effective shift rotation or continuous quality assurance); acceptability and impact, including public opinion; operator confidence in results leading to decision-making; crowd control and behavioural issues

and compliant *versus* enforced use; robust operating procedures including actions on an indication and levels of response.

The above list demonstrates that the portability of instrumentation adheres extremely well to the spirit of Green Analytical Chemistry.

If the greenness of a method is associated with portability, the key issue in green spectroscopy is how to make analytical instruments portable. Laboratory instruments are frequently large in size and as analytical methods typically involve a sample preparation step, a time- and labour-consuming sample collection step is usually needed to transport samples to a laboratory for analysis. Sample collection can often be problematic. For instance, when analyzing polluted soil, samples have to be collected in many locations according to a time consuming and laborious sampling plan, to avoid missing the "hot spot." Usually the samples must be treated with specific reagents and stored in sample containers under certain conditions to maintain their integrity before analysis. Furthermore, in many cases, there are difficulties in sampling, such as hazardous polluted sites or precious cultural relics and archaeological objects. FAC can not only eliminate sample transportation, but can also greatly shorten the analysis time or even provide real-time results, thereby affording rapid warning and accurate feedback. The development of portable instruments for field analysis is welcome because they are convenient to transport, compact, and require small amounts of sample, reagent, and energy during the analytical process. These requirements accord well with the principles of Green Analytical Chemistry. As a method of instrumental analysis, chromatography is not well suited to portability because of the need for eluents for analysis. There are, however, publications on field portable gas[53,54] and liquid chromatographs.[55] As we will demonstrate in Chapter 5.2.2.4, capillary electrophoresis is a promising technology for developing field instruments. Mass and optical spectroscopy are more conducive to portability. In the following paragraphs, portable versions of some well known spectroscopic methods, such as ion mobility, mass, laser ablation and X-ray emission spectroscopy, are described in detail.

4.4.1.1 Ion Mobility and Mass Spectroscopy. Electrophoresis in the gas phase, or ion mobility spectrometry (IMS), is well known as a basis for airport security systems,[56] for the detection of explosives,[57] and it is acquiring popularity as a field analysis method, especially for military applications for detecting chemical warfare agents.[58] The speed at which separations occur (typically in the order of 10 s of milliseconds) combined with its ease of use, high sensitivity, and highly compact design have

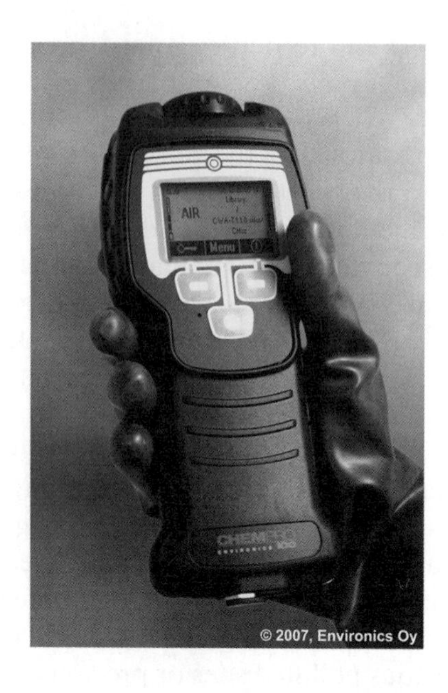

(a)

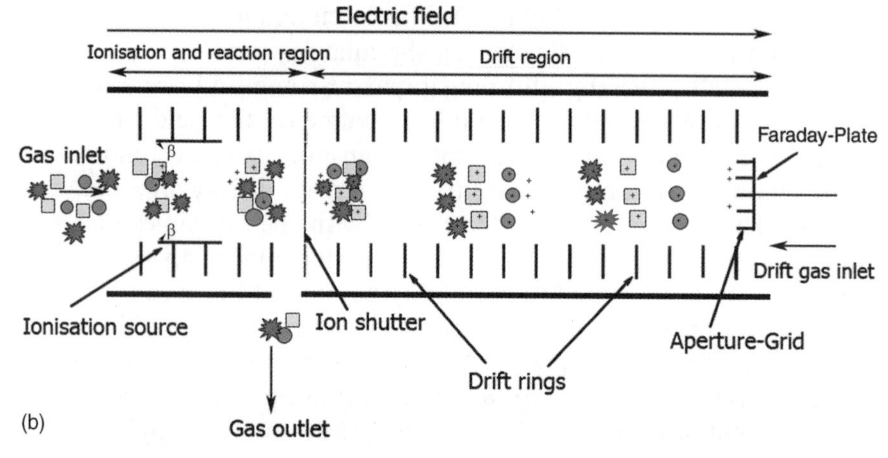

(b)

Figure 4.8 Ion mobility spectrometer. ((a) Picture of a portable IMS is reproduced with kind permission of Environics Oy;[59] (b) the schematics of an IMS are adapted from ref. 60).

encouraged the routine use of IMS as a commercial product for the field detection of the analytes mentioned above.[59] As shown in Figure 4.8, the principle of ion mobility spectrometry is very simple and the instrument can be made compact and convenient. In its simplest form, an IMS system measures how fast a particular ion moves through a given atmosphere in a uniform electric field.[60] The IMS system comprises a metallic tube consisting of two compartments – an ionization region and a drift region. Sample gas enters the ionization compartment and is ionized by various means: by corona discharge or electrospray, atmospheric pressure photoionization or a radioactive source, *e.g.* a small piece of ^{63}Ni or ^{241}Am, similar to the one used in ionization smoke detectors. Ionized analytes enter the drift region as follows: in specified intervals, a sample of the ions is let into the drift chamber; the gating mechanism is based on a charged electrode. Ions migrate against the flow of drift gas in the electric field generated by the drift ring electrodes. This electric field drives the ions through the drift tube where they interact with the neutral drift molecules contained within the system. Separation of chemical species is achieved based on ion mobility (a parameter that is dependent on ion mass, size and shape) whereupon they arrive at the detector for measurement. Analytes are detected on a Faraday plate.

A disadvantage of IMS is its poor resolution and broadening of the ion flight time distribution, caused by ions continually colliding with air molecules. For this reason, the rate of false alarms is high when other chemical substances are present in the background. In addition, if radioactive elements are used as an intense source for ionization at atmospheric pressure, this can give rise to environmental issues. An advantage of IMS is that it does not need a vacuum pump; therefore, it can be small in weight and size. When the inclusion of a vacuum pump into a field portable analytical instrument can be tolerated, FAC techniques may also include mass spectrometry.

Since mass spectroscopy is an established technique which has been well documented elsewhere,[61] we will not describe it here. The importance of MS has risen tremendously due to its implementation in proteomics. From the standpoint of sample consumption, MS can be considered a green method. However, the mass spectrometer is a bulky instrument that consumes a large amount of energy and is expensive to buy and maintain. Nevertheless, even MS can be miniaturized. For example, a palm portable mass spectrometer measuring $8.2 \times 7.7 \times 24.5\,cm^3$ and weighing 1.48 kg with an average power consumption of 5 W has been developed.[62] Its size and performance results point to its usefulness as a personal mobile chemical analyzer in the field (portable MSs are shown in Figure 4.9). The most recent advancements in miniature MS design have been reviewed by Ouyang *et al.*[63] The authors point out that these small mass spectrometers are not useful if their performance is too severely compromised. However, the combination of two existing technologies – the miniature mass spectrometer and desorption electrospray ionization – could lead to a new instrument that allows for the direct analysis of almost any type of sample in the ambient environment (such as the detection of cocaine on money).[64]

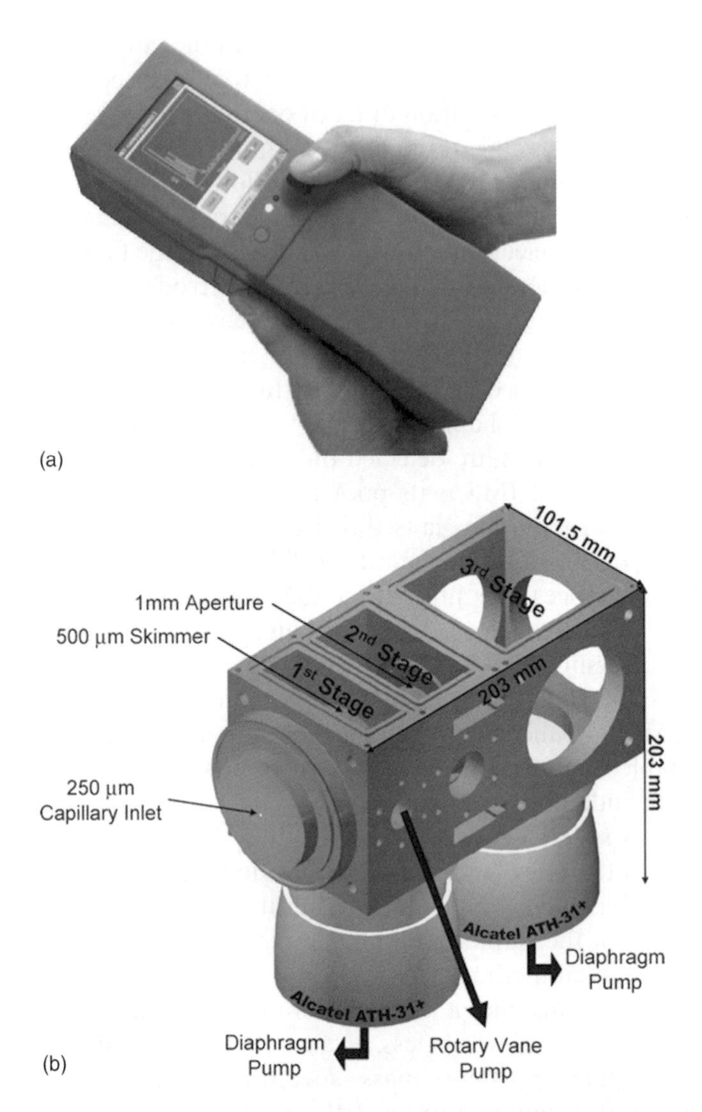

Figure 4.9 (a) A palm mass-spectrometer (reproduced with kind permission of Elsevier),[62] and (b) a portable mass spectrometer standing on two vacuum pumps (reproduced with kind permission of the American Chemical Society).[64]

4.4.1.2 Portable X-Ray Fluorescence (PXRF) for Multiple Metals. X-ray fluorescence is an emission of characteristic X-radiation from an atom, as a result of the interaction of electromagnetic (primary X-ray) radiation with its inner orbital electrons. Measuring X-ray fluorescence *in situ* is a typical example of a direct analytical procedure, where a sample preparation step is not necessary and may therefore be considered friendly to the environment. Portable XRF is one of the instrumental techniques that has developed rapidly in recent years, largely because of advances in miniaturization and semiconductor detector technology. The modern generation of instruments can truly be regarded as hand-held devices. Examples of such instrumentation are systems designed for extraterrestrial measurements, such as the PXRF on Beagle 2 (the Mars landing device for the European Space Agency's Mars Express mission). With a mass of just 280 g, this PXRF was designed for the measurement of rock and soil on the surface of Mars.[65] Like other XRF instrumentation, PXRF consists of an excitation source, sample positioning facility, detector and pulse processing and analysis facility. PXRF instruments are fast, non-destructive, and they provide instantaneous multi-element analysis results. They have high throughput, accuracy, and outstanding precision. In addition, these instruments are simple to operate by any technician (see Figure 4.10).

In situ analysis provided by PXRF has special meaning in the context of Green Analytical Chemistry. Since the analyzer is in contact with the surface of the sample, no sample preparation is necessary; therefore, the technique is particularly valuable in situations where it is not possible or

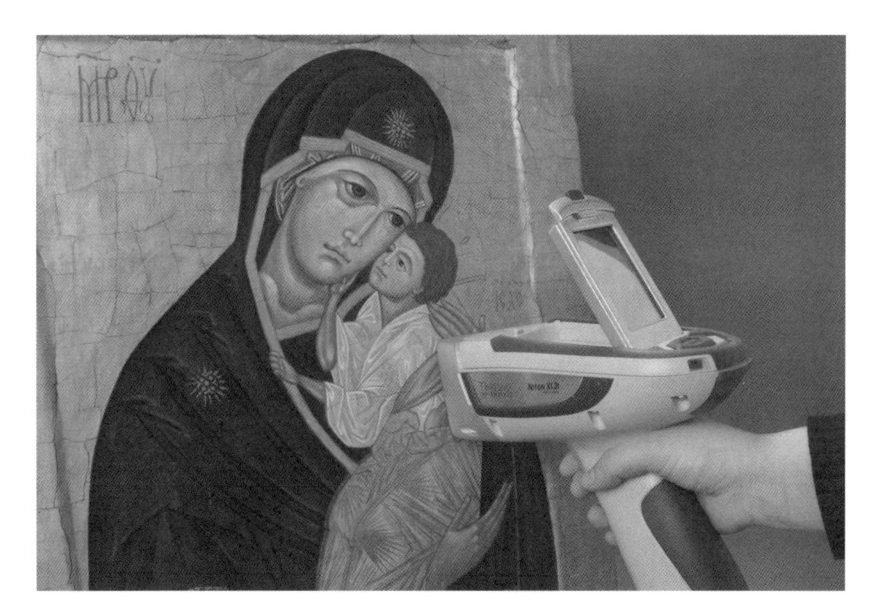

Figure 4.10 Using handheld XRF for screening. (Thermo Scientific Niton XL3t analyzers, reproduced with kind permission of Thermo Scientific).

desirable to remove the test object. Such objects could be unique and/or valuable archaeological artefacts, museum pieces and works of art, where it is not acceptable to remove a sample for conventional laboratory analysis because of the curators' duty of care or security issues. Similarly, test objects may be an integral part of either monumental stonework or a building. Clearly, in those cases the only acceptable technique has to be both portable and non-destructive. Those are the properties of PXRF which very few available analytical methods could rival.

Ironically, because they are inherently "green", the usage of PXRF instruments raises some green policy-related issues. These concern the health and safety of the operator. "XRF" and "hand-held" operation seem to be incompatible concepts, because of the risk associated with the exposure of the operator or others to ionizing radiation. However, modern instruments incorporate a range of safety features, including key operation, relevant interlocks, and contact sensors, all designed to minimize the risk of significant exposure.[65] It is essential that any operator be fully trained in these features.

The publication *Portable X-ray Fluorescence Spectrometry: Capabilities for in situ Analysis*[65] provides extensive information on PXRF applications, which we will not discuss in detail here. The book edited by Potts and West offers a general introduction to PXRF technique and applications, including the assessment of contaminated land, surfaces, coatings and paints, workplace monitoring, metal and alloy sorting, geochemical prospecting, archaeological investigations, museum samples and works of art, and extraterrestrial analysis.

4.4.1.3 Laser-Induced Breakdown Spectroscopy. Laser-induced breakdown spectroscopy (LIBS) is a mode of atomic spectroscopy for simultaneous multi-element analysis. It utilises a highly energetic laser pulse as the excitation source. The laser is focused onto the material of interest and when the laser is discharged it ablates a very small amount of material, in the range of nanograms to picograms, which generates a plasma plume with temperatures in excess of 100 000 K. Because all elements emit light of characteristic frequencies when excited to sufficiently high temperatures, LIBS can (in principle) detect all elements.[66] Besides the laser, LIBS devices also include a spectrograph and detector (see Figure 4.11). The instrument consists of a pulsed laser, optics for focusing the laser energy onto a sample surface, optics for collecting the light produced during the LIBS reaction and delivering it to a spectrometer system for resolution of the light

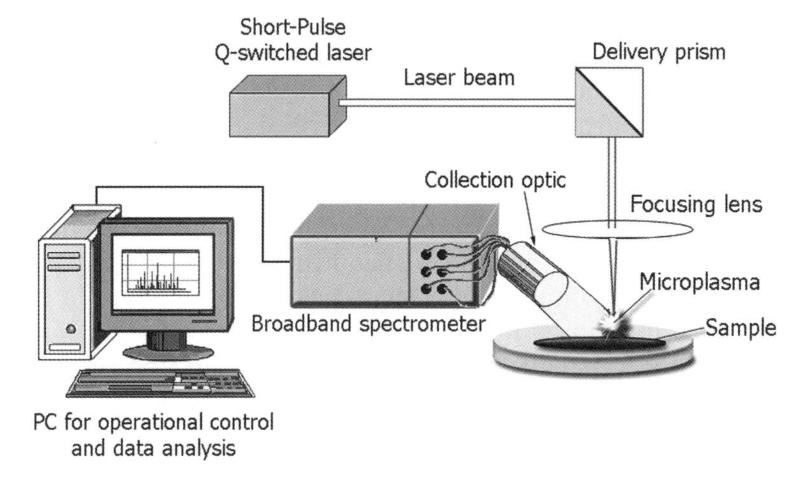

Figure 4.11 Schematic drawing of a LIBS.[66]

spectrum, and a computer for system control and data processing and analysis.

Laser-induced breakdown spectroscopy has several advantages from the point of view of Green Analytical Chemistry:

- it is considered essentially non-destructive or minimally-destructive (due to the small amount of material consumed);
- there is almost no heating of the specimen surrounding the ablation site, because an average power density of <1 W is radiated onto the specimen;
- sample preparation is typically minimized;
- depth profiling of a specimen is possible by repeatedly discharging the laser in the same position, and effectively going deeper into the specimen with each shot;
- it is a very rapid technique that gives results within seconds, making it particularly useful for on-line industrial monitoring;
- only optical access to the specimen is required, thus it is non-invasive, non-contact and can even be used as a standoff analytical technique when coupled to the appropriate telescopic apparatus.

Laser-induced breakdown spectroscopy is not a new analytical technique – the earliest application studies date from the 1980s. The simplicity of the technique and recent technological advances have led to the development of a man-portable LIBS sensor system. Compared with X-ray fluorescence spectrometers, portable LIBS systems are more sensitive, faster, can detect a wider range of elements, and do not utilise ionizing radiation to excite the sample (as X-ray fluorescence does), which is both penetrating and potentially carcinogenic. However, as the adjective "man-portable" indicates, the instrument is not "hand-held" like most of the portable instruments described in this book.

Recent man-portable LIBS applications include the determination of lead in road sediments,[67] copper determination in soil samples[68] and the *in situ* characterization of karstic formations (speleothems).[69] Identification analysis of the speleothems' alteration layers and depth profiles of strontium and calcium have been reported. In a completely different application, LIBS was evaluated for its terrorist attack detection capability, *i.e.* the detection of hazardous biological powders on indoor office surfaces and wipe materials.[70] An identification of pure unknown powders was performed by comparing them with a library of spectra containing biological agent surrogates and confusant materials, such as dusts, diesel, soot, natural and artificial sweeteners, and drink powders, using linear correlation analysis. *Bacillus subtilis* spores on wipe materials and office surfaces were successfully identified.

4.4.2 Fluorescence Excitation Emission Matrix Spectroscopy

Fluorescence excitation emission matrix (EEM) spectroscopy has long been recognized as a powerful method for complex mixture analysis, such as the characterization of dissolved organic material (DOM) in seawater, river water or in the contents of a bioreactor. The measurement principle of emission-excitation matrix measurements is multichannel fluorescence spectroscopy. The sample is irradiated sequentially with light of various wavelengths. For each excitation wavelength the fluorescence emission spectrum is measured. Assuming that the emission spectra are recorded digitally, as is common in contemporary instrumentation, the result is a two-dimensional matrix containing detailed information about the sample. The term *matrix* in this case means a mathematical object, not to be confused with the sample environment. The EEM spectrometer is a non-invasive sensor system that eliminates interference with the sample. Sampling and off-line analysis is therefore unnecessary, saving manpower, expense, and reducing the risk of contamination. Being non-invasive and informative, it is not surprising that EEM has found applications in the areas where the sample is complex, *e.g.* in bioprocess monitoring and environmental analysis. Another advantage is that in EEM spectrometers, sample pretreatment is usually reduced to sample dilution. This is characteristic of a green analytical method – traditionally time-consuming sample pre-treatment is minimized or eliminated completely. In addition, EEM spectrometer-sensors can facilitate probing of an analyte in its native environment.

Although every contemporary fluorescent spectrometer can record EEM spectra, EEM fluorescence measurements are frequently performed by a robust instrument like the BioView® sensor (Delta Light and Optics, Denmark),[71] designed for industrial applications. It uses two independently rotating filter wheels with 16 different filters each, with excitation of 270–550 nm and emission of 310–590 nm. The filters can be individually selected according to the process requirements. A version that contains up to 20 individual filters was also used in some of the experiments. Thus, the two-dimensional landscape of the measured matrix allows for the simultaneous detection of various fluorophors. In this instrument, the recording of a whole spectrum takes 90 seconds with a resolution of 20 nm. The BioView® sensor is packaged in a high-grade steel box for protection against harsh environments (*e.g.* high temperature, moisture) and electromagnetic interference.

In a paper by Hart and JiJi,[72] a design for single-measurement EEM spectroscopy is simplified even further. It is based on an array of light-emitting diodes (LED). The LED array is focused onto a sample cuvette, which creates spatially separated excitation spots. Fluorescence from analytes in solution is collected at right angles by another lens, which projects the fluorescent spots onto the entrance of a spectrograph with a CCD camera for detection. The broad emission spectrum of LEDs permits continuous coverage over a large excitation range with a limited number of LEDs, allowing excitation of all analytes with absorption within the LEDs excitation range. Successful implementation of low excitation resolution LED-EEM spectroscopy demonstrates that it is not always necessary to have high excitation wavelength resolution. As the authors demonstrate, the method performs well irrespective of the overlapping excitation wavelength regions for even heavily overlapped excitation ranges. Each analyte may then be excited uniquely by each LED, thereby conserving the multiple characteristics of the data common to all EEM methods. This knowledge makes possible many compromises in the development of a spectroscopic sensor system, such as reduced size, cost and complexity, without sacrificing analytical performance. Surprisingly, the genuinely benign features of EEM spectroscopy – simple construction conducive to field portability in conjunction with the elimination of sample preparation – have been hardly noticed so far by the research community: only Nahorniak and Booksh have mentioned EEM spectroscopy in this regard, albeit not in relation to Green Analytical Chemistry.[73]

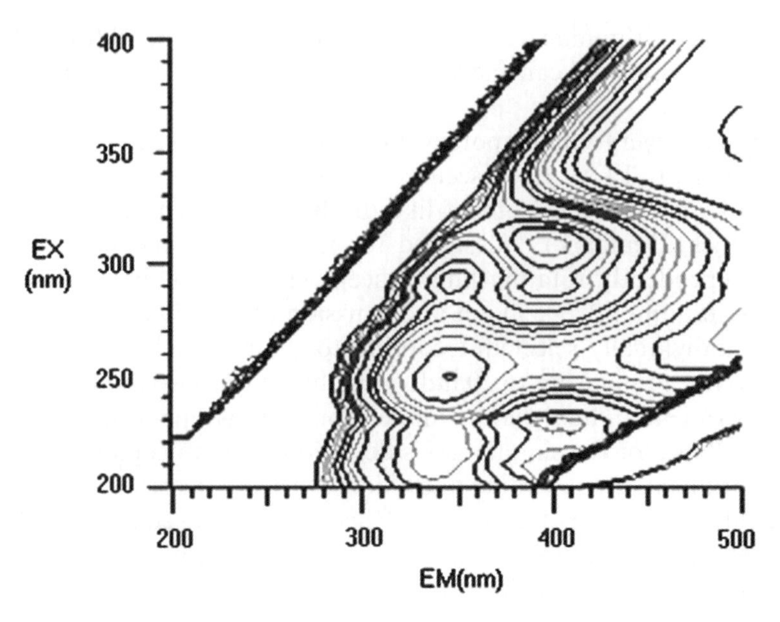

Figure 4.12 A contour plot of an EEM spectrum of human serum albumin complex with warfarin in aqueous buffer.

Typical EEM spectrometer output is usually represented as a 2D fluorescence contour map (see Figure 4.12). Note the absence of the upper triangular part of the data plane (Stokes' rule) and the diagonal feature in the right lower corner (diffraction harmonics) that is an instrumental artefact and unrelated to the sample. These signals should be avoided by selecting a suitable wavelength since they are unrelated to analyte concentrations.

The use of EEM fluorescence sensors in bioprocess monitoring is of great importance in biotechnology. With two-dimensional fluorescence spectroscopy, all fluorophores such as proteins, coenzymes and vitamins, can be simultaneously detected qualitatively and quantitatively in intra- or extracellular samples. As we saw above, the construction of EEM spectrometers makes them suitable for on-line, *in situ* measurement and they have been used in monitoring growth and metabolic changes. In addition to biotechnology, EEM spectroscopy can be used in many other analytical situations. Table 4.1 contains approximately one hundred examples, which give some idea of the possibilities of this environmentally benign method whose green capabilities, as was mentioned above, are not yet fully recognized. The data in Table 4.1 was collected up to February 2009 and is by no means exhaustive.[74–169]

Table 4.1 Applications of EEM fluorescence spectroscopy.

Topic	*Objects*	*Ref.*
Environmental science and technology	Sewage-impacted rivers	74
	Investigation and monitoring of contaminated sites	75
	Adulteration of petrol by kerosene	76
	PAH compounds present in petrol, diesel, kerosene and 2T oil; PAH in water samples; PAH detection in complex matrices	77,78
	Diesel fuel contamination; diesel-kerosene mixtures: A comparative study; Quantitative analysis of petrol-kerosene mixtures; transformer oil ageing	79–82
	River waters impacted by tissue mill effluent	83
	Trihalomethane formation potential in the Tama River, Japan	84
	Natural organic matter of a river; Amazon basin rivers; dissolved organic matter (DOM); characterization of DOM; fresh and decomposed DOM; DOM fractions formed during composting of winery and distillery residues; colored DOM in Taihu Lake; DOM in the water of a coal-mining area; DOM in marine microalgaes growth process; assessing the dynamics of DOM in coastal environments; chromophoric DOM in the Baltic Sea	85–97
	Kinetics of metal-fulvic acid complexation	98
	Soil matrix effects; indole-3-Acetic Acid in soil	99,100
	Water-soluble organic compounds in atmospheric aerosols	101
	Monitoring of chemical oxygen demand as an organic indicator in waste water and treated effluent	102
	Humic substances and fulvic and humic acids from varied origins;	103–105
	A conceptual groundwater flow model with implications for subsidence hazards: an example from Co. Durham, UK	106
	Quantification of enrofloxacine in poultry feeding water	107
	Freshwater colloids and particles	108
	Characterization of natural organic matter	109
	Organic matter dynamics during in-vessel composting of an aged coal-tar contaminated soil	110
	EEM spectroscopy as a probe of complex chemical environments	111
	Effects of organic removal by traditional purification process	112
	Properties of the effluents from ozonated biofilters	113

Table 4.1 Continued.

Topic	*Objects*	*Ref.*
	Water soluble soil organic matter as basis for the determination of conditional metal binding parameters	114
	Determination of a water integrated organic pollution index	115
	Hydrolysis of carbaryl in tap water and river	116
	EEM on Nylon Membranes. Determination of Benzo[a]pyrene and Dibenz[a,h]anthracene at Parts-Per-Trillion Levels in the Presence of the Remaining EPA PAH Priority Pollutants	117
Bio- and pharmaceutical analysis	Determination of various biochemical analytes: carbendazim, fuberidazole and thiabendazole; ciprofloxacin in the presence of enrofloxacin; N-phenylanthranilic acid derivatives; verapamil drug;	118–121
	Determination of analytes in human plasma: doxorubicin; riboflavin; fluoroquinolone antibiotics; honokiol and magnolol in human plasma and complex magnoliae officinalis;	122–125
	Tissue spectroscopy	126
	Determination of analytes in human urine: methotrexate and leucovorin; flufenamic and meclofenamic acids; daunomycin; amoxicillin; reserpine; sulpiride; monohydroxy-polycyclic aromatic hydrocarbons;	127–133
	Human placental extract used as a wound healer: multicomponent drugs	134
	Single-nucleotide polymorphism analysis	135
	Interactions of berberine and daunorubicin with DNA	136
	Determination of analytes 6-methylcoumarin and 7-methoxycoumarin in cosmetics; 6 testosterone propionate	137,138
	Ethanol production at aerobic Saccharomyces cerevisiae fed-batch cultivations. Tryptophan, nicotinamide adenine dinucleotide phosphate (NAD(P)H) and riboflavins detection for on-line yeast biomass estimation during the growth phase	139,140
	On-line monitoring of spectroscopic-detectable substrates L-phenylalanine-7-amido-4-methylcoumarine (L-PheAMC) and D-PheAFC in supercritical carbon dioxide	141
Food	Characterization and classification of olive oils; commercial Spanish olive oils; olive-pomace oil; protected denomination extra virgin olive oils; French virgin olive oil; possible artifacts in EEM spectroscopy of Spanish olive oils; malonaldehyde in olive oils	142–151

Table 4.1 Continued.

Topic	*Objects*	*Ref.*
	Various analytes in different food matrixes: carbendazim in bananas; pesticides in honey; thiabendazole in orange extract; danofloxacin in milk	152–155
Photochemistry	Photochemically induced excitation-emission kinetic fluorescence spectra	156
	Photocatalytic degradation for increased selectivity of polycyclic aromatic hydrocarbon analyses; photocatalytic degradation of PAH; Determination of phenol during a TiO2-photocatalytic degradation process; phenol and its di-hydroxyderivative mixtures - application to the quantitative monitoring of phenol photodegradation	157–160
	Spin-coated polymeric thin films	161
	The influence of visible light and inorganic pigments on EEM spectra of egg-, casein- and collagen-based painting media	162
Equipment and software development	Light-emitting diode EEM fluorescence spectroscopy	163
	Practical aspects of PARAFAC modeling of EEM data	164
	Implementation of the PARAFAC method for near-real-time calibration	165
	Handling of Rayleigh scatter in PARAFAC modeling of EEM data	166
	Alternating penalty trilinear decomposition algorithm for second-order calibration with application to interference-free analysis of EEM data	167
	Second-order advantage achieved by unfolded-partial least-squares/residual bilinearization modeling of EEM data presenting inner filter effects	168
	Fourier fluorescence spectrometer for EEM measurement	169

Table 4.1 shows that there has been a rapid growth of interest in EEM fluorescence spectroscopy in recent years. About half of the applications are concerned with environmental samples in general, and with the analysis of dissolved organic material in particular. Besides environmental analysis, there are publications on bioanalysis and the determination of analytes in complex matrices like urine and plasma. In food chemistry, the applications are mostly concerned with the analysis of olive oils. In connection with the latter, we draw the reader's attention to the publications that concern how the accuracy of such measurements may be affected by artefacts in EEM measurements.[149,150]

In most of the published applications of EEM, multi-way chemometric analysis is an essential part of the analytical procedure, which means that the simplification and robustness of the instrumentation – so well suited to the principles and spirit of Green Chemistry – does not degrade the analytical procedure since sophisticated computational procedures can compensate. In Chapter 6 we will demonstrate how chemometric data processing increases the capability of EEM spectroscopy without any increase in its ecological footprint. Some proponents of multi-way methods enthusiastically advocate the avoidance of separation methods in cases where critical information is found in the patterns of many variables. Bro[170] has given examples of what he calls mathematical chromatography, because it allows relative concentrations and spectra of individual components to be extracted from a group of EEM spectra by a chemometric technique known as PARAFAC analysis. However, he emphasizes that in order to use PARAFAC models, it is important to have an idea of the data structure beforehand. For example, mathematical chromatography could be relevant to quality control in the pharmaceutical industry, because concentrations of compounds present in a substance can be determined very quickly, in time to have a positive impact on quality control.

4.4.3 Chemical Analysis with Consumer Electronics

4.4.3.1 Computer Screen Photoassisted Technique. Although scientists' efforts and widespread usage have reduced the complexity and cost of analytical instruments, the application of various methods has been hindered by the relatively high cost of the instrumentation. On the other hand, despite the high cost of optical instruments, rapid technological progress in fields such as consumer electronics has produced low-cost advanced optical equipment such as digital scanners, video cameras, and screens that possess useful properties that could be exploited in the field beyond the use for which they were initially intended, *e.g.* in routine chemical analysis. Since the "analytical instrument" already exists, the "ecological footprint" of the methods utilizing consumer electronics does not contribute much further to environmental pollution (the amounts of chemicals involved are in the micloliters) and can be classified as "green". The first demonstration of this method was made by Suslick and co-workers[171] when they showed how a simple flatbed digital scanner could be converted into a powerful analytical instrument for the detection of subtle colour changes in chemical dyes to detect volatile compounds. More recently, scientists from Linköping University, Sweden, have demonstrated the usefulness of personal computers for various colorimetric assays. Filippini *et al.* showed how a combination of a computer monitor and an inexpensive webcam can be used as a spectrophotometer.[172] This last technique, named a computer screen photoassisted technique (CSPT) by Fillipini *et al.* is based on the fact that a computer screen can easily be order programmed to display millions of colours, combining three narrow band emission profiles. The light emitted from a computer screen is not monochromatic, but rather, a combination of three polychromatic primary colours that excite the human perception of red, green and blue (RGB).

Compared with the use of a digital scanner, the colour computer display allows the sample to be probed with a variable combination of wavelengths instead of using the white light from scanners. This provides a spectroscopic measurement combining the absorbance and fluorescence of samples. Although this form of "spectroscopy" may seem robust and limited to the visible part of the spectrum, due to the huge availability of portable computers, digital cameras and cellular phones which are all now supplied with a colour screen, camera and extended computation capabilities, the application of the CSPT concept could give rise to a ubiquitous analytical capacity. This technique may be especially useful in developing countries with limited recourse and access to high-tech products.

The experimental set-up for computer screen/web camera spectroscopy is strikingly simple and can be set up by virtually anyone. In CSPT measurement, the web camera captures the image of the colorimetric assay under screen illumination (see Figure 4.13). In one possible set-up, an area of 130 mm × 130 mm on the upper right corner of a LCD screen (*e.g.* Dell 1905FP) operating at 1280 × 1024, 32 bits resolution, normal intensity and contrast settings was used as a light source. During measurement, a sequence of colours was programmed to be displayed in this area and to illuminate the assays at a rate of 1 colour s^{-1}. The illuminating sequence used a 50 colour set. A mirror at 45° to the screen surface guided the light from the screen to illuminate a sampler holder from underneath. A microtiter plate, for example, can be located on the sample holder. The holder also provides a light shield from external illumination. The holder dimensions are fitted to the illumination area and microtiter plate. The sample holder is attached to the frame of the computer and keeps the microtiter plate in a horizontal position with the camera about 10 cm above it and focused on the assay. The image of the assay under these illuminations is captured using a web camera (*e.g.* in the publication by Malik *et al.*[173] the Logitech QuickCamPro4000 is used with a CCD detector of 640 × 480 pixels and 24 bits color resolution). The acquisition is synchronized with the illumination at a rate of 1 frame s^{-1}. The experiment can be controlled by a program written in Matlab 7, for example, using its image acquisition toolbox.

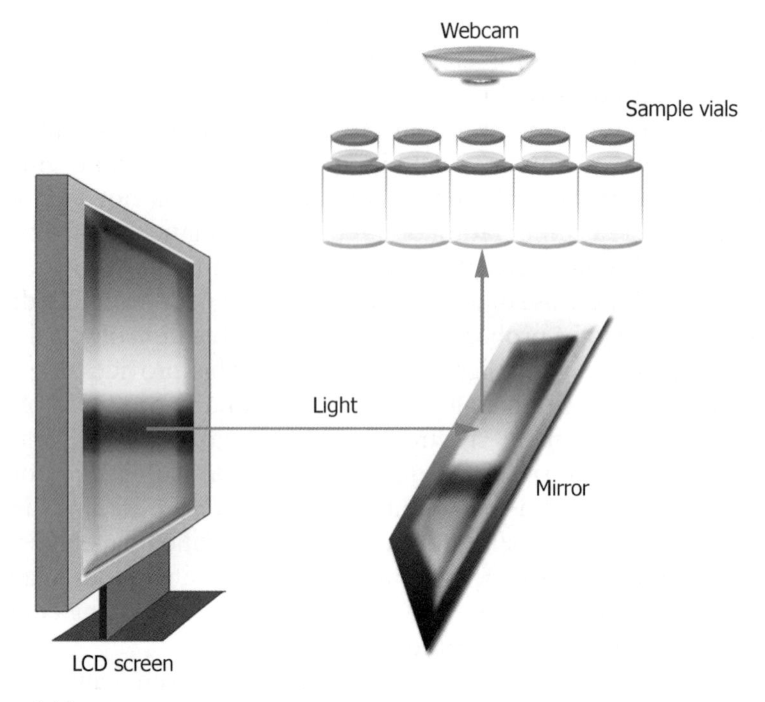

Figure 4.13 CSPT setup.

The illuminating set-up represented in Figure 4.13 generates screen colours by combining primary spectral radiances that induce the perception of red, green and blue colours [$R(\lambda)$; $G(\lambda)$; $B(\lambda)$]. Any other illuminating radiance $c(j, \lambda)$ can be composed according to the formula

$$c(j, \lambda) = r_j R(\lambda) + g_j G(\lambda) + b_j B(\lambda)$$

Colours are specified by triplets of weights r_j; g_j; b_j. The measured CSPT signal is the result of the modulation of these sources by any substance transmittance $T_i(\lambda)$ integrated under the red, green and blue camera filters [$Fr(\lambda)$; $Fg(\lambda)$; $Fb(\lambda)$]

$$\begin{aligned} Ir_i(j) &= \int_\lambda c(j, \lambda) T_i(\lambda) Fr(\lambda) D(\lambda) d\lambda \\ Ig_i(j) &= \int_\lambda c(j, \lambda) T_i(\lambda) Fg(\lambda) D(\lambda) d\lambda \\ Ib_i(j) &= \int_\lambda c(j, \lambda) T_i(\lambda) Fb(\lambda) D(\lambda) d\lambda \end{aligned} \tag{1}$$

where $D(\lambda)$ is the spectral response of the detector. Despite the complicated form of Equation (1), it is clearly obvious (and Equation 1 demonstrates this as well) that the key element of the success of CSPT methods is the existence of a wide and structured visible absorption spectrum of a sample.

The intensities of each webcam filter channel are collected in a single fingerprint of the sample. The described fingerprints are too complex for direct inspection, and a multivariate technique such as principal component analysis, (PCA as described in Chapter 6.1), is used to summarize the information. PCA represents each fingerprint as a single point in a two-dimensional principal components space that usually preserves more than 80–90% of the original information. PCA is well suited to the analysis of CSPT data. It has been demonstrated that CSPT could be useful for fingerprinting (for classification purposes) a huge number of samples from different origins or locations. Various aspects of the use of computer screens as light sources combined with chemometric data processing have been enthusiastically studied in many publications by the Linköping group. They demonstrated recently that even the separation of emitted light from transmitted light for the fingerprinting of fluorescent substances is possible using CSPT.[174] In this case, the omnidirectional property of fluorescent emission is used to separate it from the background, by means of a simple optical arrangement compatible with CSPT. This enhances the sample classification capability and makes it possible at sub μM concentrations. The method can be improved using an embedded local reference that is simultaneously measured in the tests. The achieved performance (based on 580 classifications covering all the ranges of the assay, using synthetic samples) yielded 97% correct determinations compared with 90% for colorimetric determinations.[175] The performance of CSPT, defined as its ability to distinguish different colour substances, depends on the colour of the evaluated substances and the particular illuminating sequences used for the measurements. The study of CSPT performance for sets of substances of similar colours identified specific substance transmittances, which is very favourable in terms of classification capability.[176] Furthermore, the issue of spurious spatial distribution of light overlapping on the image was examined.[177] The effect of spatial variability can be minimized by the selection of colours composing an illuminating sequence. The approach was tested for the classification of different colour substances, and it showed improvements of up to 53%, when compared with randomly chosen colours.

Thus, CSPT can serve as a spectrometric instrument for various analytical methods such as EEM spectroscopy and SPR. As we saw in Chapter 4.4.2, the ability of EEM to record complex and highly selective signatures of fluorescent substances typically requires dedicated instrumentation. Computer screen/web camera technology indicates that more widespread application scenarios (through ubiquitous CSPT platforms) are possible. It has been demonstrated that CSPT fingerprinting of fluorescent substances can provide a detailed signature of the EEM of such substances.[178] Angle and spectra resolved SPR images of gold and silver thin films with protein deposits can also be recorded using CSPT. The screen provides multiple-angle illumination, p-polarized light and controlled spectral radiances to excite surface plasmons. The sensitivity and resolution of the method, determined in air and solution, are $0.145\,nm\,pixel^{-1}$, 0.523 nm, 5×10^{-3} RIU $degree^{-1}$ and 6×10^{-4} RIU, respectively. These are encouraging results at the proof-of-concept stage considering the ubiquity of the instrumentation.[179]

Despite the fact that at first sight the development of CSPT technology looks like a secondary school science project, its potential lies in its availability to an extremely wide audience. The use of computer screens as controlled light sources and web cameras as image detectors is an omnipresent alternative for colorimetric quick tests in homes or primary care units.[180] The fact that CSPT uses a globally distributed and familiar infrastructure for instrumentation and that it is able to serve multiple purposes with the same set-up, makes it an attractive candidate for home testing. Its potential to support the global monitoring of environmental and sanitary parameters, using Internet geography browsers, is significant.[181] Since CSPT platforms require only standard computers and web cameras as a measuring set-up, application of these kinds of assays outside specialized laboratories becomes possible. One can envision various sensing strips on paper, plastic or glass substrates being sold in general stores for lay testing of various clinical or environmental parameters. The situation could be similar to home pregnancy tests in which sophisticated multiple immunoassays are involved in the adsorbent chemistry of the strip, but its usage is plain and simple. In CSPT, an additional opportunity is provided by software that could come with the test.

There is, therefore, an enormous challenge to Green Analytical Chemistry to develop such test strips on glass or other substrates to take advantage of opportunities afforded by the CSPT method. Some steps in this direction are described below. The capacity of CSPT for the characterization of colorimetric assays and its ability to retain key spectral features has been described.[182,183] The detection and classification of fluorescent dyes,[184] the possibility of spectral fingerprinting of representative fluorescent substances for prospective assays and the characterization of opaque colour samples using reflected light have been demonstrated.[185,186] Rapid quantitative determinations of creatinine, potassium and glucose – all important parameters in routine medical diagnostics – have been established.[187]

CSPT can be used in bioanalysis. An interesting application of CSPT is biosensing the hormones melatonin and alpha-melanocyte-stimulating hormone (alpha-MSH) as lightening or darkening stimuli, respectively. Melanophores are pigmented cells in lower vertebrates capable of quick color changes and thereby suitable as whole cell biosensors. In the frog dermis skin layer, the large and dark pigmented melanophore surrounds a core of other pigmented cells.[188] The distinctive generation of biochemical response patterns of eight different substances, using an assay based on pigment-containing cells was demonstrated. *Xenopus laevis* melanophores, transfected with human beta(2)-adrenergic receptor, were seeded in a 96-well microplate and used to generate individual biochemical images through a two transient measuring protocol that contributes to highlighting the response signatures of the agents. The assays – evaluated both with a standard microplate reader and with CSPT – yielded similar results.[189] Furthermore, CSPT was used for the evaluation of a prospective enzyme-linked immunosorbent assay (ELISA). An anti-neutrophil cytoplasmic antibodies test, typically used for diagnosing patients suffering from chronic inflammatory disorders of the skin, joints, blood vessels and other tissues, was comparatively tested with a standard microplate reader and CSPT, which yielded equivalent results at a fraction of the instrumental cost.[190] The capability of CSPT for spectral fingerprinting of a photoactive polythiophene derivative used as a pH reporter, poly(3-[(*S*)-5-amino-5-carboxyl-3-oxapentyl]-2.5-thiophenylene) hydrochloride (POWT), is demonstrated. POWT is part of a family of industrial scalable materials, well established in organic electronics and biomedical applications. The ability of CSPT for substance classification, corroborated by principal component analysis (PCA), successfully compares with standard spectroscopy.[191] In another study, CSPT has been used to detect the attachment of complementary DNA strands to a complex of single DNA strands and a POWT. The complex is a highly sequence-specific indicator based on non-covalent coupling of DNA to a water-soluble, zwitterionic, electroactive and photoactive polymer able to produce a combined absorption-emission signal readable by CSPT. The observed CSPT signal retains key spectral features of the spectrum of the complex, distinguishing the DNA attachment, as well as other stimuli, such as pH regulation at concentrations of

30 μM POWT and 15 μM DNA.[192] When CSPT is used for examining a disposable microstructured substrate, the fluorescent fingerprinting of 100 μm elements packed at 46 elements mm^{-1} is possible.[193,194]

4.4.3.2 Computer Optical Disk Drives for Chemical Analysis. In addition to computers, optoelectronic consumer products – optical disk drives – are another class of widely employed devices in the office and home. Due to their dramatic cost advantage over analytical instruments, which are produced only in small quantities, optical disk drives are attracting significant attention for optical sensor applications because of their robustness and ease of operation. Compact disk-based, high-density microarrays were explored many years ago by Kido *et al.*[195] The results of this research indicate that compact disk-based microarray technology can yield qualitative and quantitative results and has potential for simultaneous multianalyte analyses. The use of conventional computer optical disk drives for chemical and biological sensing has been reported, and the ability to extract an analog signal from a drive and to use that signal for the quantitative detection of optical changes in sensor films deposited on conventional CD and DVD optical disks has been demonstrated by several groups of researchers, such as Lange *et al.*[198] and Potyrailo *et al.*[196–200] Figure 4.14 illustrates the concept of analytical detection using a conventional CD/DVD drive in a laptop computer that can be employed for quantitative chemical analysis. The drive has 650 and 780 nm lasers to read DVDs and CDs, respectively, a Si photodiode detector, and a laser tracking system to scan across the disk surface. In the "lab-on-DVD" system, an analog signal from the photodiode is extracted during the reading of an optical disk, and used for quantitative detection of the changes in the optical properties of sensing films deposited on the read surface of the optical disk.

In the absence of a sensor film, laser light is transmitted through the surface of the optical disk, reflected from the disk's reflective data layer, and returned to the photodiode detector. When a sensor film is applied to the read side of a CD or DVD disk, the laser light travels through the sensor film twice, as shown in Figure 4.14. If the optical properties of the sensor film vary, this will cause a change in the amount of light detected by the laser pick-up head photodiode detector and allow quantization of analytes embedded in the sensor film. Numerous possibilities exist for the use of conventional sensor materials as well as nanomaterials for the realization of these optical phenomena as demonstrated in Figure 4.15.

In addition, the optical system of an optical disk drive provides polarization and phase control of the light that reaches the detector. These

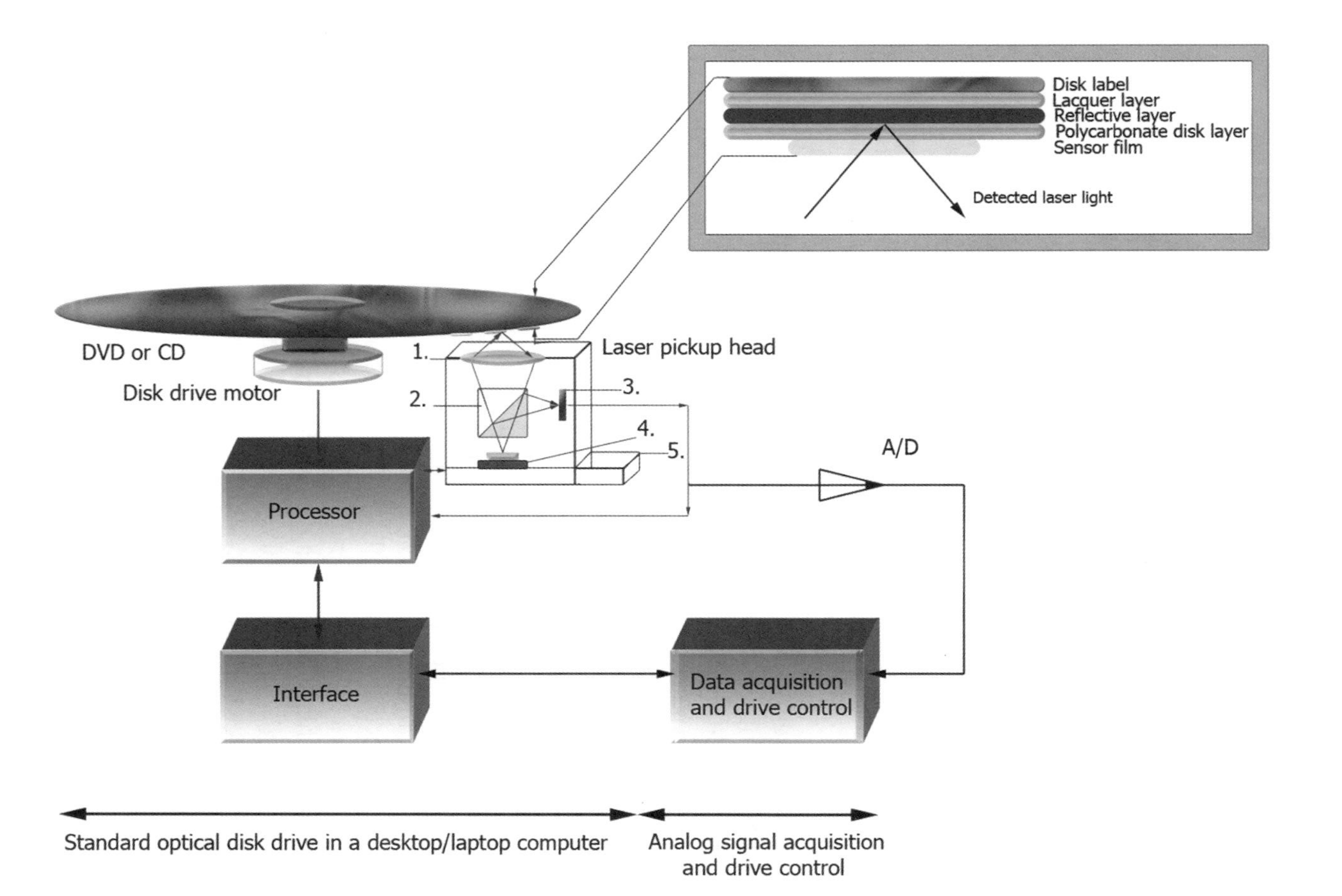

Figure 4.14 Schematic of a conventional optical disk drive, and the methodology for obtaining an analog signal from a photodiode detector and for controlling the optical disk drive; 1 – lens; 2 – beam splitter; 3 – photodiode; 4 – laser diode; 5 – tracking drive.

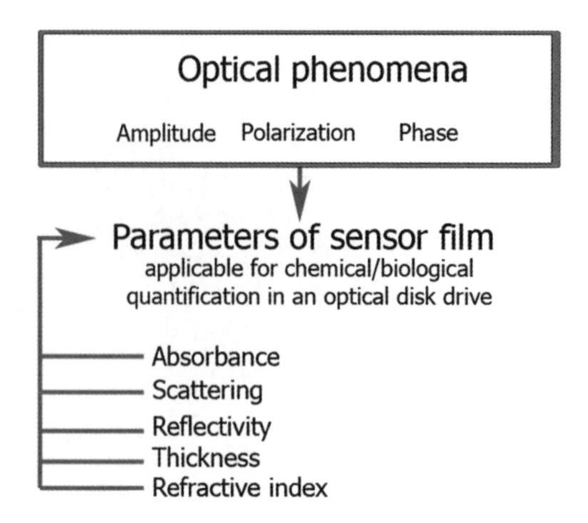

Figure 4.15 Basics of quantitative chemical and biological analysis using optical disk drives, based on the optical phenomena and parameters of sensor films involved in signal generation.[199]

features are used for the rejection of ambient light and light scatter from scratches and other imperfections on the disk surface, but they also provide an additional opportunity for chemical and biological quantification based on several optical phenomena that can be produced in sensing films.

To demonstrate the application of optical disk drives in chemical analysis, Potyrailo *et al.*[199] used a straightforward procedure to create polymeric films on DVD that are sensitive to certain ions. As can be expected, reagent-based sensing films are complex multicomponent compositions. The key formulation ingredients include: a chemically sensitive reagent, a polymer matrix, auxiliary additives and a common solvent. The optimization of composed sensing materials is a cumbersome process because it typically involves the evaluation of numerous polymeric matrices and multiple additives at multiple concentrations. For example, films for sensing calcium ions were produced by incorporating a Ca^{2+}-sensitive dye (Xylidyl Blue) in a polymer matrix (poly-(2-hydroxyethyl)methacrylate). For the fabrication of films, the indicator, pH modifier and polymer were dissolved in 1-methoxy-2-propanol and manually applied to the DVDs through a 100-μm-thick tape mask with 3×4-mm openings. After the solvent evaporated, the sensor films adhered to the DVD surface and were ready for testing for Ca^{2+}. In the tests, 20–40 μL water samples with varied concentrations of Ca^{2+} were manually pipetted onto the sensor films and removed with pressurized house nitrogen gas after 2 min of exposure. Figure 4.16 shows a conventional DVD with an array of sensing films for the

Figure 4.16 Super audio CDs with screen-printed sensing regions, with optimized concentration of the chlorine-sensitive cyanine reagent immobilized into sensing films, for detection of chlorine in water.[200] (Reproduced with kind permission of Elsevier).

measurement of multiple chemicals in water (*e.g.* calcium, chlorine, pH, *etc.*). In another application, a determination of chlorine in water was performed using sensing films that contained 1',1'-diethyl-4,4'-carbocyanine iodide dye. The calculated detection limit for chlorine determination (at S/N = 3) was 200 ppb.[198] For the detection of relative humidity in air, sensing films containing rhodamine 800 dye in Nafion polymer (Sigma Aldrich, Milwaukee, WI) were used. A reversible signal change was observed in the absorption of the sensing film.

As shown in Figure 4.14, the light beam is not precisely focused onto the sensor film in this set-up, which reduces the total number of sensor film points that can be spotted on the disk, thus reducing the spatial resolution of the assay. This issue has recently been addressed by Morais *et al.* using another type of compact disk: low reflectivity compact disks, (L-CDs), as analytical platforms in combination with a commercial CD player for developing immunoassays in microformat.[201,202] This approach uses a gold layer of L-CD to reflect 30% of the light of the laser beam CD drive ($\lambda = 780$ nm); the remainder is transmitted through the disk. The reflected light allows the pick-up head of the CD drive to follow the data track of the disk, which means that the entire disk can be read. The transmitted light was detected by a planar photodiode that was incorporated into the CD drive by the authors as follows: a planar photodiode,[203] (25.4 mm long and 5.04 mm width; with a spectral range of 400–1100 nm), was attached to the upper part of the CD reader, 2 mm above the L-CD, along the line that is scanned by the laser beam during the reading process. The position of the photodiode in combination with the radial and rotational movement of the laser and disk, respectively, allows for coverage of almost the entire disk area, the surface being large enough to simultaneously detect 8 separate assays per disk, 2560 spots per disk. This set-up produces the best spatial resolution in terms of laser focusing. The electrical signal generated by the photodiode is digitalized by a data acquisition board, stored in memory and then deconvoluted into an image. The immunoreaction is detected simply by absorbance: in the absence of an analytical response, approximately 70% of the laser light is transmitted through the surface of the L-CD and detected as a background signal. In the case of a positive signal, the optical properties of the disk are modified causing a variation in the light intensity detected by the photodiode, which is further related to the analyte concentration. The gold surface of the disk was spin-coated with a polystyrene solution dispensed with a pipette near the internal radius of the gold surface, evenly distributing the polymer solution over the disk. The coated CD was then heated in an oven at 60 °C for 30 min in order to remove any remaining solvent. This produces a solid polymeric film that forms an analytical platform to perform microimmunoassays. Other polymers can be attached to the disk surface using this spin-coating approach in order to improve the physical adsorption of the biomolecules and, consequently, assay performance. The polymers can be functionalised prior to the spin-coating process in order to develop covalent and oriented immobilization of bioreceptors such as protein and DNA.

Furthermore, the disk can be coated with several layers as was done by Gopinath *et al.*[204] In their project, a "BioDVD" was prepared as a multilayered disk containing five-layer structures that included a phase-change material layer (see Figure 4.17). The multilayered structure of the BioDVD was fabricated by a radio frequency magnetron sputtering on a polycarbonate disk substrate. The operating principle of the BioDVD with phase-change film relies on interference with the laser light when it passes the multilayer system: two gold films (one thick and one thin) generate a standing wave between them. The sandwiched dielectric film and the phase-changing film thicknesses are designed to optimize the interferometric response, making it highly sensitive to a tiny refractive index change on the upper thin gold film. Once a molecule binds to the

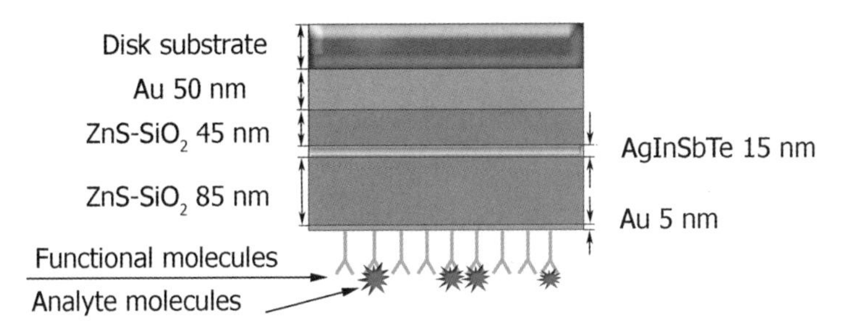

Figure 4.17 The multilayered structure of the BioDVD.[204]

surface, the optical standing waves generated in the structure are suppressed, leading to a change in reflectivity. On the outer gold surface of the DVD, a DNA oligomer was immobilized *via* a thiol-linker as an anchor for analyzing various biomolecular interactions, including nucleic acid hybridization, RNA-protein interactions and RNA small-ligand interactions. The sensitivity of the BioDVD appeared to be comparable with other known methods, such as surface plasmon resonance.

Tamarit-Lopez *et al.*[201] speculate that the simplicity of operation and management, versatility, low cost and portability of the CD player could make it a very competitive detector system in comparison with, for example, commercial microarray scanners. Moreover, the CD array methodology offers advantages with respect to conventional plate immunoassay, because CDs demonstrate high-throughput analytical capacity as well as considerable immunoreagent savings (between two and 10 000 times for coating conjugate and primary antibody, respectively). In work by Morais *et al.*, [202] the analytical capability of the methodology was demonstrated as a proof of concept through an analysis of a neurotoxic compound (2560 spots per disk), reaching a 0.08 $\mu g\,L^{-1}$ limit of detection. Low concentrations of compounds commonly used as pesticides were detected in a total assay time of 60 min, with a limit of detection ranging from 0.02 to 0.62 $\mu g\,L^{-1}$ for fenoprop (2,4,5-TP), chlorpyriphos and metolachlor. These numbers demonstrate the enormous potential of compact disks in combination with CD players, as easy-to-operate and portable devices for lab-on-a-disk analytical applications. The use of a planar photodiode was later abandoned and detection was performed by the pick-up head of the CD player only. The quantitative immunoanalysis, in microarray on CD format, of a cancer marker (alpha-fetoprotein (AFP)) and a selective herbicide (atrazine) on four types of audio-video disk was presented by Morais *et al.*[205] Enzyme or gold nanoparticle-labeled antibodies were used as tracers, forming a precipitate on the sensing disk surface that was proportional to the darkness of the immunoreaction product. The detection limit for AFP ($8.0\,\mu g\,L^{-1}$) was under the threshold needed to detect non-seminomatous testicular cancer, and for atrazine ($0.04\,\mu g\,L^{-1}$), it was below the maximum E.U. residue limit for drinking water. Later, the same group of authors developed multiplexed microimmunoassays on DVD for five compounds.[206] The multiplexed assay achieved detection limits (IC10) of 0.06, 0.25, 0.37, 0.16 and $0.10\,\mu g\,L^{-1}$, sensitivities of (IC50) 0.54, 1.54, 2.62, 2.02, and $5.9\,\mu g\,L^{-1}$, and dynamic ranges of 2 orders of magnitude for atrazine, chlorpyrifos, metolachlor, sulfathiazole, and tetracycline, respectively. Compared with reference chromatographic methods, immunoassay on DVD demonstrated its potential for high-throughput multiplexed screening applications. Analytes of different chemical natures (pesticides and antibiotics) were directly quantified without sample treatment or pre-concentration in a total time of 30 min, with similar sensitivity and selectivity to the ELISA plate format using the same immunoreagents.

A CD was used by Alexandre and co-workers[207] as a support for constructing DNA microarrays for genomic analysis. In a project by La Clair *et al.*,[208] a procedure was developed to attach ligands to the reading face of a CD by activating the terminus of its polycarbonate component. Displays were generated on the surface of a CD by printing tracks of ligands on the disk with an inkjet printer. Recognition between surface-expressed ligands and biomolecules was screened by an error determination routine. Li *et al.* demonstrated that UV/ozone treatment of polycarbonate sheets produces a hydrophilic surface with a high density of reactive carboxylic acid groups [$(4.8 \pm 0.2) \times 10^{-10}\,mol\,cm^{-2}$] in less than 10 min at ambient conditions, facilitating covalent immobilization of DNA probes *via* both passive and flow-through procedures.[209] No significant aging or physical damage to the substrate was observed, which means that the activation protocol is mild and efficient. Prepared CDs were used for performing three different types of biochemical recognition reactions (biotin–streptavidin binding, DNA hybridization, and protein–protein interaction).[210] If well correlated with the optical darkness of the binding sites (after signal enhancement by gold nanoparticle-promoted autometallography), the reading error levels of pre-recorded audio files can serve as a quantitative measure of biochemical interaction.

The described methodology using CDs highlights the huge potential of chemical analysis and immunoassay methods employing standard audio-video disk surfaces in combination with a CD/DVD drive for environmental or clinical analysis, drug detection, or high-throughput multi-residue screening applications. The slight modification of a common CD reader has to be undertaken – it was necessary to record analog signals using data acquisition boards. As in the case of CSPT described above, the methodology implementing CD readers is cost-effective, sensitive, and suitable for point-of-need or diagnostics in resource-poor settings that require low-priced and easy-to-use ubiquitous technologies. As in the case of CSPT, the essence of the assay is in preparation of the sensing film on the substrate. For immunoassay, the film preparation could be a lengthy process involving the application of various solutions (conjugate, sample, enhancer, *etc.*) to the disk. Comparing CSPT with DVD-reader based methodologies, the latter seems a bit more complicated than the former, since lab-on-a-disk applications require hardware modification of the original consumer product: recording the analog signal from the laser readout head and digitizing it for the computer.

REFERENCES

1. L. H. Keith, L. U. Gron and J. L. Young, *Chem. Rev.*, 2007, **107**, 2695.
2. D. Raynie, presented at the 13th Annual Green Chemistry and Engineering Conference, 2009, ACS, Abstract 51, 34
3. J. Ruzicka and E. H. Hansen, *Flow Injection Analysis*, J. Wiley and Sons, 1981.
4. B. Karlberg and G. E. Pacey, *Flow Injection Analysis: A Practical Guide*, Elsevier, 1989.
5. S. Armenta, S. Garrigues and M. de la Guardia, *TrAC*, 2008, **27**, 479.
6. N. W. Barnett, C. E. Lenehan and S. W. Lewis, *TrAC*, 1999, **18**, 346.
7. B. F. Reis, M. F. Gine′, E. A. G. Zagatto, J. L. F. C. Lima and R. A. Lapa, *Anal. Chim. Acta*, 1994, **293**, 129.
8. http://www.globalfia.com/tutorial7.html
9. C. Fernandez, A. C. L. Conceicao, R. Rial-Otero, C. Vaz and J. L. Capelo, *Anal. Chem.*, 2006, **78**, 2494.
10. Z. Song, S. Hou and N. Zhang, *J. Agric. Food Chem.*, 2002, **50**, 4468.
11. H. Liu, Y. Hao, J. Ren, P. He and Y. Fang, *Luminescence*, 2007, **22**, 302.
12. S. Gil, I. Lavilla and C. Bendicho, *Spectrochim. Acta, Part B*, 2007, **62**, 69.
13. P. R. M. Correia, R. C. Siloto, A. Cavicchioli, P. V. Oliveira and F. R. P. Rocha, *Chem. Educ.*, 2004, **9**, 242.

14. T. Leelasattarathkula, S. Liawruangratha, M. Rayanakorna, B. Liawruangrathc, W. Oungpipatb and N. Youngvise, *Talanta*, 2007, **72**, 126.
15. S. Kruanetr, S. Liawruangrath and N. Youngvises, *Talanta*, 2007, **73**, 46.
16. J. G. March and B. M. Simonet, *Talanta*, 2007, **73**, 232.
17. C. M. C. Infante, J. C. Masini and F. R. P. Rocha, *Anal. Bioanal. Chem.*, 2008, **391**, 2931.
18. L. S. G. Teixeira and F. R. P. Rocha, *Talanta*, 2007, **71**, 1507.
19. M. L. C. Passos, A. M. Santos, A. I. Pereira, J. R. Santos, A. J. C. Santos, M. L. M. F. S. Saraiva and J. L. F. C. Lima, *Talanta*, 2009, **79**, 1094.
20. P. Gründler, *Chemical Sensors: An Introduction for Scientists and Engineers*, Springer, 2007.
21. P. A. Lieberzeit and F. L. Dickert, *Anal. Bioanal. Chem.*, 2007, **387**, 237.
22. W. Wang, S. T. He, S. Z. Li and Y. Pan, *IEEE Sensors*, 2006, **1–3**, 690.
23. C. Coimbatore, S. M. Presley, J. Boyd, E. J. Marsland and G. P. Cobb, in *Advances in Biological and Chemical Terrorism Countermeasures*, ed. R. J. Kendall, S. M. Presley, G. P. Austin and P. N. Smith, CRC Press, 2008.
24. T. Alizadeh and S. Zeynali, *Sens. Actuators, B*, 2008, **129**, 412.
25. A. T. Nimal, U. Mittal, M. Singh, M. Khaneja, G. K. Kannan, J. C. Kapoor, V. Dubey, P. K. Gutch, G. Lal, K. D. Vyas and D. C. Gupta, *Sens. Actuators, B*, 2009, **135**, 399.
26. J. C. Kapoor and G. K. Kannan, *Defence Sci. J.*, 2007, **57**, 797.
27. J. Lozano, M. J. Fernández, J. L. Fontecha, M. Aleixandre, J. P. Santos, I. Sayago, T. Arroyo, J. M. Cabellos, F. J. Gutiérrez and M. C. Horrillo, *Sens. Actuators, B*, 2006, **120**, 166.
28. H. W. Chang and J. S. Shih, *Sens. Actuators, B*, 2007, **121**, 522.
29. R. C. Stevens, S. D. Soelberg, S. Near and C. E. Furlong, *Anal. Chem.*, 2008, **80**, 6747.
30. T. M. Chinowsky, S. D. Soelberg, P. Baker, N. R. Swanson, P. Kauffman, A. Mactutis, M. S. Grow, R. Atmar, S. S. Yee and C. E. Furlong, *Biosens. Bioelectron.*, 2007, **22**, 2268.
31. S. D. Soelberg, T. Chinowsky, G. Geiss, C. B. Spinelli, R. Stevens, S. Near, P. Kauffman, S. Yee and C. E. Furlong, *J. Ind. Microbiol. Biotechnol.*, 2005, **32**, 669.
32. A. N. Naimushin, S. D. Soelberg, D. K. Nguyen, L. Dunlap, D. Bartholomew, J. Elkind, J. Melendez and C. E. Furlong, *Biosens. Bioelectron.*, 2002, **17**, 573.

33. T. M. Chinowsky, J. G. Quinn, D. U. Bartholomew, R. Kaiser and J. L. Elkind, *Sens. Actuators, B*, 2003, **91**, 266.
34. A. N. Naimushin, C. B. Spinelli, S. D. Soelberg, T. Mann, R. C. Stevens, T. Chinowsky, P. Kauffman, S. Yee and C. E. Furlong, *Sens. Actuators, B*, 2005, **104**, 237.
35. R. C. Stevens, S. D. Soelberg, B. T. L. Eberhart, S. Spencer, J. C. Wekell, T. M. Chinowsky, V. L. Trainer and C. E. Furlong, *Harmful Algae*, 2007, **6**, 166.
36. T. M. Chinowsky, S. D. Soelberg, P. Baker, N. R. Swanson, P. Kauffman, A. Mactutis, M. S. Grow, R. Atmar, S. S. Yee and C. E. Furlong, *Biosens. Bioelectron.*, 2007, **22**, 2268.
37. R. C. Stevens, S. D. Soelberg, S. Near and C. E. Furlong, *Anal. Chem.*, 2008, **80**, 6747.
38. S. D. Soelberg, R. C. Stevens, A. P. Limaye and C. E. Furlong, *Anal. Chem.*, 2009, **81**, 2357.
39. J. Wang, *Acc. Chem. Res.*, 2002, **35**, 811.
40. J. Wang and J. Lu, *Electrochem. Comm.*, 2000, **2**, 390.
41. R. D. Gardner, A. H. Zhou and N. A. Zufelt, *Sens. Actuators, B*, 2009, **136**, 177.
42. O. T. Guenat, J. F. Dufour, P. D. van der Wal, W. E. Morf, N. F. de Rooij and M. Koudelka-Hep, *Sens. Actuators, B*, 2005, **105**, 65.
43. S. Y. Liu, Y. P. Chen, F. Fang, J. Xu, G. P. Sheng, H. Q. Yu, G. Liu and Y. C. Tian, *Environ. Sci. Technol.*, 2009, **43**, 1160.
44. D. Lowinsohn, H. E. M. Peres, L. Kosminsky, T. R. L. C. Paixao, T. L. Ferreira, F. J. Ramirez-Fernandez and M. Bertotti, *Sens. Actuators, B*, 2006, **113**, 80.
45. T. G. Anjos and C. E. W. Hahn, *Sens. Actuators, B*, 2008, **135**, 224.
46. K. Nahoko, S. Akiyoshi, N. Tobias and T. Keiichi, *Trans. Inst. Electric. Eng. Japan*, 2007, **127**, 217.
47. A. Mugweru, B. L. Clark and M. V. Pishko, *J. Diabetes Sci. Technol.*, 2007, **1**, 366.
48. P. Ciosek, K. Zawadzki, D. Stadnik, P. Bembnowicz, L. Golonka and W. Wróblewski, *J. Solid State Electrochem.*, 2009, **13**, 129.
49. A. R. Rahman, G. Justin and A. Guiseppi-Elie, *Biomed. Microdevices*, 2009, **11**, 75.
50. G. Plaza, K. Ulfig and A. J. Tien, *Polish J. Environ. Stud.*, 2000, **9**, 231.
51. Y. He, L. Tang, X. Wu and X. Hou, *Appl. Spectrosc. Rev.*, 2007, **42**, 119.
52. D. E. P. Turl and D. R. W. Wood, *Analyst*, 2008, **133**, 558.

53. J. A. Contreras, J. A. Murray, S. E. Tolley, J. L. Oliphant, H. D. Tolley, S. A. Lammert, E. D. Lee, D. W. Later and M. L. Lee, *J. Am. Soc. Mass Spectrom.*, 2008, **19**, 1425.
54. H. Q. Lin, Q. Ye, C. H. Deng and X. M. Zhang, *J. Chromatogr. A*, 2008, **1198**, 34.
55. M. A. Nelson, A. Gates, M. Dodlinger and D. S. Hage, *Anal. Chem.*, 2004, **76**, 805.
56. J. Yinon, *TrAC*, 2002, **21**, 292.
57. J. M. Perr, K. G. Furton and J. R. Almirall, *J. Separation Sci.*, 2005, **28**, 177.
58. S. Zimmermann, S. Barth, W. K. M. Baether and J. Ringer, *Anal. Chem.*, 2008, **80**, 6671.
59. http://www.environics.fi/index.php?page = chempro
60. J. I. Baumbach, *Anal. Bioanal. Chem.*, 2006, **384**, 1059.
61. J. H. Gross, *Mass Spectrometry: A Textbook*, Springer, Berlin, 2004.
62. M. Yang, T.-Y. Kim, H.-C. Hwang, S.-K. Yi and D.-H. Kim, *J. Am. Soc. Mass Spectrom.*, 2008, **19**, 1442.
63. Z. Ouyang, R. J. Noll and R. G. Cooks, *Anal. Chem.*, 2009, **81**, 2421–2425.
64. B. C. Laughlin, C. C. Mulligan and R. G. Cooks, *Anal. Chem.*, 2005, **77**, 2928.
65. P. J. J Potts and M. West, ed., *Portable X-Ray Fluorescence Spectrometry: Capabilities for In Situ Analysis*, The Royal Society of Chemistry, 2008.
66. R. S. Harmon, F. C. DeLucia, C. E. McManus, N. J. McMillan, T. F. Jenkins, M. E. Walsh and A. Miziolek, *Appl. Geochem.*, 2006, **21**, 730.
67. J. Cunat, F. J. Fortes and J. J. Laserna, *Anal. Chim. Acta*, 2009, **633**, 38.
68. E. C. Ferreira, D. M. B. P. Milori, E. J. Ferreira, R. M. Da Silva and L. Martin-Neto, *Spectrochim. Acta, Part B*, 2008, **63**, 1216.
69. J. Cunat, F. J. Fortes, L. M. Cabalin, F. Carrasco, M. D. Simon and J. J. Laserna, *Appl. Spectrom.*, 2008, **62**, 1250.
70. C. A. Munson, J. L. Gottfried, E. G. Snyder, F. C. De Lucia, B. Gullett and A. W. Miziolek, *Appl. Optics*, 2008, **47**, G48.
71. C. Lindemann and E. Poulsen, *GIT Labor-Fachzeitschrift*, 2003, **09**, S.920.
72. S. J. Hart and R. D. Ji.Ji, *Analyst*, 2002, **127**, 1693.
73. M. L. Nahorniak and K. S. Booksh, *Analyst*, 2006, **131**, 1308.
74. A. Baker, *Environ. Sci. Technol.*, 2001, **35**, 948.
75. U. Panne, R. Dusing and R. Niessner, *Field Screening Europe 2001. Proceedings of the Second International Conference on*

Strategies and Techniques for the Investigation and Monitoring of Contaminated Sites, Ed. W. Breh, J. Gottlieb, H. Hötzl, F. Kern, T. Liesch and R. Niessner, Kluwer Acad. Publ., 2002, p. 169.
76. D. Patra and A. K. Mishra, *Appl. Spectrosc.*, 2001, **55**, 338.
77. D. Patra, L. K. Sireesha and A. K. Mishra, *Indian J. Chem., Sect. A*, 2001, **40**, 374.
78. M. L. Nahorniak and K. S. Booksh, *Analyst*, 2006, **131**, 1308.
79. D. Patra and A. K. Mishra, *Anal. Chim. Acta*, 2002, **454**, 209.
80. O. Divya and A. K. Mishra, *Anal. Chim. Acta*, 2007, **592**, 82.
81. O. Divya and A. K. Mishra, *Appl. Spectrosc.*, 2008, **62**, 753.
82. S. Deepa, R. Sarathi and A. K. Mishra, *Talanta*, 2006, **70**, 811.
83. A. Baker, *Environ. Sci. Technol.*, 2002, **36**, 1377.
84. F. Nakajima, M. Hanabusa and H. Furumai, *presented at the 3rd World Water Congess: Drinking Water Treatment*, 2002, **2**, 481.
85. P. G. Coble, *Marine Chem.*, 1996, **51**, 325.
86. A. Baker, *Environ. Sci. Technol.*, 2001, **35**, 948.
87. F. C. Wu, R. D. Evans and P. J. Dillon, *Environ. Sci. Technol.*, 2003, **37**, 3687.
88. N. Patel-Sorrentino, S. Mounier and J. Y. Benaim, *Water Res.*, 2002, **36**, 2571.
89. W. Chen, P. Westerhoff, J. A. Leenheer and K. Booksh, *Environ. Sci. Technol.*, 2003, **37**, 5701.
90. P. Q. Fu, C. Q. Liu and F. C. Wu, *Spectrosc. Spectral Anal.*, 2005, **25**, 2024.
91. J. F. Hunt and T. Ohno, *J. Agric. Food Chem.*, 2007, **55**, 2121.
92. F. C. Marhuenda-Egea, E. Martinez-Sabater, J. Jorda, R. Moral, M. A. Bustamante, C. Paredes and M. D. Perez-Murcia, *Chemosphere*, 2007, **68**, 301.
93. Z. G. Wang, W. Q. Liu, N. J. Zhao, H. B. Li, Y. J. Zhang, W. C. Si-Ma and J. G. Liu, *J. Environ. Sci. China*, 2007, **19**, 787.
94. C. Yang, N. N. Zhong, Y. L. Shi, F. Y. Wang and D. Y. Chen, *Spectrosc. Spectral Anal.*, 2008, **28**, 174.
95. P. Kowalczuk, J. Ston-Egiert, W. J. Cooper, R. F. Whitehead and M. J. Durako, *Marine Chem.*, 2005, **96**, 273.
96. B. W. Ren, W. H. Zhao, J. T. Wang and L. Wang, *Spectrosc. Spectral Anal.*, 2008, **28**, 1130.
97. Y. Yamashita, R. Jaffe, N. Maie and E. Tanoue, *Limnol. Oceanogr.*, 2008, **53**, 1900.
98. F. C. Wu, R. B. Mills, R. D. Evans and P. J. Dillon, *Anal. Chem.*, 2004, **76**, 110.
99. M. L. Kram and A. A. Keller, *Soil Sediment Contamin.*, 2004, **13**, 119.

100. Y. N. Li, H. L. Wu, S. H. Zhu, J. F. Nie, Y. J. Yu, X. M. Wang and R. Q. Yu, *Anal. Sci.*, 2009, **25**, 83.
101. R. M. B. O. Duarte, C. A. Pio and A. C. Duarte, *J. Atmos. Chem.*, 2004, **48**, 157.
102. S. Lee and K. H. Ahn, *Water Sci. Technol.*, 2004, **50**, 57.
103. M. M. D. Sierra, M. Giovanela, E. Parlanti and E. J. Soriano-Sierra, *Chemosphere*, 2005, **58**, 715.
104. M. C. G. Antunes and J. C. G. E. da Silva, *Anal. Chim. Acta*, 2005, **546**, 52.
105. Z. Q. He, T. Ohno, F. C. Wu, D. C. Olk, C. W. Honeycutt and M. Olanya, *Soil Sci. Soc. Am. J.*, 2008, **72**, 1248.
106. J. Lamont-Black, A. Baker, P. L. Younger and A. H. Cooper, *Environ. Geol.*, 2005, **48**, 320.
107. D. Gimenez, L. A. Sarabia and M. C. Ortiz, *Analyst*, 2005, **130**, 1639.
108. J. R. Lead, A. DeMomi, G. Goula and A. Baker, *Anal. Chem.*, 2006, **78**, 3609.
109. R. D. Holbrook, P. C. DeRose, S. D. Leigh, A. L. Rukhin and N. A. Heckert, *Appl. Spectr.*, 2006, **60**, 791.
110. B. Antizar-Ladislao, J. Lopez-Real and A. J. Beck, *Chemosphere*, 2006, **64**, 839.
111. N. J. Reilly, T. W. Schmidt and S. H. Kable, *J. Phys. Chem. A*, 2006, **110**, 12355.
112. E. M. Ouyang, X. H. Mang and W. Wang, *Spectrosc. Spectral Anal.*, 2007, **27**, 1373.
113. W. L. Lai, L. F. Chen, S. W. Liao, S. L. Hsu, L. H. Tseng and C. L. Miaw, *Water Air Soil Pollut.*, 2007, **186**, 43.
114. T. Ohno, A. Amirbahman and R. Bro, *Environ. Sci. Technol.*, 2008, **42**, 186.
115. Z. G. Wang, W. Q. Liu, Y. J. Zhang, H. B. Li, N. J. Zhao, J. G. Liu, W. C. Sima and L. S. Yang, *Spectrosc. Spectral Anal.*, 2007, **27**, 2514.
116. S. H. Zhu, H. L. Wu, A. L. Xia, Q. J. Han, Y. Zhang and R. Q. Yu, *Talanta*, 2008, **74**, 1579.
117. S. A. Bortolato, J. A. Arancibia and G. M. Escandar, *Anal. Chem.*, 2008, **80**, 8276.
118. M. J. Rodriguez-Cuesta, R. Boque, F. X. Rius, D. P. Zamora, M. M. Galera and A. G. Frenich, *Anal. Chim. Acta*, 2003, **491**, 47.
119. D. Gimenez, L. Sarabia and M. C. Ortiz, *Anal. Chim. Acta*, 2005, **544**, 327.
120. A. M. de la Pena, A. E. Mansilla, N. M. Diez, D. B. Gil, A. C. Olivieri and G. M. Escandar, *Appl. Spectrosc.*, 2006, **60**, 330.

121. J. M. M. Leitfio, J. C. G. E. da Silva, A. J. Giron and A. M. de la Pena, *J. Fluoresc.*, 2008, **18**, 1065.
122. M. G. Trevisan and R. J. Poppi, *Anal. Chim. Acta*, 2003, **493**, 69.
123. A. Niazi, A. Yazdanipour, J. Ghasemi and A. Abbasi, *J. Chin. Chem. Soc.*, 2006, **53**, 503.
124. D. M. Fang, H. L. Wu, Y. H. Ding, L. Q. Hu, A. L. Xia and R. Q. Yu, *Talanta*, 2006, **70**, 58.
125. J. D. Jiang, H. L. Wu, A. L. Xia, S. H. Zhu, D. S. Liu, H. F. Zhang and R. Q. Yu, *Chem. J. Chin. Universities*, 2008, **29**, 71.
126. R. S. DaCosta, H. Andersson and B. C. Wilson, *Photochem. Photobiol.*, 2003, **78**, 384.
127. A. C. Olivieri, J. A. Arancibia, A. M. de la Pena, I. Duran-Meras and A. E. Mansilla, *Anal. Chem.*, 2004, **76**, 5657.
128. A. M. de la Pena, N. M. Diez, D. B. Gil, A. C. Olivieri and G. Escandar, *Anal. Chim. Acta*, 2006, **569**, 250.
129. A. L. Xia, H. L. Wu, D. M. Fang, Y. J. Ding, L. Q. Hu and R. Q. Yu, *Anal. Sci.*, 2006, **22**, 1189.
130. A. Garcia-Reiriz, P. C. Damiani and A. C. Olivieri, *Anal. Chim. Acta*, 2007, **588**, 192.
131. Q. J. Han, H. L. Wu, J. F. Nie, A. L. Xia, S. H. Zhu, Y. Zhang and R. Q. Yu, *Chem. J. Chin. Univ.*, 2007, **28**, 827.
132. J. F. Nie, H. L. Wu, A. L. Xia, S. H. Zhu, Y. C. Bian, S. F. Li and R. Q. Yu, *Anal. Sci.*, 2007, **23**, 1377.
133. K. Vatsavai, H. C. Goicoechea and A. D. Campiglia, *Anal. Biochem.*, 2008, **376**, 213.
134. P. Datta and D. Bhattacharyya, *J. Pharm. Biomed. Anal.*, 2004, **36**, 211.
135. Y. Xu and R. G. Brereton, *Anal. Bioanal. Chem.*, 2007, **388**, 655.
136. A. L. Xia, H. L. Wu, S. F. Li, S. H. Zhu, Y. Zhang, Q. J. Han and R. Q. Yu, *Talanta*, 2007, **73**, 606.
137. J. F. Nie, H. L. Wu, S. H. Zhu, Q. J. Han, H. Y. Fu, S. F. Li and R. Q. Yu, *Talanta*, 2008, **75**, 1260.
138. J. F. Nie, H. L. Wu, X. M. Wang, Y. Zhang, S. H. Zhu and R. Q. Yu, *Anal. Chim. Acta*, 2006, **628**, 24.
139. K. Hantelmann, M. Kollecker, D. Hüll, B. Hitzmann and T. Scheper, *J. Biotechnol.*, 2006, **121**, 410.
140. S. Hisiger and M. Jolicoeur, *J. Biotechnol.*, 2005, **117**, 325.
141. T. Knttel, H. Meyer and T. S Scheper, *Anal. Chem.*, 2005, **77**, 6184.
142. F. Guimet, J. Ferre, R. Boque and F. X. Rius, *Anal. Chim. Acta*, 2004, **515**, 75.
143. F. Guimet, R. Boque and J. Ferre, *J. Agric. Food Chem.*, 2004, **52**, 6673.

144. F. Guimet, J. Ferre and R. Boque, *Anal. Chim. Acta*, 2005, **544**, 143.
145. N. Dupuy, Y. LeDreau, D. Ollivier, J. Artaud, C. Pinatel and J. Kister, *J. Agric. Food Chem.*, 2005, **53**, 9361.
146. F. Guimet, J. Ferre, R. Boque, M. Vidal and J. Garcia, *J. Agric. Food Chem.*, 2005, **53**, 9319.
147. F. Guimet, R. Boque and J. Ferre, *Grasas Y Aceites*, 2005, **56**, 292.
148. F. Guimet, R. Boque and J. Ferre, *Chemometrics Intell. Lab. Syst.*, 2006, **81**, 94.
149. M. Zandomeneghi and G. Zandomeneghi, *J. Agric. Food Chem.*, 2005, **53**, 5829.
150. M. Zandomeneghi, L. Carbonaro and G. Zandomeneghi, *J. Agric. Food Chem.*, 2006, **54**, 5214.
151. A. Garcia-Relriz, P. C. Damiani, A. C. Olivieri, F. Canada-Canada and A. M. de la Pena, *Anal. Chem.*, 2008, **80**, 7248.
152. S. H. Zhu, H. L. Wu, A. L. Xia, Q. J. Han and Y. Zhang, *Anal. Sci.*, 2007, **23**, 1173.
153. S. H. Zhu, H. L. Wu, B. R. Li, A. L. Xia, Q. J. Han, Y. Zhang, Y. C. Bian and R. Q. Yu, *Anal. Chim. Acta*, 2008, **619**, 165.
154. X. M. Wang, H. L. Wu, J. F. Nie, Y. N. Li, Y. J. Yu and R. Q. Yu, *Sci. China, Ser. B*, 2008, **51**, 729–735.
155. F. Canada-Canada, A. Espinosa-Mansilla, A. M. de la Pena, A. J. Giron and D. Gonzalez-Gromez, *Food Chem.*, 2009, **113**, 1260.
156. M. L. Nahorniak, G. A. Cooper, Y. C. Kim and K. S. Booksh, *Analyst*, 2005, **130**, 85.
157. Y. C. Kim, J. A. Jordan, M. L. Nahorniak and K. S. Booksh, *Anal. Chem.*, 2005, **77**, 7679.
158. M. V. Bosco, M. P. Callao and M. S. Larrechi, *Anal. Chim. Acta*, 2006, **576**, 184.
159. M. V. Bosco, M. Garrido and M. S. Larrechi, *Anal. Chim. Acta*, 2006, **559**, 240.
160. M. V. Bosco, M. P. Callao and M. S. Larrechi, *Talanta*, 2007, **72**, 800.
161. S. Nishiyama, M. Tajima and Y. Yoshida, *J. Photopolym. Sci. Technol.*, 2006, **19**, 21.
162. A. Nevin, D. Anglos, S. Cather and A. Burnstock, *Appl. Phys. A: Mater. Sci. Process.*, 2008, **92**, 69.
163. S. J. Hart and R. D. JiJi, *Analyst*, 2002, **127**, 1693.
164. C. M. Andersen and R. Bro, *J. Chemometrics*, 2003, **17**, 200.
165. M. L. Nahorniak and K. S. Booksh, *J. Chemometrics*, 2003, **17**, 608.
166. A. Rinnan and C. M. Andersen, *Chemometrics Intell. Lab. Syst.*, 2005, **76**, 91.

167. A. L. Xia, H. L. Wu, D. M. Fang, Y. J. Ding, L. Q. Hu and R. Q. Yu, *J. Chemometrics*, 2005, **19**, 65.
168. D. B. Gil, A. M. de la Pena, J. A. Arancibia, M. Escandar and A. C. Olivieri, *Anal. Chem.*, 2006, **78**, 8051.
169. L. Peng, J. A. Gardecki, B. E. Bouma and G. J. Tearney, *Optics Express*, 2008, **16**, 10493.
170. R. Bro, *Tensor Decompositions Solving Fundamental Problems in Chemistry, presented at the 2008 SIAM Annual Meeting*, San Diego, CA, July 7–11 2008.
171. N. Rakow and K. Suslick, *Nature*, 2000, **406**, 710.
172. D. Filippini, S. Svensson and I. Lundström, *Chem. Commun.*, 2003, **2**, 240.
173. M. A. Malik, E. Gatto, S. Macken, C. DiNatale, R. Paolesse, A. D'Amico, I. Lundstrom and D. Filippini, *Anal. Chim. Acta*, 2009, **635**, 196.
174. J. W. P. Bakker, D. Filippini and I. Lundström, *Sens. Actuators, B*, 2005, **110**, 190.
175. D. Filippini and I. Lundstrom, *Analyst*, 2006, **131**, 111.
176. D. Filippini and I. Lundstrom, *Anal. Chim. Acta*, 2006, **557**, 393.
177. J. Barka, D. Filippini and I. Lundstrom, *Sens. Actuators, B*, 2006, **120**, 79.
178. M. A. Malik, E. Gatto, S. Macken, C. DiNatale, R. Paolesse, A. D'Amico, I. Lundström and D. Filippini, *Anal. Chim. Acta*, 2009, **635**, 196.
179. D. Filippini, F. Winquist and I. Lundstrom, *Anal. Chim. Acta*, 2008, **625**(2), 207.
180. D. Filippini and I. Lundstrom, *Analyst*, 2006, **131**, 118.
181. D. Filippini, C. DiNatale, R. Paolesse, A. D'Amico and I. Lundstrom, *Sens. Actuators, B*, 2007, **121**, 93.
182. D. Filippini and I. Lundstrom, *Anal. Chim. Acta*, 2004, **521**, 237.
183. D. Filippini, J. Manzano and I. Lundstrom, *Sens. Actuators, B*, 2004, **103**, 158.
184. J. W. P. Bakker, D. Filippini and I. Lundstrom, in Spectral Imaging: Instrumentation, Applications, and Analysis III ed. G. H. Bearman, A. Mahade van Jansen and R. M. Leveson, Book series: *Proceedings Of The Society Of Photo-Optical Instrumentation Engineers*, 2005, **5694**, pp. 9–15
185. D. Filippini, J. Bakker and I. Lundstrom, *Sens. Actuators, B*, 2005, **106**, 302.
186. D. Filippini, G. Comina and I. Lundstrom, *Sens. Actuators, B*, 2005, **107**, 580.

187. R. Bjorklund, D. Filippini and I. Lundstrom, *Sens. Actuators, B*, 2008, **134**, 199.
188. T. P. M. Andersson, D. Filippini, A. Suska, T. L. Johansson, S. P. S. Svensson and I. Lundstrom, *Biosens. Bioelectron.*, 2005, **21**, 111.
189. A. Suska, D. Filippini, T. P. M. Andersson and I. Lundstrom, *Biosens. Bioelectron.*, 2005, **21**, 727.
190. D. Filippini, K. Tejle and I. Lundstrom, *Biosens. Bioelectron.*, 2005, **21**, 266.
191. D. Filippini, P. Asberg, P. Nilsson, O. Inganas and I. Lundstrom, *Proc. IEEE Sensors*, 2004, **1–3**, 1377.
192. D. Filippini, P. Asberg, P. Nilsson, O. Inganas and I. Lundstrom, *Sens. Actuators, B*, 2006, **113**, 410.
193. S. Macken, C. Di Natale, R. Paolesse, A. D'Amico, I. Lundstrom and D. Fillippini, *Transducers '07 and Eurosensors XXI, Digest Of Technical Papers*, 2007, **1–2**, U1068–U1069.
194. S. Macken, C. DiNatale, R. Paolesse, A. D'Amico, I. Lundstrom and D. Filippini, *Anal. Chim. Acta*, 2009, **632**, 143.
195. H. Kido, A. Maquieira and B. D. Hammock, *Anal. Chim. Acta*, 2000, **411**, 1.
196. S. A. Lange, G. Roth, S. Wittermann, T. Lacoste, A. Vetter, J. Grassle, S. Kopta, M. Kolleck, B. Breitinger, M. Wick, J. K. H. Horber, S. Dubel and A. Bernard, *Angew. Chem., Int. Ed.*, 2006, **45**, 270.
197. R. A. Potyrailo, W. G. Morris, A. M. Leach, L. Hassib, K. Krishnan, C. Surman, R. Wroczynski, S. Boyette, C. Xiao, P. Shrikhande, A. Agree and T. Cecconie, *Appl. Optics*, 2007, **46**, 7007.
198. R. A. Potyrailo, W. G. Morris, A. M. Leach, T. M. Sivavec, M. B. Wisnudel, K. Krishnan, C. Surman, L. Hassib, R. Wroczynski, S. Boyette, C. B. Xiao, A. Agree and T. Cecconie, *Am. Lab.*, 2007, **39**, 32.
199. R. A. Potyrailo, W. G. Morris, A. M. Leach, T. M. Sivavec, M. B. Wisnudel and S. Boyette, *Anal. Chem.*, 2006, **78**, 5893.
200. R. A. Potyrailo, W. G. Morris, R. Wroczynski, L. Hassib, P. Miller, B. Dworken, A. M. Leach, S. Boyette and C. Xiao, *Sens. Actuators, B*, 2009, **136**, 203.
201. J. Tamarit-Lopez, S. Morais, R. Puchades and A. Maquieira, *Anal. Chim. Acta*, 2008, **609**, 120.
202. S. Morais, J. Carrascosa, D. Mira, R. Puchades and A. Maquieira, *Anal. Chem.*, 2007, **79**, 7628.
203. SLSD-71N6 from Silonex (Montreal, Canada)

204. S. C. B. Gopinath, K. Awazu, J. Tominaga and P. K. R. Kumar, *ACS Nano*, 2008, **2**, 1885.
205. S. Morais, J. Tamarit-López, J. Carrascosa, R. Puchades and Á. Maquieira, *Anal. Bioanal. Chem.*, 2008, **391**, 2837.
206. S. Morais, L. A. Tortajada-Genaro, T. Arnandis-Chover, R. Puchades and A. Maquieira, *Anal. Chem.*, 2009, **81**, 5646.
207. I. Alexandre, Y. Houbion, J. Collet, S. Hamels, J. Denarteau, J. -L. Gala and J. Remade, *Biotechniques*, 2002, **33**, 435.
208. J. J. La Clair and M. D. Burkart, *Org. Biomol. Chem.*, 2003, **1**, 3244.
209. Y. Li, Z. Wang, L. M. L Ou and H.-Z. Yu, *Anal. Chem.*, 2007, **79**, 426.
210. Y. Li, L. M. L. Ou and H. -Z. Yu, *Anal. Chem.*, 2008, **80**, 8216.

CHAPTER 5

Separation Methods in Analytical Chemistry

Chromatography is now one of the most common analytical techniques for the separation and analysis of mixtures. Chromatography is usually divided into gas and liquid variants depending on the physical state of the mobile phase. Liquid chromatography was first discovered by Michail Tsvett in 1903, and another widely used separation method, electrophoresis, was developed two decades later (mainly by Arne Tiselius). These methods are still in the active development phase compared with gas chromatography, which was reported by Archer J. P. Martin and Richard L. M. Snyge in 1945. It might seem surprising that such a complicated technology has replaced other known separation methods, *i.e.* extraction or distillation, but the reason is the high efficiency of the process, or the high number of theoretical plates, which determines the number of compounds that a particular column can separate.

5.1 GAS PHASE SEPARATIONS

The basic challenge of gas chromatography (GC), in the context of Green Analytical Chemistry, is the waste that the method produces. Gas chromatography is still a relatively green technique, because the eluents employed are usually helium and hydrogen, which are gases that are harmless to the atmosphere. Nevertheless, the sample preparation methods used in GC should be revisited in the light of Green Chemistry

Green Analytical Chemistry
By Mihkel Koel and Mihkel Kaljurand

Published by the Royal Society of Chemistry, www.rsc.org

principles. Namieśnik has presented various approaches to the implementation of Green Chemistry principles in gas chromatography:[1]

- the use, as far as it is possible, of so-called direct chromatographic analysis, which permits the determination of analytes in a sample without any pre-treatment or sample preparation;
- labour and energy consumption, *e.g.* reducing sample preparation time when direct chromatographic analysis is not possible, and eliminating or reducing the amount of solvent in the sample preparation steps prior to the final chromatographic analysis;
- conducting all operations involving the use of solvents in a hermetic system;
- a reduction in matrix interference;
- a reduction in chromatographic run-time and the need for reanalysis;
- an integration of the various steps in the analytical procedure, *i.e.* by using hyphenated techniques.

Eliminating or minimizing the amount of solvent in sample preparation techniques before the final chromatographic analysis is highly recommended. Concern over the disposal of toxic solvents has led to cleaner extraction methods. Such methods are commonly described as solventless sample preparation techniques[2] in contrast to the usual liquid–liquid or liquid–solid extraction using an organic solvent (often followed by clean-up and pre-concentration steps). There are three known solventless approaches: gas extraction, membrane-based techniques and supercritical fluid extraction. Techniques using gas and supercritical fluids for the extraction of many pollutants are also very popular in gas chromatography.

The U.S. Environmental Protection Agency recommends dynamic gas extraction, *i.e.* purging gas through the solution under investigation and trapping it on a suitable sorbent, (*i.e.* Tenax). This is the widely accepted and preferred method for the routine analysis of volatile organic compounds in water. Static headspace micro-extraction is also becoming very popular, mainly because it does not require such sophisticated instruments as the purge technique. Gas extraction is the most widely used method for isolating volatile pollutants from different matrices, mainly because it can be considered a pro-ecological (solvent-free) method of isolation and enrichment. Gas extraction provides the required sensitivity (up to ppt level) and can be automated by combining it with GC. Thermal desorption of pollutants collected on a sorbent is a standard method for the measurement of volatile organic compounds in the workplace or environmental air.[3] The advantages of thermal desorption *versus* conventional solvent extraction include: an improvement in detection limits (three orders); no chromatographic interference from solvent or solvent impurities; enhanced sample throughput; and a lower cost-per-analysis. Another advantage is that thermal desorption is a very straightforward gas extraction process. In light of these advantages, it can be said that thermal desorption meets all the requirements of Green Chemistry for chromatographic analysis.

In addition to sample preparation, the importance of fast gas chromatography for greening gas phase separations is quickly gaining attention. High-speed (fast) gas chromatography reduces overall analysis time, leading to significant savings in time and money.[4,5] Fast chromatography can be especially attractive for laboratories where many routine samples are analyzed every day. It can also be advantageous in situations where quick results are needed. However, increasing the speed of analysis requires modification of commercially available instrumentation. Faster separation can be achieved by decreasing the inner diameter of the capillary columns; reducing the thermal mass of the column thermostat for fast temperature programming; applying shorter columns; working at turbulent flow using either a vacuum outlet operation; or working above optimal carrier gas velocities.

5.2 LIQUID PHASE SEPARATIONS

5.2.1 Liquid Green Chromatography

Green chromatography is a new concept in HPLC that attempts to overcome the disadvantages – from the point of view of Green Chemistry – of traditional liquid chromatography. Since the aim of Green Analytical Chemistry is, first of all, to reduce the use of toxic solvents, green chromatography attempts to either:

- reduce the amount of toxic solvents *via* miniaturization, or
- use less toxic solvents in combination with HPLC at elevated temperatures. The latter approach is advocated mainly by Pat Sanda (Research Institute for Chromatography, Kortrik, Belgium) and Selerity Technologies, Inc., (Salt Lake City, UT).

5.2.1.1 Micro-Scale HPLC. Micro-scale HPLC has been developed mainly by instrumentation companies, and the following description is for illustration, rather than promotion of any of the companies. Microscale refers to HPLC that uses columns, the inner diameter of which is

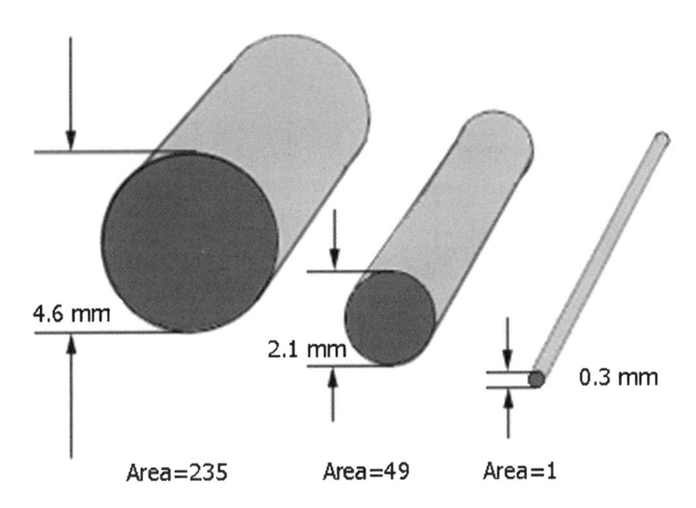

Figure 5.1 Comparison of relative cross-sectional areas of common and μ-columns.

less than 500 μm, commonly 300 microns. In contrast, typical inner diameters of conventional HPLC columns are 2.0–4.6 mm. The diameter determines the cross-sectional area of a column, which in turn determines the mobile phase volumetric flow rate and, eventually, the amount of solvent consumed. It is instructive to compare the cross-sectional areas of these columns (see Figure 5.1).

The typical conventional HPLC flow rate is 1 ml min^{-1} on a 4.6 mm column. Assuming the same linear velocity on common HPLC columns and micro-scale columns (diameter 300 μm), the volumetric flow rate is 4.3 μL min^{-1} (a reduction in flow rate of 99.6%). Micro-column HPLC systems were developed in response to the need to couple HPLC to the mass spectrometer. Their design has specific requirements; however, it is not easy to deliver eluents with a low μL min^{-1} flow rate. Flow-splitting systems (see Figure 5.2), added in an attempt to adapt conventional-scale pumps to the capillary flow rates required for coupling HPLC to MS, reduce gradient formation accuracy, add connections and complexity, and waste litres of solvent every day. The systems use split ratios up to 1000 : 1, without eluent flow monitoring in the column. It is difficult to imagine how such splitting systems could fulfil the requirements of Green Chemistry. Moreover, these systems cannot adjust flow rate when the backpressure changes. Still, because there is an urgent need to develop analytical tools for proteomics, eluent splitting is widely used, due to the need to interface micro-columns with electrospray ionization mass spectrometers. Electrospray ionisation occurs at μL min^{-1} flow rates, and available HPLC pumps cannot deliver stable flows lower

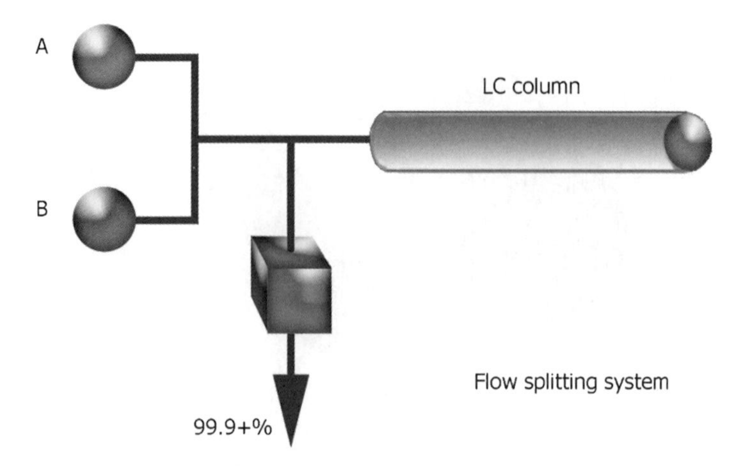

Figure 5.2 Flow-splitting in micro-HPLC; A and B eluent pumps.

than 50 μL min^{-1}. Thus, in order to interface HPLC with MS, the column flow and ESI flow must be matched and splitting becomes inevitable.

Innovations in pressure generation by delivering flows at μL min^{-1} rates would make it possible to avoid splitting. In Eksigent HPLC systems,[6] pressure is generated by connecting laboratory air or nitrogen to a pneumatic amplifier that produces a one hundred-fold amplification in pressure. The incorporated microfluidic flow controller regulates this pressure to generate the required flow rate. Flow meters in each mobile phase continuously monitor flow rate and feed a proportional signal back to a microprocessor. The system allows precise, rapid control of nanoscale and capillary flow rates. Two microfluidic flow control systems combine to form a binary gradient LC system. The microprocessor sends a voltage signal to the controller at the pressure source for each mobile phase. The signal is proportional to the pressure required in each mobile phase to achieve the desired flow rate or gradient. For example, 100 psi (0.7 MPa) incoming air pressure from the laboratory air system can be used to produce a hydraulic pressure range upwards of 10 000 psi (70 MPa).

This pressure-generating system provides precise gradients at nL min^{-1} rates without flow-splitting, and it can also respond extremely rapidly to set-point changes.

To take advantage of the inherent benefits of micro-scale HPLC, the instrument must have very little dispersion (longitudinal diffusion) of the analyte zone when it moves along the column from injection to the detector. In general, total dispersion in chromatography, σ^2_{TOT}, comprises several components, according to the formula

$$\sigma^2_{TOT} = \sigma^2_{column} + \sigma^2_{incector} + \sigma^2_{tubing} + \sigma^2_{fittings} + \sigma^2_{detector} \quad (1)$$

In micro-column HPLC, zone dispersion is small and total dispersion comes mainly from the extra-column volume. This causes the analyte bands to widen. All parts of Equation (1) must be considered and the system properly designed, but the main issues are the injection and detection of the sample.

A conventional-scale HPLC injection volume of 10 µL becomes a 40 nL injection in a capillary system. Conventional auto-samplers cannot deliver precise 40 nL injections; Eksigent solved this problem by developing an electronically controlled injection valve coupled with a microfluidic flow control delivery system to deliver precise nanoliter injection volumes. The main idea is to use only part of an ordinary HPLC sampling loop. A standard auto-sampler fills a sample loop (typically 250 nL) with sample. The valve is then toggled from load to inject and back to the load position, based on integrating the flow rate over time to reach the desired injection volume. This solution provides high-precision, software-selectable injection volume control, and compatibility with some popular auto-sampler systems.

Tubing and fitting without dead volumes can now be easily achieved using commercially available zero-volume connectors. Tubing and fitting also includes the design of micro-scale mixing. The advantages of micro-LC over conventional HPLC disappear unless there is a proportional reduction in fluid mixing volume. Thus, in a micro-fluidic HPLC system, a simple short length of capillary provides complete mixing and negligible delay volume. Eksigent's micro-scale mixer has a volume of just 300 nL, and a mixing delay of only 2 seconds at 10 µL min^{-1}. This rapid mixing virtually eliminates the gradient delays inherent in conventional systems at the start of the process and at re-equilibrations. Rapid "deterministic" mixing leads to extremely precise gradients, rapid system response and high gradient linearity.

Micro-scale UV detection also requires a new design. Scaling down a conventional UV absorbance detection system to capillary levels results in increased dispersion and reduced detection sensitivity. Eksigent systems utilize a micro-fluidic-based, non-dispersive UV absorbance flow cell. The detection system brings the dynamic range and linearity of conventional detection systems into the capillary format. Furthermore, it offers fully dispersed array-based detection from 200 nm to 380 nm, providing full spectral information for each peak.

Ultimately, nanoflow high-performance liquid chromatography can be performed in a micro-fluidic format as well. The centrepiece of Agilent's new HPLC-Chip[7] technology is a reusable micro-fluidic polymer chip. It integrates the sample enrichment and separation columns of a nanoflow LC system directly onto the polymer chip, with the intricate connections and spray tip used in electrospray mass spectrometry. The technology eliminates 50% of the traditional fittings and connections typically required in a nanoflow LC-MS system, thus reducing the possibility of leaks and dead volumes. The second component of HPLC-Chip technology is the HPLC-Chip-MS interface. A chip is inserted into the interface, which is mounted on an Agilent mass spectrometer. The design configuration guarantees that the electrospray tip is in the optimal position for mass analysis when the chip is inserted. Replacement of the chip is simple and can be completed in a few seconds, as opposed to the much longer time it takes to change nanoLC columns. The HPLC-Chip interface will be available as a standard module in the Agilent 1100 Series LC system.

In recent years, interest in micro-column LC has increased considerably. Researchers have recognized the ability of micro-column HPLC to work with small sample sizes, small volumetric flow rates, and to facilitate coupling with multidimensional LC systems and directly with MS detectors. Certainly, the latter properties (rather than a green worldview) have been the main driving force for the developments in micro-column chromatography. However, the list of routine applications of micro-column HPLC is limited, with only slightly more than 100 publications (judging by a Web of Science search in the summer of 2009) for applications in various areas of bioanalytical and environmental analysis. Most HPLC equipment vendors provide micro-columns either in capillary, packed or monolithic format.

5.2.1.2 Elevated Temperature and Temperature Programming in HPLC. Chromatographers frequently overlook temperature as an optimization parameter in HPLC, but it can play an important role in selectivity since it affects nearly all of the physical parameters that play a role in liquid chromatographic separation. However, temperature as a way to tune separation and shorten analysis time in HPLC has only recently attracted the attention of chromatographers. Since temperature is an instrumental setting, it is much easier to adjust during the method development stage than, for example, the buffer pH or mobile phase composition. Still, most HPLC separations have been performed at temperatures ranging from 18–30 °C, *i.e.* at room temperature. If the column temperature is higher, the term "high-temperature HPLC" is frequently used in the literature. In this book, the authors prefer to use the term "elevated temperature", because many applications of interest to chromatographers reviewed in this chapter are at temperatures in the 40–200 °C range, which are certainly elevated with respect to ambient temperature, but would not normally be considered high. A number of reviews have been published recently which summarize the developments and applications of elevated-temperature HPLC.[8–12] A special issue of the *Journal of Separation Science* (2001) was also dedicated to the role of temperature in LC.[13] The earlier lack of interest in this topic has obviously been due to the lack of dedicated equipment for temperature-controlled LC. In recent years the situation has improved considerably. The work of Pat Sandra's group and companies like Pfizer and Selerity have contributed considerably to elevated-temperature HPLC. In several publications they have advocated the use of elevated temperature as a way to make HPLC greener compared with conventional liquid chromatography.[14–16] The following discussion is based on these publications.

The advantages of applying elevated temperature in LC separations are many. First of all, temperature affects the viscosity of the liquid, which is a measure of resistance of the eluent to flow. The viscosity of a liquid will decrease with higher temperature. Viscosity influences the backpressure of a chromatographic column, which is necessary to create the flow of eluent through the column. This is expressed as a proportionality coefficient, according to the well-known Hagen–Poiseuille equation (for empty tubing):

$$\Delta P = \eta \frac{128l}{\pi d^4} F \tag{2}$$

where ΔP is the pressure drop over tubing, l and d are the length and diameter of the tubing, η is the so called dynamic viscosity of the eluent, and F is the volumetric flow rate. The most obvious advantage of reduced backpressure at increased temperature is the gain in speed of analysis. For example, the viscosity of acetonitrile–water mixtures is decreased by up to a factor of 4, as the temperature is increased from 30 °C to 120 °C (ref. 17). Due to the decreased retention, the amount of organic modifier in the mobile phase can be reduced significantly, and in some cases eliminated completely, resulting in pure aqueous mobile phases. The reduced backpressure also allows ethanol (which is commonly avoided in HPLC separations because of its higher viscosity compared to methanol) to replace methanol or acetonitrile. Water and ethanol are both non-toxic liquids, and analysis using these substances can be considered green chromatography. Therefore, one of the most important principles of Green Chemistry – replacement of harmful solvents – can be implemented in chromatographic practice. In addition, increasing the temperature can also lead to improved peak symmetry by reducing secondary interactions with the silica support.

The influence of temperature on chromatographic retention can be expressed in thermodynamic parameters. Increasing the temperature increases the enthalpy of solute transfer from the mobile phase to the stationary phase.[18,19] The retention factor, k, of an analyte is related to its equilibrium constant, K, between the stationary and mobile phases through the volume ratio of stationary to mobile phase, β, as follows: $k = K\beta$. The retention factor is a function of Gibbs free energy change:

$$\ln \frac{k}{\beta} = -\frac{\Delta H}{\mathrm{RT}} + \frac{\Delta S}{\mathrm{R}} \tag{3}$$

where ΔH is the enthalpy and ΔS is the entropy of the sorption process of a molecule to the stationary phase. T is the absolute temperature (in Kelvin), and R is the gas constant (8.31441 J K^{-1} mol^{-1}). As shown in Equation (3), if $\Delta H < 0$ (which is normally the case), an increase in temperature will cause a decrease in retention. This is valid in nearly all reverse-phase separations. As a result, linear plots with positive slopes can usually be obtained.

The temperature effect can be seen on such chromatographic parameters as efficiency resolution and selectivity. At elevated temperatures, the solute transfer from the mobile phase to the stationary phase is more efficient. This leads to a flatter van Deemter curve and makes use of higher flow rates without hampering efficiency. As shown in Figure 5.3, as temperature increases, H_{min} slightly increases in value and shifts to higher flow rates. Combining high temperature and increased flow rates results in higher efficiency per time unit. As a result, implementing high temperatures enables a chromatographer to work in the "flattened" region of the van Deemter curve, which means that higher flow rates can be used without hampering efficiency. Indeed, it has been proved that an increased flow rate is required to fully benefit from increased temperature.[18]

Low backpressure due to decreased solvent viscosity at elevated temperatures allows the use of longer columns and increased flow rates, to increase efficiency and resolution without significantly increasing the analysis time. An example of this approach is Vanhoenacker and Sandra's analysis of a mixture of a pharmaceutical compound and its related impurities (see Figure 5.4).[16]

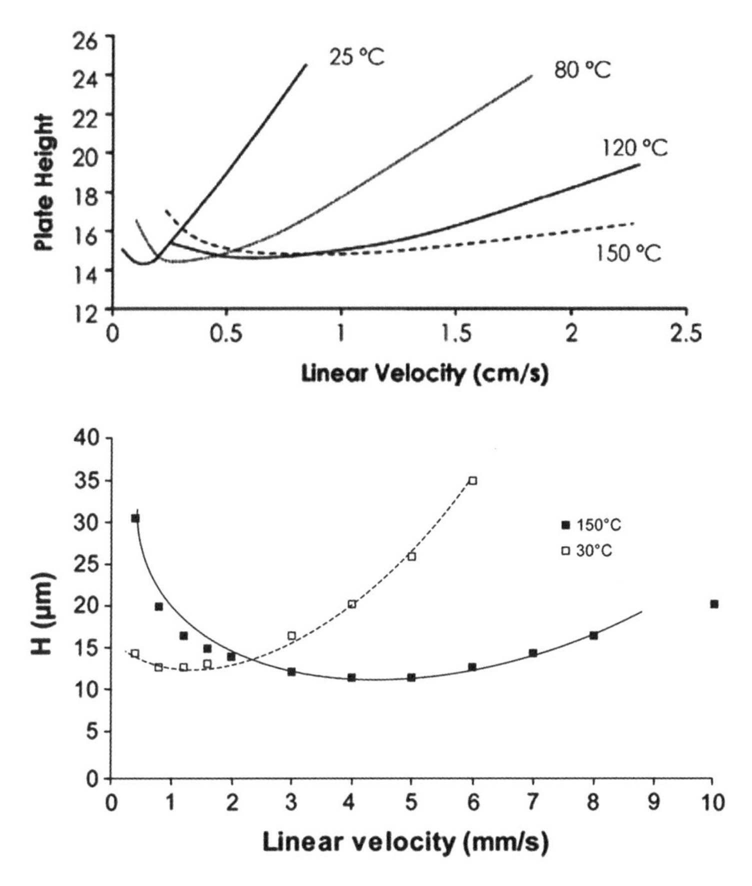

Figure 5.3 (Top): plate height (H) *versus* linear velocity at various temperatures.[16] (Reproduced with kind permission of Springer Science.) (Bottom): experimental demonstration for propylparaben. Conditions: column = Blaze 200 (150 mm × 4.6 mm i.d.; particle diameter 5 μm); mobile phase = water–ACN (60 : 40); and detection = 210 nm.[15] (Reproduced with kind permission of Wiley-VCH Verlag GmbH & Co. KGaA.)

The mixture was analyzed on a Zorbax StableBond C18 column at 40 °C (column length 150 mm; 3 mm ID; particle diameter 3.5 μm) and at 90 °C (column length 600 mm – four 150 mm columns were coupled in series), using gradient elution with mobile phase A: formic acid in water, and mobile phase B: formic acid in acetonitrile. Because of the reduced backpressure and the higher value of the optimum linear velocity for high-temperature separations, a higher flow rate was used. As a result, significantly higher resolution was obtained within a shorter analysis time.

Since the temperature can be set by an instrument, it is one of the easiest and most straightforward parameters for tuning

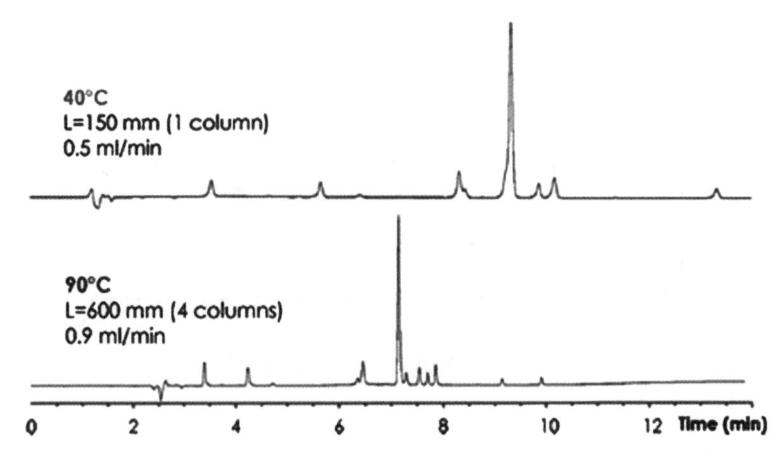

Figure 5.4 Analysis of a pharmaceutical mixture at (top): 40 °C and (bottom): 90 °C.[16] (Reproduced with kind permission of Springer Science.)

chromatographic selectivity. Controlling temperature on an instrument should be more reproducible than adapting the composition or pH of a buffered mobile phase. Tuning selectivity by temperature could be especially useful for polar and ionisable compounds, since ionization equilibrium is temperature dependent.[16] The potential role of temperature in method development is still underexploited. A drastic change in selectivity for the analysis of octylphenol ethoxylate oligomers by RP LC was recently demonstrated.[20] The elution order of the oligomers could be reversed by simply changing the temperature, using the same column and mobile phase composition. Another representative application was the separation of a mixture of ten phenylurea and ten triazine pesticides, demonstrated by Vanhoenacker and Sandra.[16] They showed that it was impossible to separate all 20 compounds if they were analyzed on a C18 column with only a solvent gradient. Temperatures between 40 and 90 °C with 10 °C intervals were tested and significant selectivity changes were noted, but complete resolution was impossible under isothermal conditions. Temperature-programmed HPLC at temperatures up to 200 °C allows the user to perform a temperature gradient to alter retention and selectivity, instead of using a solvent gradient. Increasing the temperature reduces water polarity, causing it to behave like a moderately polar organic solvent during the separation process. Many separations requiring a binary solvent gradient can thus be performed isocratically using a thermal gradient.

When a moderate temperature gradient was combined with the solvent gradient, the separation of all the polar pesticides was achieved.

This clearly demonstrates the potential of a column thermostat that enables temperature programming. Another example of green chromatography was presented,[15] in which the application of high-temperature HPLC produced a relatively fast separation of naphthylamine isomers on a cyanopropyl column. The mobile phase in this case consisted of water and ethanol. Since gradient analysis was performed, the mobile phase can be considered (in their words): "to evolve from a wine-like to a strong spirit strength".

At elevated temperatures, it appears possible to avoid organic compounds entirely. This is because the hydrogen-bonding effects of water are reduced if the temperature is increased, thus making it less polar. By using a backpressure regulator, as in supercritical fluid chromatography, it is possible to use superheated water at temperatures up to 250 °C as the mobile phase.[21] Sub-critical water, at 140 °C, approaches the characteristics of an organic solvent in terms of its dielectric constant. It also has greatly improved solubilization properties for hydrophobic analytes, as compared with water at ambient conditions. This allows reversed-phase HPLC to be performed using a mobile phase that contains no organic co-solvents, thereby providing safety and a completely green environment for LC. Water at sub-critical temperatures behaves like a moderately polar organic solvent, such as methanol, during the separation process. Many separations requiring a binary solvent gradient can thus be performed isocratically. Eliminating the use of solvent gradients will not only ease some difficulties in the method development process, but also lessen the issues when transferring methods from development laboratories to quality control laboratories.

An analysis of steroids on a coated zirconium oxide column at 200 °C (ref. 22) provides an example of using water as a mobile phase. The use of water as a green eluent that is environmentally friendly and available at reduced cost has been reviewed by several groups.[23,24] A number of studies have demonstrated the environmental and cost benefits of using pure water as a mobile phase.[21,25,26] For example, for aniline and phenol, k values in sub-critical water separation at 150 °C are similar to those achieved by using methanol–water (43:57) or acetonitrile–water (40:60).[21] Water also has the advantage of enabling UV detection to 190 nm (ref. 27–29) and even flame ionization detection.[30–33]

Because hydrolysis is enhanced at high temperatures, the use of water in the mobile phase for reversed-phase-type separations requires temperature-resistant stationary phase materials to prevent the loss of bonded phase from the silica support.[34] Considerable effort has been devoted to developing stationary phases that are robust enough to tolerate temperatures of 200 °C or higher. Silica-based stationary phases

Table 5.1 Commercially available LC columns for high temperature operation.

Base material	*Column (manufacturer)*	*Available stationary phases*	T_{max} (°C)
Silica, sterically protected	Zorbax StableBond (Agilent Technologies)	C18	90
		C8, C3, CN, Phenyl	80
Silica, polydentate	Blaze 200 (Selerity Technologies)	C18	200
	Cogent (MicroSolv Technologies)	C18, cholesterol	100
Silica, hybrid particle technology	Xbridge (Waters Corporation)	C18, C6 Phenyl	100
Silica, encapsulated	Pathfinder (Shant Laboratories)	C18	150
Zirconium oxide	Zirchrom (Zirchrom Separations)	PBD, PS	150
		Carb, Diamondbond-C18	200
	Discovery Zirconia (Supelco)	PBD, PS	150
		Carbon, Carbon C18	200
Polystyrene/divinylbenzene	PLRP-S (Polymer Laboratories)	PS/DVB	150
	PRP-1 (Hamilton)	PS/DVB	150
	Jordi RP (Jordi Associates)	DVB, C18-DVB	150
Graphitized carbon	Hypercarb (Thermo Electron)	Graphitized carbon	200

usually are stable at temperatures up to 60 °C and even higher in some cases (see Table 5.1). A new type of silica-based phase that is stable at temperatures up to 200 °C under certain reversed-phase conditions has recently been introduced. Stationary phases with the highest temperature stability are based on materials other than silica, *e.g.* graphitized carbon, zirconium oxide,[35,36] polystyrene/divinylbenzene (DVB) and other polymeric phases. An overview of commercially available stationary phases is presented in Table 5.1.[16]

In addition to new column materials, high-temperature HPLC requires the redesign of conventional HPLC equipment. The column must be supplied with a thermostat and the mobile phase preheated before it enters the column and cooled after it leaves the column. The thermal mismatch between the column and the mobile phase produces a temperature gradient across the column radius. The cool mobile phase entering the heated column will warm up faster along the walls than in the centre of the column if the mobile phase is not preheated.

The warmer mobile phase in that region will flow faster than in the column centre and lead to band broadening, poor peak shape and split peaks. The negative impact on peak shape can render the advantages of high temperature useless, even at temperatures as low as 80 °C with 4.6 mm i.d. columns. Therefore, performing LC at elevated temperatures requires accurate control of the column temperature and the incoming mobile phase. "Thermal mismatch" band broadening is eliminated if the mobile phase is preheated. The temperature of the incoming mobile phase should be within ±6 °C of the oven/column temperature to minimize band broadening by radial temperature gradients.[37] This is not difficult to achieve when using packed capillary columns at high temperatures. However, heating aqueous mobile phases to significantly high temperatures with typical analytical flow rates of several hundred microlitres to several millilitres per minute is more difficult.

In the past, there was no dedicated equipment for high-temperature HPLC. Chromatographers who wanted to investigate high-temperature HPLC used gas chromatograph ovens and long pieces of stainless steel tubing to heat the HPLC columns and to preheat the incoming mobile phase, respectively. The column effluent was subsequently cooled by a water or ice bath. Extra-column dispersion due to the preheating and cooling tubing volumes often leads to reduced efficiency in these systems. Recently, instrumentation for elevated-temperature and temperature-programmed HPLC with conventional column dimensions became available from Selerity Technologies[38] (see Figure 5.5).

This system enables elevated-temperature and temperature-programmed LC with conventional columns and HPLC equipment, and actively heats the incoming mobile phase to the same temperature as the column/oven temperature with a low-volume, Peltier-based pre-heater unit. The instrument also has an integrated post-column cooling/thermostatic unit. Post-column or post-detection backpressure regulation is

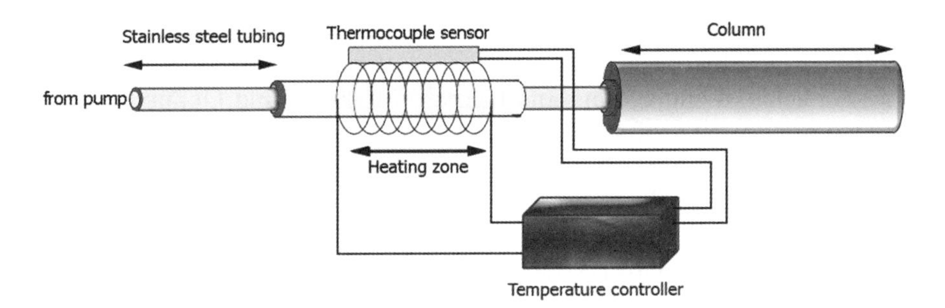

Figure 5.5 A mobile phase preheating system.

sometimes required to maintain the column effluent in the liquid state at high temperatures. 10 to 30 bar backpressure will enable operation well above the normal boiling point.

A traditional flame ionization detector (FID) design with a grounded flame jet can be used for superheated water eluent. This detector – which is well known in gas chromatography – can tolerate a stable flow (up to 100 μL min^{-1}) of superheated water coming directly into the flame jet. It is capable of detecting analytes at low nanogram levels and it operates at temperatures up to 400 °C. However, column backpressure must be maintained to minimize detector spiking when using an FID as a HPLC detector. The Selerity column-to-detector interface[38] contains an all-metal design column end restriction from sintered metal beads that elutes superheated water to the detector in aerosol form. The pressure drop region remains short (2–5 mm) and is positioned in the heated region of the FID.

The application of high temperatures enables the use of columns packed with sub-2 μm particles at high flow rates. Therefore, it is possible to exploit the advantages of these columns using conventional HPLC equipment rather than the ultra-performance liquid chromatography that has recently become popular, which requires special instrumentation and dedicated columns.

Column temperature has a strong effect on retention, and not only because of the properties of the mobile phase components. It follows from Equation (3) that the other chromatographic parameters (*e.g.* composition of the mobile and stationary phases) as well as the properties of the analytes are affected by temperature through various mechanisms.[39] The relative retention of two solutes is sensitive to changes in the conformation of the stationary phase as the temperature is varied. Therefore, temperature-dependent stationary phases can be synthesized. For example, retention can be modulated by a temperature-responsive polymer with reversible hydrophilic–hydrophobic conformation.[40,41] Temperature-responsive stationary phases also contribute to green chromatography. Sensitive polymers with external physical, chemical and electrical stimuli are termed "intelligent materials" and have been used in various fields of engineering and medicine. Kikuchi and Okano describe several applications of surfaces and interfaces modified with stimuli-responsive polymers for stimuli-responsive surface property alteration, and their application in the separation sciences for the affinity separation of proteins.[42] Special attention is paid to the temperature-responsive polymer, poly(*N*-isopropylacrylamide) (PNIPAAm), and its derivatives, as surface modifiers for novel green chromatography, in which only an aqueous mobile phase was utilized for the separation of bioactive compounds. This polymer exhibits hydrophilic properties below 32 °C and becomes hydrophobic above that temperature. Several effects of bioactive compounds on separation were investigated and discussed; they include the effects of temperature-responsive hydrophilic–hydrophobic changes, copolymer composition, graft polymer molecular architecture and the incorporation of charged groups. The technique has superior chromatographic separation characteristics for reducing organic wastes and costs; therefore, it is an environmentally friendly separation tool. In another study, HPLC columns were packed with PNIPAAm attached to an aminopropyl silica gel.[43] The temperature-responsive properties of the coupled phase were demonstrated by using only water as a mobile phase; an increase in retention was observed with rising temperature. Mixtures of compounds covering a wide polarity range and including phenones, alkylbenzenes, phenols, alkylated benzoic acids, anilines, sulfonamides and carbamates were analyzed, and the retention, peak shapes and plate counts were compared under identical conditions. For retained solutes, an increase in retention as a function of temperature between 25 and 55 °C was observed, and this was higher for analytes containing a longer hydrophobic chain. Compounds with similar hydrophobic chains but additional polar functions showed increased retention and improved peak shapes, suggesting a mixed mode interaction mechanism also exists at temperatures well above the transition temperature of the polymer. Weak acids and bases can be analyzed by pH adjustment. This was demonstrated for mixtures of benzoic acid derivatives and sulfamide drugs. A carbamate pesticide mixture was analyzed at 55 °C with water (pH 5.5) as the mobile phase, and electrospray ionization MS detection.

As mentioned above, the usual aim of elevated-temperature chromatography is to avoid using multicomponent mobile phases. If such phases are still required, one must take into account that changing the column temperature also affects retention through its effect on mobile phase composition. Melander, *et al.*[44] propose a relationship between the logarithmic retention factor, mobile phase composition and column temperature to describe the retention behaviour of numerous polar and non-polar compounds. They concluded that the retention data have a reciprocal dependence on absolute temperature, T, as shown in the following equation:

$$\log k = A_1 \varphi (1 - \frac{T_c}{T}) + \frac{A_2}{T} + A_3 \tag{4}$$

where φ is the volume fraction organic co-solvent of the hydro-organic mobile phase; T_c is the so-called compensation temperature, which is usually found to be constant in reverse phase chromatography, and A_1, A_2 and A_3 are parameters that vary with solutes and stationary phases. Equation (4) can be used to develop software to rapidly optimize both temperature and mobile phase composition. For a particular sample, four initial experiments can be carried out at two different temperatures, using either isocratic or gradient elution. On the basis of Equation (4), the corresponding data allow the resolution map to be calculated as a function of both temperature and either mobile phase composition in isocratic mode, or gradient slope in gradient elution mode.[45,46]

Even analyte property dependence can be used in elevated-temperature chromatography. Many properties of molecules are temperature-dependent, and this can be used for tuning their retention. For example:

- the relative size or shape of two molecules is different, leading to differences in their entropies;
- two molecules may have different functional groups with different temperature dependence for retention;
- an ionizable solute is partially ionized, so that the molecule exists in both ionized and neutral form.

The latter case deserves special attention since most pharmaceutical and biological substances possess ionizable carboxylic or amino functional groups. The separation of these compounds (especially the basic ones) has been a difficult challenge. The separation of basic compounds is frequently performed at low pH where the ionization of the residual silanols at the silica surface is suppressed, which reduces the peak tailing at the expense of low solute retention. The use of high temperatures can overcome many of the problems encountered at room temperature. From a theoretical study of selectivity variation with temperature, Li[47] concluded that major changes in selectivity with temperature can be expected for ionizable compounds and that buffer composition, pKa and enthalpy of ionization have a significant effect on this variation. Temperature can be a powerful tool to vary selectivity and could replace the mobile phase pH in method development. As we pointed out above, there are several reasons to use temperature rather than pH for selectivity tuning. For instance, the relationship between retention and temperature is smooth; therefore, retention modelling as a function of temperature is more reliable than as a function of pH. Temperature is also a more flexible and easily adjusted parameter than pH. Optimizing temperature rather than mobile phase pH can produce a more robust analysis. As a matter of fact, a 2 °C change in temperature can have the same effect on the solute dissociation rate as a 0.05 pH unit change.[48] Such a variation can therefore be controlled more easily.

Despite many advantages, high-temperature liquid chromatography does have serious limitations:[48]

- The instability of some solutes at high temperatures can limit the use of high-temperature HPLC for the analysis of complex mixtures. In addition, for reactive compounds, the on-column reaction depends on both the solute residence time in the column and the reaction rate in the column conditions. The rate of reaction can be higher in the column than in the free mobile phase if catalytically active sites exist at the stationary phase.
- The maximum allowable temperature is low for most silica-based columns used in reversed-phase conditions, particularly in acidic or basic buffered eluents, and should not usually exceed 60 °C. Alternative packing materials have been developed to withstand much higher temperatures. Those include metal oxide-based columns,[49,50] porous graphitic carbon and organic polymer. Unfortunately, these stationary phases are often less efficient than their silica-based counterparts, especially with polar and basic compounds. Information is still lacking about the long-term thermal stability of different chromatographic supports and, moreover, there is no universal test that could provide an objective comparison.
- One of the major challenges is the design of a chromatographic system that minimizes both thermal mismatch and extra-column band broadenings. As already addressed above, a high-temperature HPLC instrument should reduce insufficient mobile phase preheating and viscous heat dissipation across the column length because both phenomena produce longitudinal and radial thermal gradients. Radial thermal gradients decrease efficiency, whereas longitudinal thermal gradients affect solute retention factors.

5.2.1.3 Application of Green Eluent Modifiers. Green HPLC can benefit from monolithic stationary phases. Compared with particulate stationary phases, the macroporous structure of the silica rod induces low pressure drops, which, as in elevated temperature chromatography, allows the use of high flow rates, leading to a dramatic reduction in analysis duration. These types of columns can use mobile phases, such as ethanol–water mixtures, which have high viscosity. These mobile phases are environmentally friendly, avoid the use of toxic solvents such as methanol and acetonitrile, and consequently reduce the need to treat waste. Destandau and Lesellier[51] compare the eluting strength of different hydro-organic mobile phases and their efficiencies on Chromolith RP-18e. Separations of pesticides (triazines) and sunscreen molecules are presented.

Without adding any volatile organic solvents, aqueous solutions of room temperature ionic liquids (RTILs) were used as 'green' mobile phases to determine octopamine, synephrine and tyramine by liquid chromatography. The results showed that the retention and resolution of the bases and amino acids were improved when ionic liquids were used as additives. The separation mechanism using ionic liquids as additives to the mobile phase is discussed.[52] The content of adrenergic amines in several citrus herbs and extracts, such as *Fructus aurantii immaturus*, was simultaneously determined by this green chromatographic method.[53] The greenness of ionic liquid is understood in the paper to mean low volatility, which may have relevance to chromatography in connection with effuent recycling and use, (*i.e.* after vaporizing the water, the analytes may be retained in the low volatility and low volume ionic liquid phase).

5.2.1.4 Supercritical Fluid Chromatography. Supercritical fluid chromatography (SFC), that uses liquid carbon dioxide as an eluent, is the most popular alternative to traditional HPLC as a green chromatographic method. It is used mainly as a preparative chromatographic mode. Its usefulness as an analytical method is limited. For this reason, we give only a brief account of it in this book.

What is generally called supercritical fluid chromatography employs carbon dioxide above or near its critical temperature of 31 °C and pressure of 73 bar, combined with an organic modifier such as methanol or ethanol. Most of the available SFC analytical instrumentation is also LC-compatible. Pumps and mixing chambers are already at high pressure; however, pump heads must be cooled. The major alteration is the requirement to keep the UV detector at high pressure.

An advantage of packed column SFC relative to HPLC methodologies is the use of an inert, environmentally green, more volatile, carbon dioxide-based mobile phase. Carbon dioxide gas does cause global warming; however, the carbon dioxide used in SFC chromatography contributes no new chemicals to the environment. This is because SFC generally uses carbon dioxide collected as a by-product of other chemical reactions or is collected directly from the atmosphere. Moreover, the gas is usually recycled when used in very large processes. Thus, compared with HPLC, SFC provides rapid separations with little use of organic solvents. SFC mobile phases are carbon dioxide-based, but may contain up to 15% ethanol, methanol or acetonitrile. Mixed-phase SFC solvents are theoretically not as environmentally benign as single-phase CO_2, but they are significantly easier to dispose of or recycle than mixed organic–aqueous LC solvents.

Compared with gas chromatography, capillary SFC provides high-resolution chromatography at much lower temperatures. This allows a fast analysis of thermolabile compounds. The initial major growth period for SFC occurred 20 years ago in the 1980s. Capillary SFC, which was practiced at this time, originated from gas chromatography using capillary columns (50 μm i.d.) in a GC type of oven and a flame ionization detector. A pump for pure carbon dioxide was used as a pressure source to perform either pressure or density programming, and a fixed restrictor maintained pressure in the column and served as an interface between the column outlet and the laboratory atmosphere.[54] This type of SFC was viewed as an extension of GC, in which some of the thermal energy required for mobilizing solutes was replaced by solvation energy. Capillary SFC experienced explosive growth, mainly because of the novel combination of supercritical mobile phases and open tubular fused silica column technology. GC detectors such as flame ionization, electron capture, nitrogen phosphorus, and sulphur chemiluminescence were popular with SFC. From the early days until the present, the targeted application of open tubular SFC has been primarily in the petrochemical industry. Unfortunately, the physical properties of the fluids were disregarded and the associated problems were rapidly discovered. It was found that the polarity and solvating power of carbon dioxide are low and many analytes of interest were not soluble.

The other form of SFC uses packed columns, usually binary or ternary fluids, composition programming, and a UV detector. Stationary phases have much higher surface area to void volume ratios than capillaries and are therefore much more retentive. Polar modifiers (which are usually incompatible with flame ionization detection) mixed with the main fluid (CO_2) increase the solvating tendency and decrease the retention time of solutes. Unlike capillary column SFC, once modifiers are added, mobile phase composition becomes more important than carbon dioxide pressure or density in determining retention. Packed columns are usually operated near the critical temperature of the fluid, with flow control pumps and electronically controlled back pressure regulators mounted downstream of the column to obtain accurate flow rates and mobile phase composition. In other words, the combination of upstream flow control and downstream pressure control allows volumetric mixing of the main fluid and modifier and also gradient elution. Most of the components of packed column SFC, including the columns, are similar to HPLC. Therefore, packed column SFC can be viewed as an extension or subset of HPLC. In summary, there are dramatic differences between the instrumentation required for packed and capillary columns in SFC. Attempting to use inappropriate open tubular instrumentation with packed columns usually results in complete failure, whereas the opposite situation (*i.e.* packed instrumentation with capillary columns) is entirely feasible.

Packed column SFC is now widely accepted. It is more robust, more adaptable to a broader spectrum of compound classes, and thus more useful for routine operation than open tubular column SFC. Sophisticated commercial instrumentation that allows independent flow control under both pressure and composition gradient conditions has been available since 1992, and difficulties with back-pressure regulation, consistent flow rates, modifier addition, sample injection, automation, stationary phases, *etc.* have been resolved. The growing demand for environmentally friendly processes and the elimination of volatile organic solvents is driving technological development in this field. Connecting packed column SFC to mass spectrometry and ultraviolet detectors is more straightforward than it is for open tubular column SFC. SFC-MS is significantly more benign to the ion source of an MS instrument than water-based solvents.

Chromatographers consider SFC to be normal phase chromatography, without the problems usually associated with normal phase HPLC. Equilibration is very fast, usually requiring flow turnover of little more than three to five column volumes. Composition, pressure, and temperature can all be programmed with rapid recovery to initial conditions between analyses. Traces of water do not cause the variations in retention often encountered in normal phase HPLC.

The advantages of packed column SFC relative to HPLC methodologies are the following:

- lower viscosity and higher diffusivity of supercritical mobile phases relative to liquids, which lead to faster, more efficient separations per unit time and shorter turn-around time between injections;
- an inert, environmentally "green", more volatile carbon dioxide-based mobile phase for large-scale separations and energy-efficient isolation of the desired product;
- longer, stacked columns with the same or multiple phases, with total theoretical plates in excess of 100 000;
- selectivity that matches reversed phase HPLC, but is more easily adjustable;
- SFC instrumentation can be used for HPLC applications.

Reversed phase HPLC remains the first choice for working with water-soluble compounds. But for certain applications such as chiral separations and high-throughput screening, SFC has clear advantages. Because of the short separation time, SFC is very useful for the production of tracers for drug metabolism, since a primary requirement is speed of analysis. It has been proposed that it is more "time efficient" to synthesize racemic mixtures and then chromatographically separate the isomers than to develop enantiomeric syntheses of the drug candidates of interest.[55] In the last ten years, the acceptance and use of packed column SFC-mass spectrometry (SFC-MS) for drug applications has gained momentum because of its broad applicability to such areas as high-throughput analysis, purity assessment, structure characterization, and purification.[56]

SFC is rapidly becoming the separation method of choice for preparative scale multi-gram processes, as a support for preclinical development in the pharmaceutical industry. Preparative SFC can be viewed as a greener alternative to classical preparative chromatography since the intensive use of organic solvents is not required. The resulting decrease in solvent use and waste generation offers a green advantage with an economic bonus – preparative SFC can produce kilograms of purified materials. The SFC product is recovered in a more concentrated form relative to HPLC, thereby greatly reducing the amount of solvent that must be evaporated. This gives rise to considerable savings in labour, time, and energy costs. The higher SFC flow rates also contribute to higher productivity relative to HPLC.

In summary, analytical and preparative scale SFC researchers and vendors are currently advancing the technology, and it appears that new developments and a broader spectrum of applications in this field can be expected in the future. SFC makes use of elevated column outlet pressure, which can add new dimensions of flexibility and control to the chromatographic retention process. Taylor has reviewed many recent applications of SFC.[57] Because the technique is mature, the author notes the general absence of fundamental SFC-related work in the academic community, which "causes concern from the standpoint of both advancing and globally applying the science and the training of graduate students and future workers in the field".[57]

5.2.1.5 The Greening of Chromatography: Not Black-and-White. As discussed above, gas chromatography and supercritical fluid chromatography can already be considered green methods. Three approaches have been tested for greening HPLC:

- miniaturizing by means of advancements in microfluidic technology.

- using water as an eluent. In this case, either temperature programming with traditional reverse phase columns or special temperature dependent columns are used. Another approach is to use ingredients of the water mobile phase (such as ionic liquids) to improve the selectivity of the eluent.
- replacing acetonitrile with methanol/ethanol in reverse phase separations.

In light of the principles of Green Chemistry, we can conclude from the discussion in this chapter that minimizing or eliminating hazardous material production, exposure, and disposal; substituting safer materials for more harmful ones, and striving for energy efficiency, apply to chromatography, whether on the analytical or industrial scale. Without making any special effort to be green, HPLC provides environmental benefits through better use of knowledge that has long been available to chromatographers, (*i.e.* the effect of temperature on separations). Nevertheless, instruments must be improved and new equipment designed. Environmentally conscious or green chemistry has outgrown the period of purely academic interest and begun to provide significant returns on investment.[58] As we pointed out in Chapter 1 of this book, the pharmaceutical industry has a bad reputation for generating chemical waste. Many pharmaceutical companies are now considering and adopting green chromatography methods for analytical and preparative work. The leading companies are applying Green Chemistry, especially in manufacturing and processing. Moving to lower-volume operations as well as analytical and preparative chromatography is now becoming a trend (for example, see steps being taken by some major companies).[55] Analytical methods that are faster or more efficient, or that have that effect on related processes, are inherently green because they consume less energy, materials and human resources. For example, by producing higher throughput, (*e.g. via* mass spectrometry), HPLC helps busy labs get the most out of existing equipment.

However, one should not take a simplistic view of green chromatography, specifically by preferring aqueous to organic waste streams.[58] The whole process should be considered when evaluating the greenness of a particular form of chromatography. Elevated temperature chromatography consumes more energy. At the time this book was being written (summer 2009) the energy aspects of green chromatography were completely ignored. A remarkable example is a work by van der Vorst *et al.*[60] who performed an *exergetic* life cycle analysis of a chromatographic separation of enantiomers of a racemic mixture of phenyl acetic acid derivatives in order to compare preparative HPLC with preparative

SFC. The exergy of a system refers to the maximum work that is required to bring the system into equilibrium in a process.[61] If one considers instrumentation alone (in their terminology – "α system boundary"), the exergy consumption related to the preparative HPLC technique is about 25% higher than for the preparative SFC technique due to its inherently higher use of organic solvents. Considering the capacity of the plant ("β system boundary"), one must take into account exergy calculations for the physical boundaries of the production site (plant) and the resources passing those boundaries. They comprise the resources that have to be purchased by the pharmaceutical company to perform the separation and produce the racemic mixture, the cooling and industrial water, the cooling and heating medium, steam, and the expense of product storage. From this perspective, they again concluded that preparative SFC is more favourable than preparative HPLC because HPLC requires about 30% more resources as quantified in exergy. However, the cumulative exergy extracted from the natural environment to deliver all the mass and energy flows to the α and β system boundary via the overall industrial network (γ system boundary) reveals that preparative SFC requires about 34% more resources than preparative HPLC. The conclusion of the work[60] is astonishing: that the most sustainable process in terms of integral resource consumption is preparative HPLC. The authors reason that the requirement for electricity for heating and cooling and the production of liquid CO_2 disfavours the use of preparative SFC.

Although the study[60] limits itself to the preparative separation of a particular sample, its findings could be far reaching nevertheless. As shown, elevated temperature HPLC also avoids the use of harmful solvents and uses temperature (*i.e.* electricity) to achieve its purpose. Therefore, the actual greenness of elevated temperature chromatography is unproven until a thorough exergy life cycle analysis has been performed. At the moment, green chromatography has been developed by a small group of proponents. Opposition to its application on a wider scale has been based not on chromatographers' ignorance of the principles of Green Chemistry, but rather on the fact that green chromatography can not solve some important separation problems faced by the chemical industry and regulatory agencies. If green HPLC makes further strides and becomes widely accepted, its exergy calculations will progress from an academic exercise to organizational policy issues. One can envision conflicting scenarios in which green chromatography is environmentally acceptable and economically beneficial in laboratories and institutions but not sufficiently benign for the environment and society as a whole.

Recycling and disposal costs should also be considered on a case-by-case basis. Solvents that are considered innately green (such as ionic liquids) may not be all that green. Recycling is possible for a purely organic waste stream, but less feasible for mixed organic–aqueous solvents. Moreover, even 100% aqueous waste effluent cannot be flushed into municipal sewers. Companies must remove the water or burn the effluent, solutions that cost energy and carry an independent environmental impact.

Because of energy consumption, a completely "green" chromatography is probably impossible, in the sense that it is not sustainable as green chemists define the term.[58] Gas chromatography, arguably the greenest chromatographic method, consumes pure carrier gases, the extraction of which from the atmosphere results in a measurable carbon footprint. As shown above, the energy consumption necessary to run a chromatogram can be substantial for HPLC and SFC. Nevertheless, the trend towards greener chromatography is noticeable. The evolution of chromatography towards greater functionality and the adoption of green technologies do not only serve scientists, but are imperative to serve and save the environment.

In a recent overview of elevated temperature chromatography, S. Heinisch and J-L. Rocca speculate about what they call the "sense and nonsense" of this mode of HPLC.[48] They found that in addition to the attractive non-toxicity advantage of using superheated water, as discussed above, it is also possible to use flame ionization, low wavelength UV, inductively coupled plasma mass spectrometry and even nuclear magnetic resonance spectroscopy detectors using hot deuterium oxide as the mobile phase.[62–64] However, they point out some problems that can arise from the use of superheated water:[48]

- The eluent strength can be increased during the analysis only if the solvent gradient is replaced by a temperature gradient.
- Non-polar solutes can be retained even at high temperatures. The eluent strength of pure water at 150 °C is approximately the same as a mixture of water–methanol (50 : 50; v/v) at room temperature.[65]
- Hot water is a very aggressive solvent and the number of high temperature stationary phases suitable for the use of superheated water is very limited.
- Electrospray ionization mass spectrometry detection is less sensitive with a lower percentage of organic solvent.[66]
- Problems may arise when the solute has low solubility in cold water. In this case, a stronger injection solvent can be necessary, which may lead to further peak distortion after peak injection, as discussed in a recent paper.[67]

Temperature programming instead of gradient elution could be useful in a few situations. For example, for separations with micro- or nano-columns when a gradient elution is not easy to operate at very low flow-rates on a given gradient system, the use of capillary columns is usually recommended for temperature programming because of better heat transfer with small inner diameters. However, temperature programming is sometimes problematic. A wider temperature interval than ambient to 200 °C is usually required for programming. Commercially available equipment is not yet able to cover such a wide range of temperatures. The required temperature ramp is often too steep (>20 °C min^{-1}) to be applied to HPLC separations because heat transfer in liquids is very slow compared with heat transfer in gas. Heating an HPLC column made of steel (a material with very low thermal conductivity) can be too slow as well. Temperature programming is mainly used to improve isocratic isothermal analysis rather than to replace gradient elution conditions. In 1970, Snyder observed[68] that the application of temperature programming was limited compared with gradient elution. On the other hand, flow programming could accompany temperature programming to maintain optimum velocity during the analysis. As demonstrated by Sandra,[15] it was impossible to reach the complete resolution of all pairs of peaks under isothermal conditions using a temperature gradient combined with a solvent gradient to enhance selectivity, whereas the use of a moderate temperature gradient made it possible to achieve higher resolution between solute pairs. Inverse temperature programming has been applied by Andersen *et al.*[69] to improve the separation of polyethylene glycol (PEG) oligomers.

Finally, it should be noted that two lesser-known chromatographic techniques, steady-state recycle chromatography[70] and simulated moving bed chromatography,[71] also qualify as green separation methods.[58] Steady-state recycling is a discontinuous, single-column separation technique which involves recycling unresolved fractions back into the column. It combines high throughput and low solvent consumption. Simulated moving bed chromatography uses the counter-current flow of the stationary and mobile phases in continuous mode. Continuous chromatography provides two significant benefits: higher throughput (by means of smaller runs and columns) and up to a hundredfold reduction in solvent consumption.

5.2.1.6 Ultra-Performance Liquid Chromatography (UPLC™). Whereas micro-column HPLC relies on traditional HPLC column packing to attempt to reach a new quality of separation *via* reduction of the column diameter (and adapting the rest of the equipment to that diameter), ultra-performance chromatography (UPLC) – a new development in HPLC – tries to raise HPLC to a qualitatively new level *via* the radical reduction in the diameter of column packing particles. It is well known from chromatographic theory that as the column packing particle size decreases, efficiency improves correspondingly. Emphasizing the particle diameter factor in the van Deemter equation, one can write as follows:

$$H = \mathrm{a}d_p + \frac{\mathrm{b}}{u} + \mathrm{c}d_p^2 u \qquad (5)$$

where H is plate height; a, b, c are constants; u is the linear velocity of the eluent; and d_p is the particle diameter of the stationary phase.

Due to the squared dependence on the particle diameter, it follows that reducing this parameter mainly affects the third, mass-exchange term of the van Deemter equation, making it flatter around the minimum. This means that an increase in the eluent velocity in UPLC has a lesser effect on efficiency than in HPLC. Recent experimental studies have demonstrated that this is true; if the particle size decreases to less than 2.5 μm, not only is there a significant gain in efficiency, but the efficiency also does not diminish at increased linear velocities or flow rates.[72]

By using smaller particles, speed and peak capacity (the number of peaks resolved per unit of time) can be extended to new limits. Scientists at the Waters Corporation, where sub 2 μm diameter particles were first developed, called it ultra-performance liquid chromatography or (UPLC™). To maintain retention and capacity similar to HPLC, UPLC uses a novel porous particle that can withstand high pressures. In order to provide the kind of enhanced mechanical stability UPLC requires, a bridged ethane hybrid technology called "Acquity UPLC BEH" was developed by Waters chemists.[73] The 1.7 μm diameter particles developed at Waters derive their enhanced mechanical stability from bridging the methyl groups in the silica matrix, resulting in column packing that can withstand the rigors of both high pressure and high pH. Packing 1.7 μm particles into columns requires a smoother interior surface for the column hardware, and re-designing the end frits to retain the small particles and resist clogging. HPLC instrument technology also had to be redeveloped to take advantage of the increased speed, superior resolution and sensitivity afforded by smaller particles.

UPLC's excellent performance is mainly reported by the company producing UPLC equipment, but their findings are confirmed by independent studies as well. A comparison with common

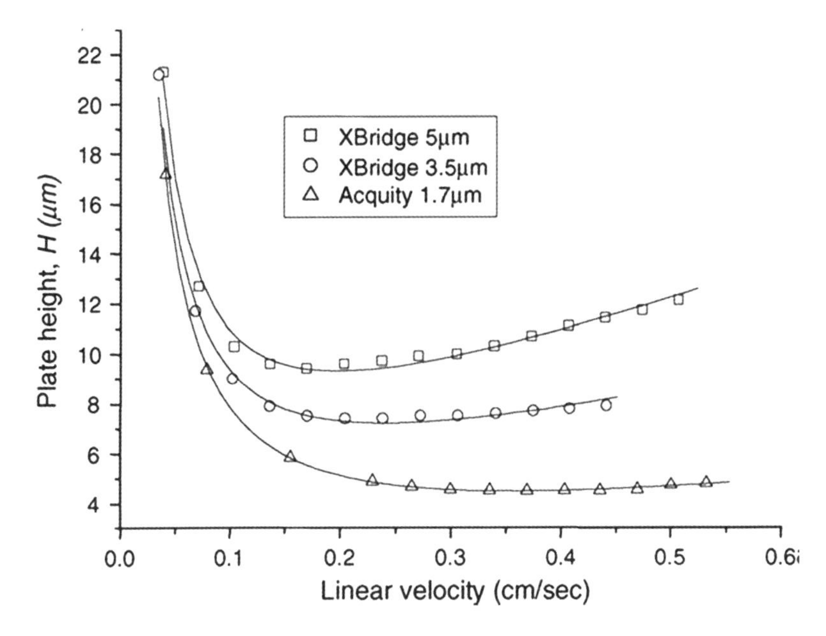

Figure 5.6 van Deempter plots for columns with different stationary phase particle sizes. The plots were obtained for acetophenone on Acquity and XBridge columns.[72] (Reproduced with kind permission of Elsevier.)

chromatographic columns (ref. 72; see Figure 5.6) reveals that the highest optimal velocity is observed for the 1.7 μm particle Acquity BEH C18 column (0.37 cm s^{-1} at a flow rate of 0.5 mL min^{-1}) compared with XBridge C18 columns using an acetonitrile–water (30 : 70) mobile phase. This is in agreement with Equation (5), which predicts that favourable mass transfer characteristics should result in a shift of the optimal velocity to higher values when the particle size is reduced (a reduction in the last term of the van Deemter equation). Also in accordance with the theory, the curve appears "flatter" at higher velocities, therefore increasing the velocity above the optimal value leads to a relatively small decrease in efficiency.

In other findings of the study,[72] a slightly higher-than-expected band broadening (C-term measured from experimental Knox plots) was ascribed to residual temperature effect under non-ideal adiabatic conditions, lower packing efficiency and extra column band broadening. The combination of high optimal flow rates and shorter column lengths allows a gain in speed by a factor of approximately 4.3 and 3.5 for 5 μm and 3.5 μm particles, respectively, without sacrificing efficiency. From "Poppe's kinetic plots"[74] it is evident that the use of UPLC has advantages over HPLC with regard to the speed of analyses requiring up to

~80 000 theoretical plates. As the authors of the paper claim,[72] this makes it possible to increase the speed of current HPLC analyses, and also to use longer columns to provide increased efficiency within an acceptable time (within practical limitations, a retention time of less than 100 min for $k = 10$ and column lengths below 60 cm).

In most chromatographic analysis, chromatographers must sacrifice resolution for speed. Because of the enhanced mass-exchange rate for smaller particles, UPLC provides a way to solve this dilemma. It has been demonstrated that it is possible to scale the original HPLC method to a new UPLC column using exactly the same mobile phase composition and recalculating each gradient step, *i.e.* keeping the column volumes proportional, the gradient steps should be re-adjusted for the new flow rate in the UPLC column. Relevant instructions and formulae can be found in a publication by Swartz.[75] In Figure 5.7, a comparison is shown for the HPLC *versus* UPLC separation of 12 phthalates.[76] This study demonstrated that analysis time can be reduced three-fold and solvent consumption six-fold by using UPLC. HPLC runtime exceeds 14 minutes, as opposed to 4 minutes for UPLC. In another study, theoretical and experimental results both indicated that a minimum of 80% savings in mobile phase consumption and disposal costs can be obtained when an HPLC method using 3.5 μm particles is replaced by a UPLC method using sub-2 μm particles.[77]

The UPLC separation illustrated in Figure 5.7 was obtained with properly scaled injection volume, flow rate and gradient time. This happens without noticeably changing the appearance of the separation; if one does not look at the time scale figures, it would be difficult to distinguish between the HPLC and UPLC separations.

Faster separations can therefore lead to higher throughput and time saving when running multiple samples. However, a significant amount of time can also be consumed in developing the method. Faster, higher resolution UPLC separations can cut method development time from days to hours, or even minutes.

In conclusion, it can be said that whereas scientists have reached barriers pushing the limits of conventional HPLC, UPLC has the potential to extend and expand the utility of this widely used separation science. New Acquity UPLC technology in chemistry and instrumentation provides more information per unit of work, and begins to deliver the increases in speed, resolution and sensitivity predicted for liquid chromatography.

A new system design with advanced pump, autosampler, detector, data system and service diagnostics technology was required to implement UPLC. Achieving rapid, high peak capacity separations by taking

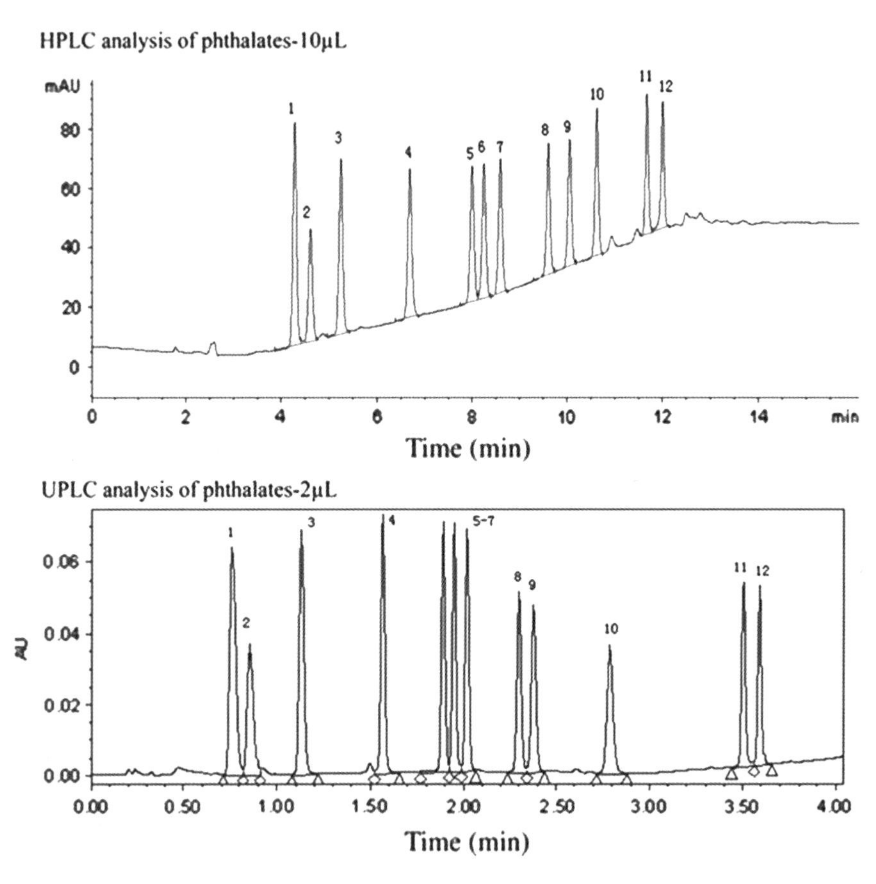

Figure 5.7 HPLC *versus* UPLC separation comparison.[76] (Reproduced with kind permission of Springer Science.)

advantage of small packing particle size requires a greater pressure range than that achievable with conventional HPLC instrumentation. The calculated pressure drop at the optimum flow rate for maximum efficiency across a 15 cm long column packed with 1.7 μm particles is approximately 1000 bar. Therefore, a pump is required that is capable of delivering solvent smoothly and reproducibly at these pressures, operates in both gradient and isocratic separation modes, and has the ability to compensate for solvent compressibility. Traditional pumps have passive check valves in which the ball that closes the valve is lifted off the seat by the force of the solvent flow. In a UPLC pump, an applied current generates a force timed with the motion of the pump plungers, which pushes a ball off its seat. Currently, the pressure limit is approximately 1000 bar, in order to take full advantage of the sub-2 μm particles.[78] High pressure makes the sample introduction process critical. The sampling process has to protect the column from extreme pressure fluctuations and be relatively pulse-free. The swept volume of the device needs to be minimal to reduce potential band spreading. Sample injection must be fast to take full advantage of the speed afforded by UPLC. The injection process uses needle-in-needle sampling for improved ruggedness and pressure-assisted sample introduction.

The Waters Corporation claims that by replacing conventional high-performance liquid chromatography (HPLC) with Acquity UPLC, acetonitrile consumption can be reduced by at least 70% without compromising productivity and performance.[79] According to the company, because the system operates at a wider velocity range, higher flow rates and greater backpressures; UPLC uses significantly less solvent, including acetonitrile. As a result, laboratory productivity and performance can be maintained or improved without the increased costs and external pressures associated with the current shortage of acetonitrile.

Light-guided flow cells of 10 mm path length have been implemented for UV detection of narrow bands in UPLC. The flow cell channel is inside a low refractive index Teflon AF tube that has total internal reflection on the walls, such as optical fibre cladding. MS detection significantly benefits from UPLC because of the increased peak concentrations resulting from reduced chromatographic dispersion at lower flow rates through columns that promote increased source ionization efficiencies. Due to the new instrument design, flow splitting is not necessary before the column – a clear indication of the greenness of the UPLC technique.

Mass spectrometric analysis of complex matrixes probably has the most to gain from UPLC technology. Increased throughput and information content has been demonstrated for *in vitro* drug metabolism experiments using UPLC coupled to a quadrupole time-of-flight (Q-TOF) mass spectrometer.[80] In one of the studies, it has been shown that UPLC offers significant advantages over conventional reversed-phase HPLC: more than twice the peak capacity, an almost ten-fold increase in speed and a three- to five-fold increase in sensitivity compared with that generated by a conventional 3.5 μm stationary phase.[81] The first functional genomic application of UPLC-MS technology was used for multivariate (see Chapter 6.1) metabolic profiling of urine from males and females of two groups of phenotypically normal strains of mice (*C57BL19J* and *Alpk:ApfCD*) and a "nude mouse" strain. Comparing this technology with conventional HPLC-MS under similar analytical conditions, UPLC-MS analysis demonstrated an improved phenotypic classification capability and a greater ability to probe differential pathway activities between strains, as a result of improved analytical sensitivity and resolution.

UPLC-MS has also been used for the analysis of priority pesticides in groundwater.[82] It has been investigated as an alternative to HPLC for the analysis of pharmaceutical development compounds,[83] and for forensic and toxicological analysis.[84] In recent years, the list of UPLC-MS applications has grown exponentially and it is difficult to review all of them. In Table 5.2 below is a list of applications of UPLC-MS in environmental chemistry and bioanalysis which gives an idea of the possibilities provided by this technique. The list is by no means exhaustive. Only the applications that have had the largest impact (in terms of citations) have been included.[85–134]

5.2.1.7 Consequences of the Acetonitrile Shortage for Chromatography in 2009. A decline in the world economy has an indirect negative impact on the performance of analytical laboratories in ways such as increases in the cost of supplies and job cuts. The economic decline at the end of 2008 has had a direct, unexpected impact on HPLC. This topic has been addressed in journals such as *Chemical and Engineering News.*[135] The popular solvent in HPLC, acetonitrile (ACN), is a by-product of the process used in the automotive industry to make acrylonitrile, a building block for acrylic fibres and acrylonitrile–butadiene–styrene resins. Demand for acrylonitrile–butadiene–styrene resins, used in cars, electronic housings, and small appliances, is decreasing around the world because of the global economic slowdown. The acrylic fibre market is also in decline because it is losing market share to polyester fibres. An acrylonitrile plant yields 2 to 4 L of acetonitrile for every 100 L of acrylonitrile produced. The status of ACN as a minor co-product of a big industry has led to the present shortage.

Acrylonitrile plants are operating less than 60% globally.[135] Moreover, there were other important reasons that contributed to the 2009 ACN shortage. Interestingly, one of them was the 2008 Summer Olympics. The industrial production that the Chinese government shut down to improve Beijing's air quality seems to have included a disproportionate amount of the country's acetonitrile production. Another reason was that the U.S. plant on the Gulf Coast was shut down during Hurricane Ike. It was also shut down during the summer because of a lightning strike.[135] At the time of the writing of this book, (the beginning of 2009), it is predicted that the ACN shortage could last well into 2009. Suppliers of laboratory chemicals have been allocating acetonitrile to existing customers or not selling it at all. Sigma–Aldrich announced that the company had prepared for the shortage by building up inventories

Table 5.2 Selected applications of UPLC-MS.

Field	*Objects of Analysis*	*Ref.*
Biomedical analysis Pharmaceutical	oligosaccharides in glycopepticles using immobilized Endo-M	85
	metabolic profiling studies on human blood plasma	86
	diastereomers of SCH 503034 mass spectrometric in monkey plasma	87
	oral contraceptive concentrations in human plasma	88
	ginsenosides in rat urine	89
	organic acid markers relevant to inherited metabolic diseases	90
	apolipoproteins in human serum	91
	azithromycin in human plasma	92
	new analogs of the marine biotoxin azaspiracid in blue mussels (*Mytilus edulis*)	93
	lercanidipine in human plasma	94
	multiresidue screening of veterinary drugs in urine	95
	amlodipine in human plasma	96
	lipophilic marine toxins	97
	high-throughput bioanalysis with simultaneous acquisition of metabolic route	98
	bacterial N-acylhomoserine lactones (AHLs)	99
	histamine and its metabolites in mice hair	100
	drug metabolites in biological samples	101
	constituents of the flower of *Trollius ledibouri Reichb.*	102
	epirubicin in human plasma	103
	phenotypic differences in the metabolic plasma profile of Zucker rats	104
	metabolomics of raw and steamed Panax notoginseng	105
	drug mixture in rat plasma	106
	solubility screening assay in drug discovery	107
	pharmacokinetics intraperitoneally administered troglitazone in mice	108
	structurally diverse drug mixtures	109
	atypical antipsychotics and some metabolites	110
	amphetamine-type substances, designer analogues, and ketamine	111
	testosterone and its metabolites in *in vitro* samples	112
Food	pesticides in fruits	113
	acrylamide in foods	114
	macrolide antibiotic residues in eggs, raw milk, and honey	115
	pesticide residues in foods	116
	(UPLC) with (ICP-MS) for fast analysis of bromine containing preservatives	117
	aflatoxins B1, G1, B2, G2 and ochratoxin A in beer	118
	heterocyclic amines in food	119
	multiresidue pesticide analysis in food	120
	mycotoxin contaminants in foods and feeds	121
	priority pesticides in baby foods	122
Environmental and forensic analysis	cannabinoids and opiates in wastewater and surface waters	123
	biologically active compounds in water	124

Table 5.2 (Continued).

Field	*Objects of Analysis*	*Ref.*
	basic/neutral pharmaceuticals and illicit drugs in surface water	125
	amphetamine-type substances and ketamine for forensic and toxicological analysis	126
	priority pesticides in groundwater	127
	multi-residue analysis of pharmaceuticals in wastewater	128
Method development	hydrogen/deuterium exchange mass spectrometry	129
	comparison of ELISA, HPLC and UPLC for ppt analysis of estrogens in water	130
	robust, high-throughput quantitative analysis of an automated metabolic stability assay, with simultaneous determination of metabolic data	131
	increasing throughput and information content for *in vitro* drug metabolism experiments using UPLC-MS	132
	UPLC-MS	133
	UPLC – TOF- MS as a tool for differential metabolic pathway profiling in functional genomic studies	134

and should be able to supply contract customers. New customers have to pay the current price[135] – six to eight times more than in August 2008.

Acetonitrile is the primary chemical solvent used in HPLC analysis and in the purification of peptides. It is also used in pharmaceutical synthesis and in the extraction of butadiene from streams of C_4 hydrocarbons. ACN availability directly affects the biotechnological and pharmaceutical industries. If the acetonitrile shortage does not end, chemists will be forced to use alternative solvents. Methanol is the most common alternative, but others include tetrahydrofuran (THF) and ethanol. Many are switching to methanol–water, which usually works, but is not always an option. Since methanol is also a toxic chemical, ethanol can be considered as an alternative. However, laboratories that are obliged to use certified methods cannot change acetonitrile without a substantial amount of documentation, because new methods must be accredited. Moreover, for some applications, no solvent works exactly the same as acetonitrile. It is a much superior solvent for HPLC compared with other solvents. At 200 nm the UV absorbance of ACN is five times less than methanol and ethanol. Methanol is 1.5 times and ethanol is up to three times more viscous than ACN. Using them in HPLC requires the application of higher pressure to the column, and back-pressure will certainly increase. Efficiency is also lower when using methanol instead of ACN. As shown in Figure 5.8, the elution strengths of methanol and ethanol are weaker than that of acetonitrile. The

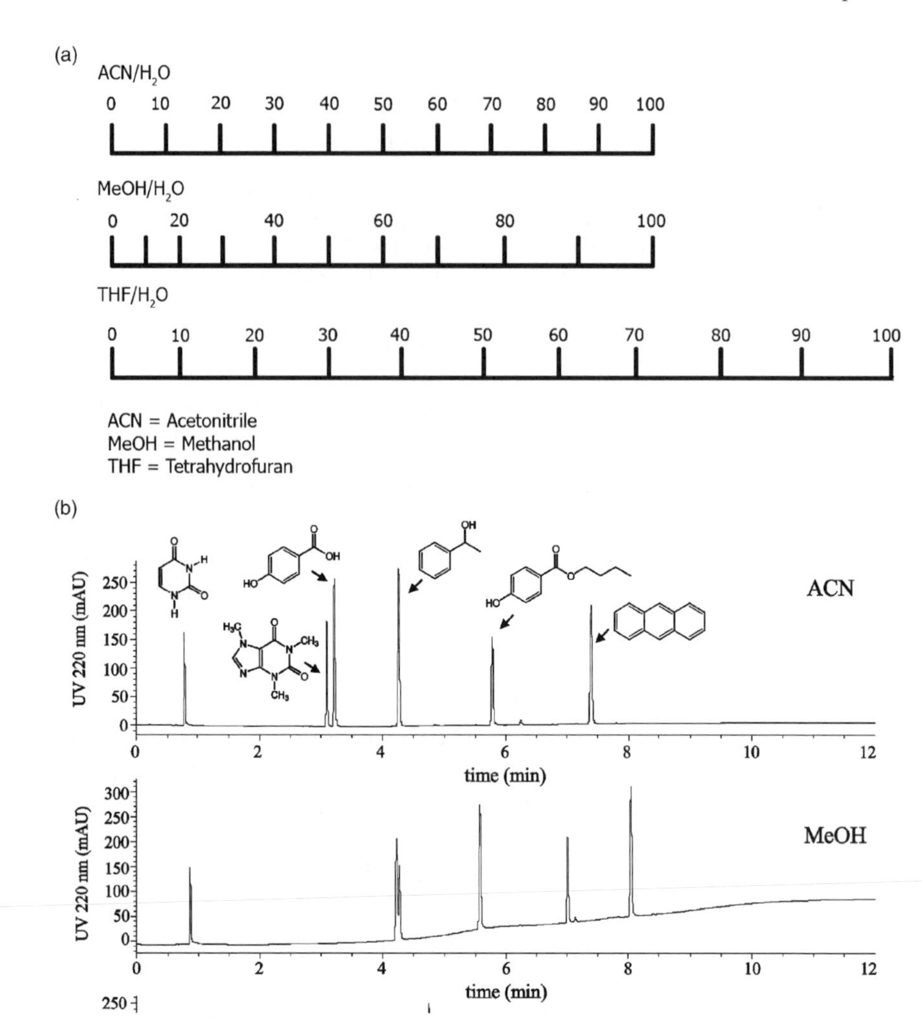

Figure 5.8 (a) Relative strength of different solvents[136] and (b) example of using methanol *versus* ACN for the separation of test mixture 2 by RP HPLC. Conditions: column = Eclipse Plus C18 (50 mm × 4.6 mm i.d.; particle diameter 1.8 μm); flow rate = 0.8 mL min^{-1}; mobile phase = aq. 0.1% H_3PO_4 from 5–95% organic in 6 min (with 6 min hold); and detection = UV 220 nm.[139] (Reproduced with kind permission of The Royal Society of Chemistry).

elution pattern may change and qualitative analysis must be performed to confirm it. The general rule is to increase the organic component by ten percentage units when replacing acetonitrile with methanol in a reversed-phase method.[136] This rule is based on the relative eluotropic strength of two solvents. Figure 5.8 shows the relative reversed-phase eluotropic strengths of acetonitrile, methanol and THF.

For example, if a method uses 50% acetonitrile, the same elution strength and retention time is achieved by using 60% methanol and 35% THF.

Discussions of Green Chemistry solutions to analytical chemistry problems are sometimes regarded as abstract and theoretical by the general public, but in regards to the ACN crisis, analytical chemistry requires that green chromatography be taken seriously. Green issues have forced large pharmaceutical companies to reconsider their practice of using various solvents.[137] Some companies have reacted already by providing their own solutions to the ACN crisis.[138]

Green solutions are generating much discussion among scientists with regard to the potential of conventional HPLC, and alternatives are being sought. The greening of chromatography could become a popular research topic. The study by Welch *et al.* serves to illustrate that point.[139] While acetonitrile undoubtedly delivers outstanding performance as an HPLC solvent, greener alternatives such as ethanol also perform reasonably well. The performance advantage of acetonitrile may be essential for the survival of a few laboratories (*e.g.* those performing high-throughput analysis or engaged in the analysis of samples with poor UV chromophores). On the other hand, many laboratories would be able to tolerate the slightly longer analysis times using ethanol. As Welch *et al.* comment: "...every bit of excess performance delivered by modern HPLC technology is probably not needed in most analytical settings, and may be contributing to the ever escalating cost and complexity of carrying out simple analyses, which is becoming increasingly difficult to justify in today's economically constrained environment".

Welch *et al.* conclude that accepting the use of ethanol as an HPLC solvent becomes rather attractive for the following reasons:[139]

- the cost of acetonitrile and the associated cost of waste disposal continue to escalate, while the cost of ethanol is decreasing;
- ethanol is universally available;
- the use of ethanol as a fuel implies that the cost and quality of ethanol should improve in the coming years;
- the environmental impact of ethanol *vs.* acetonitrile waste is much less;
- in the light of recent trends toward miniaturization of HPLC equipment, one can envision that HPLC will be used for on-site monitoring in various environments, (by doctors, dentists, farmers, and even in the home). A greener solvent than ACN would be strongly preferred in such situations.

In the previous sections, various green alternatives to conventional chromatography have been discussed. Supercritical fluid chromatography, high-temperature chromatography and the miniaturization of HPLC provide some solutions to the ACN crisis. These techniques either do not require harmful solvents or require drastically less solvent than conventional LCs. The advantages of these techniques are better quality analyses and higher productivity. Throughput is the key topic in modern bio- and pharmaceutical analysis. In contemporary laboratories, analysis has to be conducted around the clock, preferably with sample preparation included. Green chromatographic methods combined with automated sample introduction provide the opportunity to run more samples with less operator involvement. New players are also emerging on the stage. As we saw in the previous section, ultra-performance liquid chromatography is a new technique that is attracting interest in the field of green chromatography.

5.2.2 Capillary Electrophoresis

Electromigration comprises a group of analytical separation methods that are based on differences in the mobilities of charged analytes in electric fields. In this section, we will examine electromigration methods that are performed in thin capillaries, channels with i.d of less than 0.1 mm, or in droplets with μL or even nL volumes. This group includes two well-known methods – capillary zone electrophoresis (CZE or simply CE) and micellar capillary electrokinetic chromatography (MEKC). CE is an analytical technique that can easily be miniaturized, which makes it a potential green analytical technique. Electro-osmosis and electrophoresis are basic physical phenomena used in CE to provide the driving force for microfluidics in developing microminiaturized methods known as μTAS (*i.e.* micro total analysis systems). Digital microfluidics is an emerging concept based on tiny drops of liquids actuated *via* such phenomena as electrowetting and dielectrophoresis.

The phenomenon of the migration of charged species under the influence of an externally applied electric field is known as *electrophoresis*. Differences in mobility of the analytes due to their average charge, size, shape, and the properties of the electrolyte solution form the basis of an important separation method in chemistry. According to Reuss,[140] and Pertsov and Zaitseva,[141] the first electrokinetic phenomena (electrophoresis and electro-osmosis) were discovered in 1807 by Reuss (1778–1852), a professor at Moscow University. His paper, entitled *Notice sur un nouvel effet de lélectricité galvanique*, was published in 1809.

He studied the migration of colloidal clay particles and discovered that the liquid adjacent to the negatively charged surface of the wall migrated towards the negative electrode under the influence of an externally applied electric field. The theoretical aspects of this electrokinetic phenomenon (*electro-osmosis* by Reuss) were formulated in 1897 by Kohlrausch.[142] In the late 1800s and early 1900s, electrophoretic separations in U-shaped tubes were conducted by several researchers. Arne Tiselius, starting in 1925 with his PhD thesis on the development of free moving boundary electrophoresis, advanced the analytical aspects of electrophoresis. This resulted in the separation of complex protein mixtures based on differences in electrophoretic mobility.[143] In 1948, Arne Tiselius was awarded the Nobel Prize for Chemistry for his work on electrophoresis. The possibilities of performing electrophoresis in capillaries were investigated by Hjerten,[144] Everaets,[145] and Virtanen,[146] but their work did not attract much attention to capillary electrophoresis until the research of Jorgenson and Lucas. They performed the separation of fluorescent dansylated amino acids in a glass capillary with an inner diameter of 75 μm. By applying voltages up to 30 kV, they provided efficiencies of more than 40 000 within 25 minutes.[147] This efficiency, not seen in separation science before, was mainly as a result of the fact that at diameters of less than 0.1 μm, the capillary wall dissipates all the Joule heating generated in the buffer by electric current. Analyte band broadening as it moves along the capillary is determined (in most cases) by its diffusion, which is very low in liquids. After the landmark work by Jorgenson, interest in capillary electrophoresis grew rapidly.[148] Although CE was initially acclaimed for its speed and low sample volume, the technique is also useful because it is quantitative, can be automated, and will separate compounds that are difficult to process by HPLC. CE played a crucial role in determining the human genome sequence and CE is the basis for virtually all the micro-fluidics of "lab-on-a-chip" devices. CE can separate polar substances, which are notoriously difficult to analyze by HPLC. Chiral separations are another area in which the use of CE has expanded. The small sample volumes required for CE can be an advantage with a limited amount of sample. An area closely related to CE that has yet to reach its potential is capillary electrochromatography (CEC), which combines features of CE with LC by using capillaries packed with chromatographic materials.

CE instrumentation is simple (see Figure 5.9). The heart of the instrument is a thin capillary with a diameter of between 10 to 100 μm and length of 20 to 100 cm. An electric field is generated by a high voltage power supply that delivers bipolar voltage up to 30 kV. If higher voltages are required, then special precautions for isolation are required.

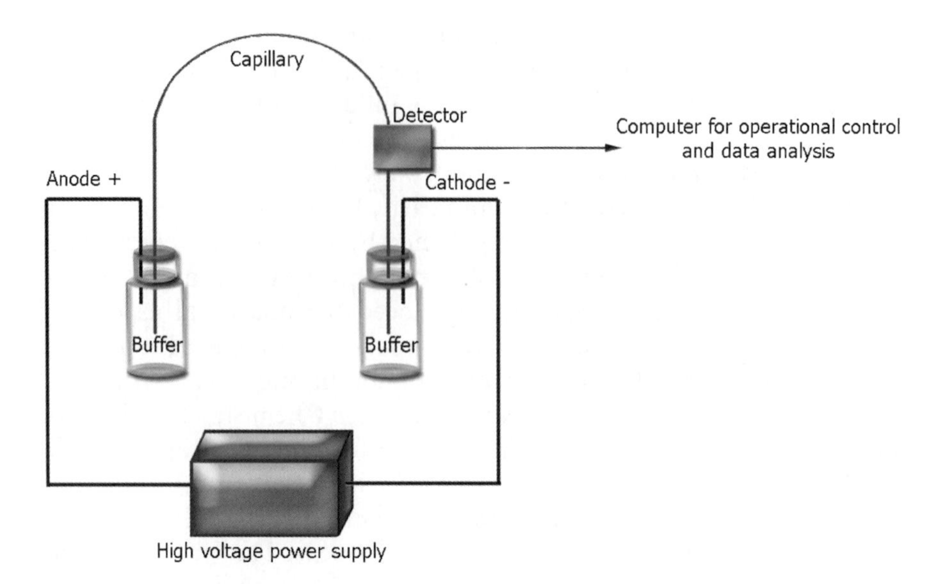

Figure 5.9 CE instrumentation.

A detector is also needed for monitoring the results of the separation processes. Absorbance of analytes is usually measured by a UV-Vis spectrophotometer, but florescence, electrochemical and mass spectrometric detectors are needed when detection limit requirements are not met by UV absorbance. Although simple in concept, CE instrumentation is complex and expensive (about 50 000 dollars) because of the high cost of the detectors and sampling systems contained in contemporary instruments.

5.2.2.1 Mobility in Capillary Electrophoresis. The simplest description of the movement of a charged analyte in an electric field can be provided by balancing the two forces that influence ion movement. The ion tends to be accelerated by the force of the electric field F_e, which is equal to $F_e = zE$, where z is the charge of the analyte and E is the strength of the electric field. However, this acceleration is balanced by the frictional resistance F_f of the environment, which for a spherical particle is equal to $F_e = 6\pi\eta rv$, where η is the viscosity of the media, r is radius of the analyte and v is the velocity of the analyte. If the two forces are equal, the analyte moves with constant *electrophoretic velocity* proportional to the strength of the electric field as follows:

$$v = \frac{z}{6\pi\eta r}E \tag{6}$$

In Equation (6) the proportionality constant

$$\mu = \frac{z}{6\pi\eta r}$$

is known as *mobility*.

It must be noted that the separation media itself moves with a certain velocity. When a voltage is applied to the capillaries used in CE, a bulk flow of electrolyte fills the capillary through the system. This process is known as *electro-osmosis* (EOF) and is a result of the surface charge on the inside of the capillary wall. Most surfaces possess a negative charge that results from the ionization of the surface or the adsorption of ionic species. In fused-silica capillaries, both processes occur and this results in the inside wall being extremely negatively charged. As the walls of the capillary are negatively charged, a layer of cations builds up near the surface to maintain the charge balance. This creates a double layer of ions near the surface and a potential difference. This is known as the *zeta* potential. When voltage is applied across the capillary, the cations forming the double layer are attracted to the cathode. They then move through the capillary, and as they are solvated, they drag the bulk solution behind them.

Equation (6) is valid for macroscopic particles moving in a continuous media. In electrophoresis, where the analyte ion moves in a media in which particle size is comparable with analyte size, this is not true. In addition, analyte ions are not spherical and the ionic radius term becomes ambiguous, which makes the value difficult to estimate. Various improvements of Equation (6) have been proposed.[149] The mobility of the analyte appears to be approximately proportional to the following ratio:

$$\mu \propto \frac{z}{\sqrt[3]{M^2}}$$

where M is molar mass and the proportionality constant has to be determined experimentally and depends on the class of substances.

If the analyte molecules cannot be charged, electrophoretic separation is impossible. However, if a charged ingredient to which analyte ions have affinity is added to the separation buffer, separation becomes possible by partitioning the analyte between the free buffer and the ingredient. Assuming the simple mass action law $A + C \rightleftarrows AC$, where A is the analyte and C is the buffer ingredient, the equilibrium constant is $K = [AC]/[A][C]$, with the brackets denoting molar concentrations of A, C and AC. The analyte now moves part of the time with the velocity of the separation media v_{EOF} and part of the time with the velocity of the separation media plus the velocity of the charged complex: $v_{EOF} + v_{[AC]}$. Taking equilibrium into account, the total migration time becomes as follows:

$$v = v_{EOF} + \frac{K[C]}{1 + K[C]} v_{[AC]}$$

The best-known CE buffer ingredient is sodium dodecylsulfate (SDS), which was proposed by Terabe.[150] SDS forms micelles; neutral analyte separation is achieved by partitioning between the buffer and the SDS micelles, *i.e.* by hydrophophicity. This is the basis of micellar electrokinetic chromatography. The mobility of analytes correlates linearly with $\log P$ values where P is the octanol : water partition ratio. Many other buffer ingredients have been proposed. Most of them implement hydrophobic interactions between analytes and buffer ingredients, but chiral selectors have also been used, as well as various affinity probes.

5.2.2.2 Detection of Analytes in Capillary Electrophoresis. Whereas the small capillary diameter implemented in CE facilitates the efficient separation of analytes, it also makes their detection difficult. Detection by UV absorbance of light is dependent on the solution path length according to Beers' law:

$$A = \varepsilon C b,$$

where A is absorbance; C is the concentration of an analyte; ε is the molar absorbivity and b is the path length that a radiation beam has to pass. The latter parameter is the capillary diameter which, as we saw above, is on the order of 10 to 75 mm. Extra-column detection would degrade the resolution to such an extent that the advantages of CE would disappear. Since the capillary has a cylindrical form, much light is lost due to scattering. Detection cells commonly used in HPLC are not suitable for CE and the detector optics must be tailored to the capillary format. Nevertheless, detection in the concentration range of ppm is usually possible if the analyte has UV-absorbing chromophores.

Detection by the fluorescence of analytes is also possible. The detection limits of analytes are approximately three orders of magnitude lower than those of UV absorbance if the analyte fluoresces. This is not a common phenomenon, and analytes must be derivatized with florescent dyes to make detection possible. Optics must also be tailored to the capillary format as in UV detection, and florescent detectors for CE

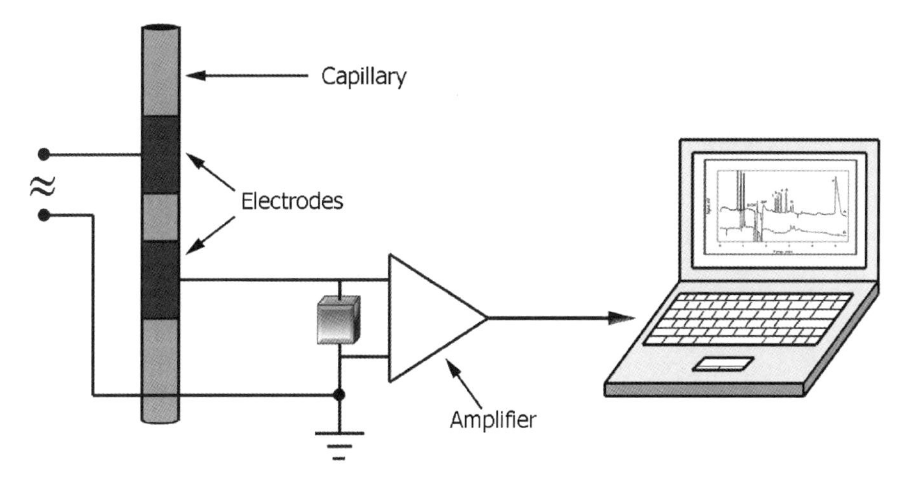

Figure 5.10 Contactless conductivity detector.

are even clumsier than UV detectors. Nevertheless, fluorescent detection is steadily gaining popularity in capillary electrophoresis, especially for bioanalytical applications where the amounts of analyte are so small that using UV detection is out of question.

Contactless conductivity detection (CCD) developed by Zeeman[151] and da Silva[152] is also gaining popularity. It measures the conductivity of the solution in the separation capillary by making use of two tubular electrodes with a small cap in between them, positioned over the capillary (see Figure 5.10).

One electrode is the actuator electrode, which applies microwave energy to the buffer solution, and the second electrode picks up changes in buffer conductivity if an analyte passes the detector gap. The detected changes are amplified and recorded. The detection limits are on the order of sub-ppm, but the selection of separation media is severely restricted, because the instrument detects changes in conductivity against the buffer conductivity. Popular common CE buffers such as phosphate and borate cannot be used; low conductivity buffer compositions like mesithylethyl sulfonate with histidine (MES-HIS) are used instead. Non-aqueous buffers can also be used.[153] Despite these restrictions, CCD can be used in many situations.[154–156] As the detector consumes little energy, and its sensor element and electronics can be made lightweight and small, CCD has good prospects for use as a portable field analyzer for environmental and military purposes.[157–159] For these reasons, it also has good prospects as a genuinely green detector. It can be used successfully in HPLC as well as in CE. One of the advantages of this detector is that neither the separation media nor the capillary need to be optically transparent. This property of the CCD is still not widely used, although a few examples can be found.[160,161]

Mass spectrometric detection – potentially the most powerful means of detecting analytes – is still in the CE development stage. The problem is the difficulty of establishing a reliable CE interface with the typical mass spectrometer, and matching the liquid flows and electric currents of both techniques.

5.2.2.3 Eluent Consumption in CE and in HPLC. The "greenness" of CE is most easily demonstrated by comparing the consumption of solvent and sample in CE with conventional HPLC. In this respect, CE can be considered a genuinely green separation method. Although the low consumption of eluent has frequently been pointed out as an advantage of CE over high-pressure liquid chromatography (HPLC), this has never been discussed in the context of Green Chemistry. As described above, capillary electrophoresis is a separation method that is performed in a narrow-bore capillary (inner diameter less than 0.1 mm) with a length of less than one meter. The inner volume of a typical capillary with an inner diameter of 75 mm and a length of 50 cm is approximately 2 μL. This is a tiny droplet of liquid that can hardly be seen by the naked eye. The driving force that carries eluent in CE is electro-osmosis, at a rate on the order of 1–5 mm s^{-1}. This means that during a typical CE analysis runtime of 10 minutes, approximately 10 μL of eluent is consumed. The CE buffer is usually a non-toxic solution of inorganic salt. Ordinary reverse-phase HPLC with a flow rate of 1 mL min^{-1} produces three orders of magnitude more toxic acetonitrile–methanol mixture. For this reason, it is surprising that CE has not been considered a genuinely green separation method. In practice, CE consumes more eluent since capillary input and output have to be located in vessels which each contain approximately 50 μL of buffer. This renders the above estimates less favourable, but still extremely attractive.

A comparison of these two methods (see Table 5.3) indicates the direction of separation method development – decreased sample size, low consumption of solvent, higher selectivity, faster analysis time and mechanically simpler instruments. The developments in CE are rapidly changing, and a growing number of new methods and instrumental solutions are published in journals and presented at conferences. At this time, CE is overcoming problems of separation instability and the lack of detector sensitivity, and more laboratories are starting to consider capillary electrophoresis (CE) as a standard procedure for the separation of complex samples.

Table 5.3 Comparison of selected parameters of liquid chromatography and CE.

	Liquid chromatography	*Capillary electrophoresis*
Injected volume	1–100 μL	1–100 nL
Flow rate of the mobile phase	1–10 mL/min	1–100 nL/min
Flow profile	Parabolic	Plug
Peak capacity (number of separated peaks)	20–30	20–100
Analysis time	10–60 min	1–20 min
Separation efficiency	> 10 000 plates	> 100 000 plates
Separation technique	High pressure	Electrical field
Robustness	Complicated pumping system	High voltage source
Solvents	Different solvents for different columns	Aquatic buffers for the same column
Level of development	Mature technique	Young, developing fast
Detection methods	Optical, electrochemical	Optical, electrochemical
Limit of detection	ng/L (ppb)	μg/L (ppm)
Portability	Not portable	Portability is possible
Energy consumption	100 W	1 W

The numbers in Table 5.3 can vary from design to design but the general trend is evident: CE consumes much less eluent and energy than HPLC, and is technically simpler, so maintenance is easier. However, HPLC is more popular in industrial and clinical laboratories. This can be attributed to the high detection limits inherent in CE, lower reproducibility, and the difficulties of interfacing CE to mass spectrometry. HPLC, as we saw in the previous section, is a routine technology – an important factor in proteomic analyses. The manufacturers of HPLC equipment are working hard to develop new columns and HPLC modes that equal the performance of CE capillaries. As we saw in Section 5.2.1.6, ultra-performance columns with particle diameters of less than 1.8 μm require an operating pressure of approximately 1000 atm, but are able to develop more than 100 000 plates per analysis. Monolithic micro columns can match the flow rates that are needed for the operation of an electrospray source in mass spectrometry.

By comparison with HPLC, CE excels with large molecules and when sample size is limited. For these reasons, it is well suited to bioanalysis, and can solve problems where HPLC has little chance of success – the Human Genome Project being the obvious example. Now that CE has been officially recognized by several regulatory agencies – the Food and Drug Administration and the Centre for Drug Evaluation and Research among them – it is finding a niche in quality control and quality assurance laboratories as well.

5.2.2.4 CE as a Method of Choice for Portable Instruments. It should be noted that the development of commercial instrumentation for CE has stagnated at the time of writing this book. There are only two main suppliers of CE instrumentation (Agilent and Beckman Coultier) and much of the research in CE has shifted to solving various problems in applied science. This has had a decelerating influence on the development of CE instruments. It might be argued that there is no longer much to invent in CE, but this is not entirely true. One of the overlooked advantages of CE is the development of portable field instruments based on the technique.

It is difficult to imagine a portable gas or liquid chromatograph, although it might be technically possible (portable gas chromatographs are used in space applications). However, portable CE can easily be constructed and there are reports of portable instrument designs from several groups. They use either a CCD detector or advanced optics based on light-emitting diodes (LED). Hauser *et al.*[157,158] developed and optimized a portable CE instrument with CCD for the sensitive field measurement of ionic compounds in environmental samples. It is powered by batteries and the high voltage modules are capable of delivering up to 15 kV at either polarity for more than one working day (see Figure 5.11). Inorganic

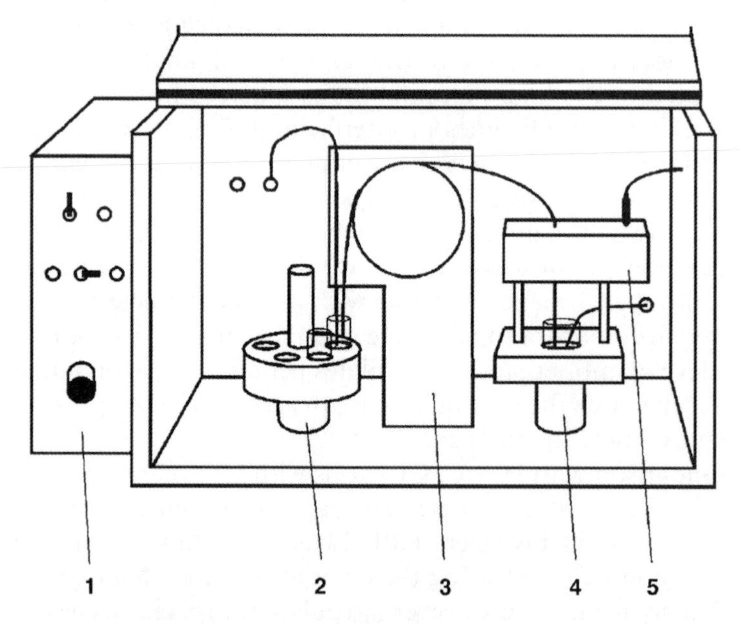

Figure 5.11 Schematic drawing of a portable CE instrument with sample buffer tray: control electronics (1), sample tray (2), capillary holder (3), vial holder (4) and detector cell (5).[158] (Reproduced with kind permission of Wiley-VCH Verlag GmbH & Co. KGaA).

cations and anions, including ions of heavy metals and arsenate, were determined with detection limits in the range of approximately 0.2 to 1 mM. The instrument was field tested in a remote region of Tasmania. Nitrite and ammonium were determined on-site at concentrations as low as 10 ppb in the presence of other common inorganic ions at concentrations that were two to three orders of magnitude higher.

In another publication, Haddad *et al.* demonstrated the use of CE for the detection of explosives in the environment.[162] They showed that it is possible to analyze blast residue at the crime scene, where it can be sampled simply by wiping hard surfaces with a wet cloth, rather than transporting the residues back to the laboratory. Paul Haddad and his team decided to use indirect photometric detection, which involves adding a UV absorbing probe or dye with the same charge as the analytes, such as the cationic dye chrysoidine, to the electrophoresis buffer and then recording the dip in absorption caused by the passage of non-absorbing analytes in front of a UV detector. They found that they could separate and detect the 12 cations at concentrations as low as 0.11 mg L^{-1} and separate and detect the 15 anions at concentrations as low as 0.24 mg L^{-1}. In both cases, the analyses took less than 10 minutes. Furthermore, in evaluating the performance of CCD, they found that the results were better than those achieved with indirect photometric detection.[163]

The aim of work by Seiman *et al.* was to develop robust operating procedures for on-site sampling and analysis.[159] A syringe is inserted into the socket, and the background electrolyte (BGE) in the cross-section of the sampler is replaced by the sample. The sample between the capillaries is inserted into the separation channel by electro-osmotic flow that carries the sample to the separation capillary, and replaces the junction between two capillaries with BGE as soon as high voltage is applied (see Figure 5.12). In this way, the manipulation of buffer vials is minimized.

This instrument has LOD for phosphonic acids of 4–8 μM and for cations of 0.3–0.5 μM, and RSD (with internal standard) of 8%. A typical electropherogram of the phosphonic acids recorded with this instrument is shown in Figure 5.13.

As we will see in the following section, electrophoresis is a key technology for micronizing analytical separation methods, most of which are based on "lab-on-a-chip" platforms. Micronization of analytical separation methods opens the way to many inexpensive point-of-care medical diagnostics. To enable "lab-on-a-chip"-based diagnostics, a compact, low-power and highly portable platform is needed. The development of portable analytical instrumentation is attractive because it

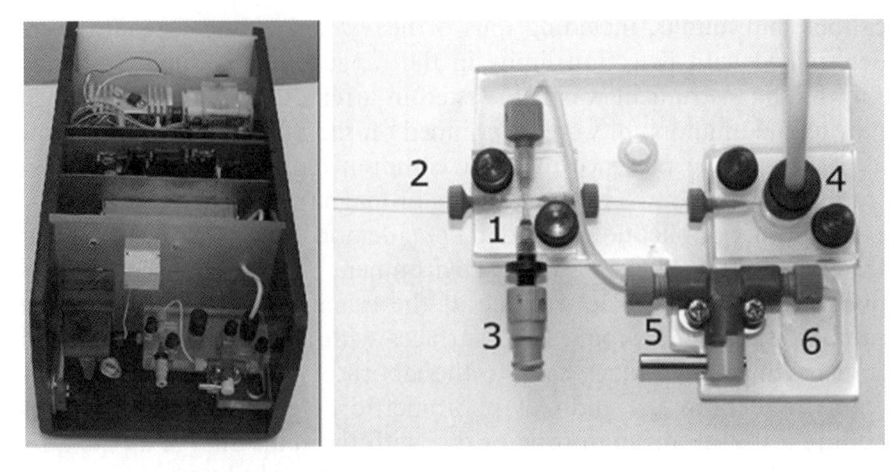

Figure 5.12 Portable CE instrument. (Left): photograph of the instrument with the upper cover removed. In the front, the chemical compartment with cross sampler and CCCD detector are visible. The electronics and high-voltage power supply are located in the middle, and the battery compartment at the back. (Right): a close-up view of the cross sampler – cross sampler (1), two aligned capillaries for separation (2), socket for syringe (3), BGE vessel with HV lead and electrode (4), shut-off valve for washing capillary (5), and waste reservoir (6).[159] (Reproduced with kind permission of Wiley-VCH Verlag GmbH & Co. KGaA).

minimizes complications arising from sample storage and transport, enables fast decisions at the sampling site, and, therefore, can be used in environmental applications, forensics and the detection of chemical warfare agents. In response to that need, many reports on portable CE analysers based on microfluidics platforms have been published in recent years.[164,165] CE-based and other microfluidic devices will be described in the following section.

5.2.3 Micronization of Separation Methods

Micronization is an important way to generate less waste and is essential when the amount of available sample is very small (less than microliters). Combinatorial chemistry has also stimulated the search for alternative separation approaches. The key to rapid and efficient synthesis is not only the parallel arrangement of reactions, but also simple work-up procedures that circumvent time-consuming and laborious purification and analysis steps.

5.2.3.1 Microfluidics. Concerns about toxic eluent consumption are relevant to microfluidics, in which the amount of eluent required is even

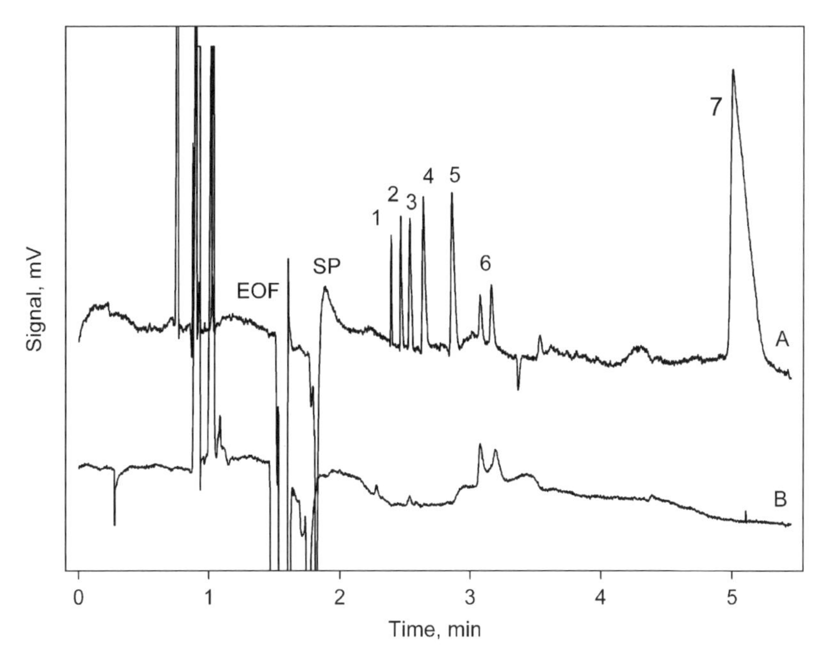

Figure 5.13 Electropherograms of sand extracts. The BGE 15 mM MES/His; 20 kV. Sample (A) contains 75 μM phosphonic acids dissolved in sand extraction water; pure sand extraction water (B); peaks: pinacolylmethylphosphonic acid (1), 1-butylphosphonic acid (2), propylphosphonic acid (3), ethylphosphonic acid (4), methylphosphonic acid (5), peaks extracted from soil (6), citric acid (7), system peak (SP) and electro-osmotic flow (EOF).[156] (Reproduced with kind permission of Wiley-VCH Verlag GmbH & Co. KGaA).

smaller than in CE. "Fluidics" means the handling of liquids and/or gases. "Micro" signifies at least one of the following features: small volumes (μL, nL or pL), small size, or low-energy consumption. Microfluidics is the science of designing, manufacturing and formulating devices and processes that handle volumes of fluid on the order of nanolitres (10^{-9} litre) or picolitres (10^{-12} litre). The devices have dimensions ranging from millimeters to micrometers. Microfluidics hardware requires a construction and design that is different from its macroscale counterpart, since it is usually not possible to scale down conventional devices and expect them to work in microfluidics applications. As the scale becomes smaller and the dimensions of a device reach a certain size, the particles of fluid, or particles suspended in the fluid, become comparable in size with the apparatus itself. This dramatically alters the behaviour of the system. Capillary action changes the way in which fluids pass through micrometer-diameter tubes, as compared with

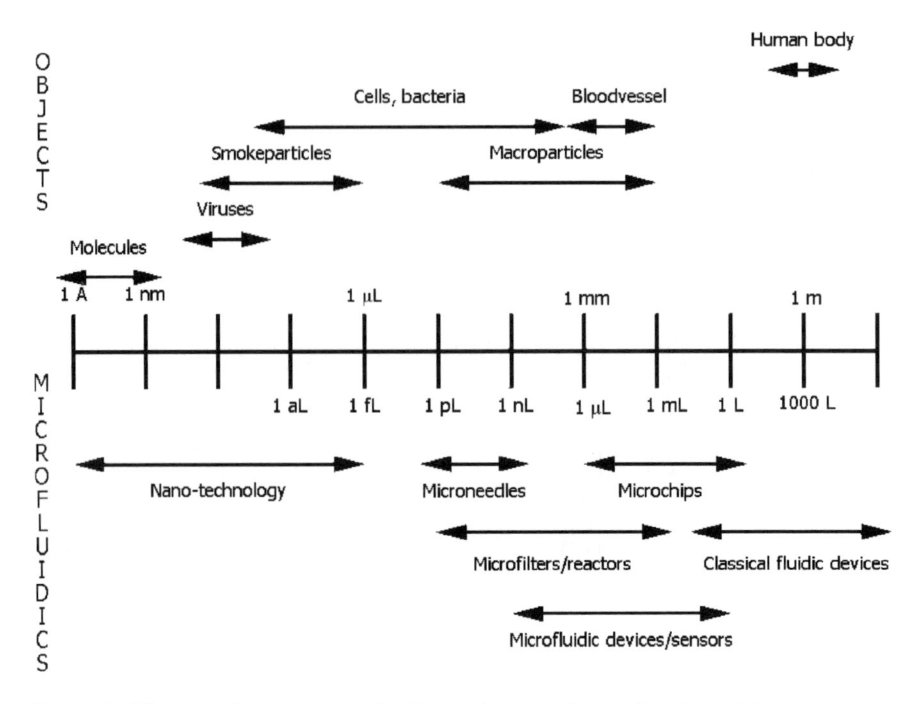

Figure 5.14 Definition of microfluidics and comparison of various objects.

macroscale channels. The dimensions of various objects and technologies are illustrated in Figure 5.14.

The volumes involved in microfluidics can be understood by visualizing the size of a one litre vessel, and then imagining fractions of this container. A cube measuring 100 mm on one side has a volume of 1 L. The volume of a thimble is 1/1000 of a litre *i.e.* 1 mL, with 1 cm sides. The size of a small grain of table sugar is a tiny cube whose height, width and depth are 1/100 of this size, or 0.1 mm. This grain has a volume of 1 nL. A volume of 1 pL is represented by a cube whose height, width and depth are 1/10 (0.1) of a 1 nL cube and requires a microscope to be seen.

Total analysis systems (TAS) can be built on microfluidics. TAS comprises all analytical procedures: sampling, processing and analysis integrated into one device. Miniaturized total analysis systems (μTAS) were first proposed for chemical sensing in 1990.[166] This led to the development of microfluidics and the vision of a "lab-on-a-chip". μTAS integrates all the steps required in chemical analysis – sampling, preprocessing, and measurement – into a single device *via* miniaturization, resulting in an improved selectivity and detection limit compared to conventional sensors. The dramatic downscaling and integration of chemical assays also hold considerable promise for faster and simpler on-site monitoring of priority pollutants, and make these analytical microsystems particularly attractive as Green Analytical Chemistry screening tools. The amount of waste generated is reduced by *ca.* 4–5 orders of magnitude, in comparison with, for example, conventional liquid chromatographic assays (*i.e.* 10 μL *vs.* 1 L per daily use). A significant amount of research has been devoted to the development of microfluidics technology and the applications of μTAS devices over the past decade.[167–170] Microfluidic systems have diverse and widespread potential applications.[171] Some examples of systems and processes that might employ this technology include: inkjet printers, blood-cell separation equipment, biochemical assays, chemical synthesis, genetic analysis, drug screening and electrochromatography. The medical industry has shown enthusiastic interest in microfluidics technology. Common analytical assays, including: polymerase chain reaction (PCR), DNA analyses and sequencing, protein separations, immunoassay, and intra- and inter-cellular analysis, have been reduced in size and fabricated in a centimeter-scale chip.[172]

Microfluidics devices, first developed in the early 1990s, were initially fabricated in silicon and glass using photolithography and etching techniques adapted from the microelectronics industry. They were precise, but expensive and inflexible.

Patterns of etch resists are defined on rigid, planar substrates such as silicon, with photolithography or electron-beam lithography, and relief structures are then created with reactive-ion or wet etching. The advantages of this methodology are high spatial resolution (about 100 nm) and parallel processing – photolithography and etching create all features in a single step. The highly developed methods of photolithography, however, have disadvantages for the fabrication of microfluidic devices. Processing must be performed in a clean room. The methods are expensive, require inconvenient processes for sealing channel structures, and provide no simple way of interconnecting channel systems (such as for sample collection, introduction, and analysis). The trend is toward the application of soft lithography, invented by Whitesides,[173] to build microdevices using fabrication methods based on printing and moulding organic materials. Soft lithography involves the following steps (see Figure 5.15):

- generating a layer of UV-curable photoresist (PR) on a silicon substrate;
- using a mask to selectively develop sections of PR (UV light hardens the PR), rinsing away undeveloped PR to create a template;

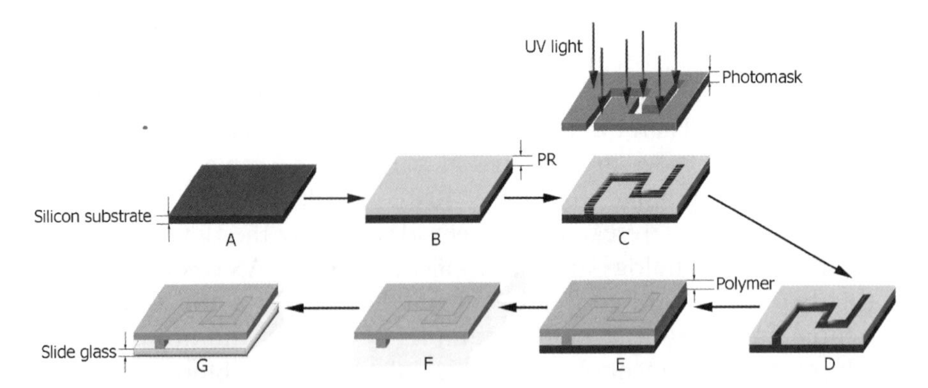

Figure 5.15 Process of soft lithography: silicone substrate (A); silicone substrate covered with photoresist PR (B); exposing photoresist through a mask (C); generating template by rinsing away uncured PR (D); pouring curable polymer onto template (E); removing cured polymer sheet from template (F); and fixing polymer sheet to a glass slide (G).

- pouring curable (by heat, UV or chemical) polymer over a template, curing, and removing from template.

Photoresist, the material mentioned above, is an important ingredient in photolithography. It is sensitive to UV radiation. One type of photoresist is based on a mixture of diazonaphthoquinone (DNQ) and a phenol–formaldehyde resin (Novolac). DNQ inhibits the dissolution of the Novolac resin in a base, however, upon exposure to light, the dissolution rate increases even beyond that of pure Novolac. The mechanism by which unexposed DNQ inhibits Novolac dissolution is not well understood, but it is believed to be related to hydrogen bonding (or more exactly, diazocoupling in the unexposed region). DNQ–Novolac resists are developed by dissolution in a basic solution (usually 0.26 N tetramethylammonium hydroxide in water). Photoresist is widely sold under the commercial name SU-8.

Two detection principles have been employed in microfluidic systems: elelctrochemical and optical. The advantages of electrochemical detection for microfluidics are: an inherent miniaturization of electrodes for detection, low power requirements, low cost and a high compatibility with advanced microfabrication and micromachining technologies.[174] As was mentioned in Section 5.2.2.4, electrochemical detection is an extremely attractive means of creating truly portable and disposable microsystems at points-of-care. The characteristic common to most of these detectors is the placement of the detector outside the separation channel, which results in effective isolation from the high separation potential. In contrast, optical detectors are still relatively large, which compromises the advantages of miniaturization.

Although there have been many successes, an important hurdle that still needs to be cleared is the connection between the microcomponents of a device and the macro-environment of the world. This part of the device is often referred to as the macro-to-micro interface,[175] interconnect,[176–179] or "world-to-chip" interface.[180–184] The difficulty results from the fact that samples and reagents are typically transferred in quantities of microliters to millilitres (or even litres) while microfluidic devices consume only nanoliters or picoliters of samples/reagents due to the reaction chambers and channels, which typically have dimensions on the order of microns. This challenge is often overlooked in research environments such as academic laboratories, but it thwarts meeting the requirements of Green Chemistry and cannot be ignored in routine applications. A good illustration is provided by the following photograph shown in Figure 5.16 (which was picked at random by a Google search on the keyword "microfluidics"): a tiny microchip is surrounded

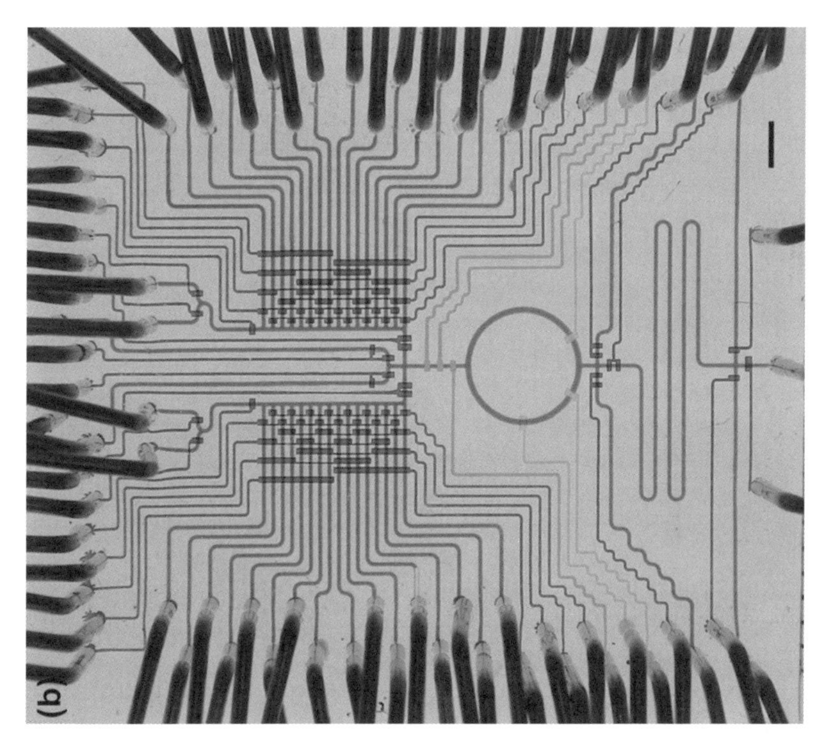

Figure 5.16 Connecting microfluidic devices to the world: an image of an actual device.

by the bundle of tubing needed for the fluid supply (pumps) and actuation (by gas) and detection.[185] In the photograph, there are nearly fifty connecting tubes, indicating the number of actuating (perhaps syringe type) pumps that are involved. Supporting equipment is not usually shown in illustrations of microfluidic devices. Although such microchips can perform many sophisticated operations, microminiaturization can perhaps be justified if the integration of the elements performing unit operations could be increased by many orders of magnitude.

The so-called "world-to-chip" interface problem has plagued microfluidics since its inception. It remains a challenge to achieve the desired practical economies of fluid manipulation in microfluidic chip systems by solving the mismatch between processed volumes on the "world" and chip scales. Microfluidics has to integrate all the components of the system on the same chip. Pumps, valves, mixers, *etc.* must be miniaturized in order to achieve an integrated system, which forces a choice between: active methods which are efficient but difficult to miniaturize and require energy sources, and passive methods which are easier to integrate but are less efficient. In biotechnology, the volume reduction from the macroscopic environment where the targets are located to the microscopic environment of the microsystem is a huge challenge. Molecules and/or particles existing in macroscopic volumes have to be concentrated in micro-volumes for biochips. Miniaturization of the whole treatment chain remains to be accomplished. Finally, the very large surface : volume ratio of miniaturized systems modifies the physical behaviour of the system, which gives rise to new problems, such as the adherence of target molecules to the solid walls, or the effect of capillary forces that may prevent the fluid from entering the microchannels.

According to some researchers, the dilemma of the "world-to-chip" interface is one of the bottle-necks in the development of μTAS.[186] It is critical for high-throughput applications where manual manipulation is not economical and the macro-to-micro interface must be developed. In a review article,[175] solutions that have appeared in the literature are presented and discussed. Although designed as lab-on-a-chip applications, these "world-to-chip" designs might be relevant to designs of similar interfaces for conventional CE as well. Such designs could aim to achieve computerized on-line sampling of small sample volumes without incurring the complications of microfabrication.

In a recent paper, two simple and economical techniques were described for computerized on-line sampling of a small amount (approximately 200 μL) of reacting media in a conventional CE capillary.[187] The first approach was similar to the "cross" injection device commonly used in labs-on-a-chip where the sample and separation channels are perpendicular on a chip. However, instead of electrokinetic loading, the sample is injected into a capillary by pressure pulse. This sampling technique has been used in microfabricated systems by Lin *et al.*[188] In the second approach, the sample is delivered as droplets (10 μL volume) into a buffer situated in a pipette tip. The falling droplet sampler was first proposed by Liu and Dasgupta,[189] and has recently been used for hyphenating flow injection analysis (FIA) to CE.[190–192] Falling droplet sampling was adopted in an FIA-CE system to avoid deterioration of separation due to hydrodynamic pressure created in the inlet, especially from higher flow rates in the FI system.[193] In this falling-drop design, a constant liquid head equal to that at the capillary outlet is maintained, avoiding Poiseuille flow. In one proposed design,[187] the construction of the falling droplet sampler was further simplified by rejecting a specially designed (FIA-CE) interface body. The liquid is kept in the pipette tip by surface tension and when a new portion of liquid is delivered, it displaces the old portion into a waste receptacle.

Despite recent achievements, "world-to-chip" interfacing is still awaiting the best solution. The high degree of integration offered by "lab-on-a-chip" devices implies that the principles of Green Chemistry must be applied to all the steps of the analytical process involved in them. In this context, the "world-to-chip" problem needs to be solved immediately. One solution has been provided by Harrison *et al.*[194] for the design of an interface that would allow microfluidic chips to sample from the "fire hose" of the external environment. The ability to continuously introduce real samples into micrometer channels would make "lab-on-a-chip" devices compatible with real-life monitoring applications.

A recent review lists microfluidics applications from over 5000 publications (as of mid-2009) covering all possible topics in chemistry and biochemistry.[195] The development of microfluidics is now passing through a decisive phase. According to a publication by Mukhopadhyay,[196] the development of microfluidics follows "the Gartner hype cycle model". According to this model, after the technological advances at the beginning of the 1990s, a peak of inflated expectations occurred that was soon followed by disillusionment when the technology was not implemented. No application of great consequence emerged. Microfluidics has created much hope and excitement among analytical chemists who are now – as mentioned above – past the period of disillusionment. For this reason, the quest for a "killer" application in microfluidics is both urgent and desperate. Becker defines a "killer" application as a product that has such highly desirable properties that it is capable of generating very large revenues with attractive margins in a comparatively short period of time.[197] Large flat-panel displays for televisions and digital photography are good examples of killer applications developed from new technology. As Becker[197] writes, microfluidics "potential killer applications are easily understandable to an audience... the idea of a universal analysis tool ...is immediately appealing...". Becker is optimistic – he expects that a killer application in microfluidics will appear within 5–10 years and predicts that such an application will probably be in the domain of diagnostics – perhaps a point-of-care device or laboratory diagnostics apparatus for a physician's office. It might also be associated with veterinary medicine or food inspection. Such a product would definitely meet the criteria of Green Analytical Chemistry. We hope that microfluidics will climb the "slope of enlightenment to the plateau of productivity" where the technology becomes mainstream.

5.2.3.2 Non-Instrumental Microfluidic Devices. The microminiaturization of instruments is only one way to make methods of analysis environmentally friendly. The "lab-on-a-chip" has the disadvantage of frequently being designed for a very specific application. If there is no widely used application, then the development of microfluidic devices is not economically profitable. Academic institutions and companies have developed an abundance of "labs-on-a-chip" for different applications, but what is really needed is a universal application that could trigger widespread use of microchips in biomedicine.

As we saw above, the typical setup for a "lab-on-a-chip" experiment consists of a small microfluidic chip which is surrounded by desktop-sized analysis instruments and power supplies. To solve this "world-to-chip" interface problem, further miniaturize instruments and integrate the elements of micro total analysis systems (μTAS), as many power-consuming and otherwise complex elements as possible must be eliminated. These elements can be replaced with passive components that operate without external power by manipulating fluids using gravity, air pressure or simple manual actions. The need to develop such simple and possibly disposable devices is motivated primarily by the need for simple point-of-care (POC) tests for developing countries, but such devices can also find application in developed countries where most medical diagnostics are performed in centralized, well-equipped laboratories. Two of those applications are home testing (pregnancy, glucose) and the detection of natural or man-made bioemergencies (bioterrorism) by first responders.

George Whitesides and his colleagues at Harvard University have achieved the most operational robustness and cost reduction with microfluidic devices so far. They have recently demonstrated that three-dimensional microfluidic devices can be made from stacked layers of ordinary paper and tape.[198] As a result of the paper's wicking ability, the devices don't require external pumps to drive the liquids through. The wicking property of paper is routinely exploited in medical tests such as those for blood glucose, pregnancy and HIV. To define the pathways in the paper-based microfluidic device, the team impregnated each layer of paper with a common photoresist and patterned them with UV light. With channels thus established on a sheet of paper, the layers of paper were alternated with layers of double-sided tape, with holes cut in the tape connecting the channels in adjacent layers of paper. The complex routing that can be achieved can wick liquid horizontally and vertically to an array of 1024 detection zones on the bottom. With reagents or antibodies placed in detection zones prior to assembly, such devices would provide highly parallel, independent assays.

The potential of paper-based devices could be put to good use in medical diagnostics in developing countries. Paper-based microfluidic devices are especially appropriate for use in distributed healthcare in the developing world and in environmental monitoring and water analysis. Paper-based 3D microfluidic devices have capabilities that are difficult to achieve using conventional open-channel microsystems made from glass or polymers. In particular, 3D paper-based devices wick fluids and distribute microliter volumes of samples from single inlet points to arrays of detection zones (with numbers in the thousands). This capability makes it possible to carry out a range of new analytical protocols simply and inexpensively (all on a piece of paper) without external pumps.

An example of a non-instrumented, microfluidics-based μTAS is the immunochromatographic strip (ICS) assay. Within a few minutes of a sample being applied to the test strip, an indicator line demonstrates the presence or absence of the analyte of interest. The test sample migrates through the filter paper matrix, which contains both fixed and impregnated antibodies. As the sample migrates, it solubilizes the tagged antibodies and initiates antibody binding to the target. The analyte of interest encounters the second set of antibodies and forms an antibody–analyte–antibody sandwich matrix with a visible signal. Because ICS relies on inexpensive reagents and components, they cost less than 2 US dollars to the end user in many cases. ICS strips require relatively little and sometimes no sample processing, and they do not require an external instrument.

The function of components that are suitable for non-instrumented, microfluidics-based disposable diagnostic devices must be based on simple physical chemical phenomena like capillary action, evaporation, endo/exothermic reaction, gravity and laminar flow in microchannels. Since a power supply is not supposed to be available, some of the energy can be supplied by the analyst's manpower. In their review, Weigl *et al.* give many examples of designing components with such elements:[199]

Pumps. Wicking and capillary action have been widely used to transport fluids for POC diagnostics. Finger-operated bellows fabricated from polydimethylsiloxane (PDMS) can also be used. Among other things, pumping can be used to drive DNA solutions through heated zones for PCR applications, contrary to conventional PCR which keeps a sample stationary while using heat sources and sinks to cycle the temperature over the necessary three-temperature profile to conduct the analysis. The viability of this approach has been demonstrated by moving fluid over a series of appropriate temperature zones separated by insulators.[200]

Mixers. In general, mixing reagents and samples in microfluidic structures is challenging because of the laminar flow of liquids in microchannels. Methods requiring external power sources (electrohydrodynamic, magnetohydrodynamic, pressure perturbations, centrifugal, electrophoretic effects, pulsed flow and an on-card stir bar stator with an oscillating field on a card reader) cannot be used in noninstrumental POC devices. Some passive methods such as transverse mixing wells, spiral microchannels, expansion vortex, channel obstacles, lamination split and recombining have been designed.

Valves. Microvalves have been developed that utilize magnetic, electrostatic, thermal, mechanical, pneumatic and piezo-electric actuators, but the complexity of manufacturing methods has largely limited the success of this approach.

Separators. The H-filter is a typical microfluidic device that is based on the parallel laminar flow of two or more miscible streams in contact with each other. The streams do not mix (because of the laminarity of the flow), but chemicals can diffuse from one stream to the other, with smaller molecules diffusing faster than larger ones. An H-filter can be used to extract unwanted components or desired components from one of several fluids being processed simultaneously.

Concentrators. based on isothermal evaporation can be constructed. Methods like chromatography, dielectrophoresis and isoelectric focusing are not appropriate for noninstrumented POC devices because of their reliance on instruments.

Precise fluid handling. can be performed by standard pipette techniques.

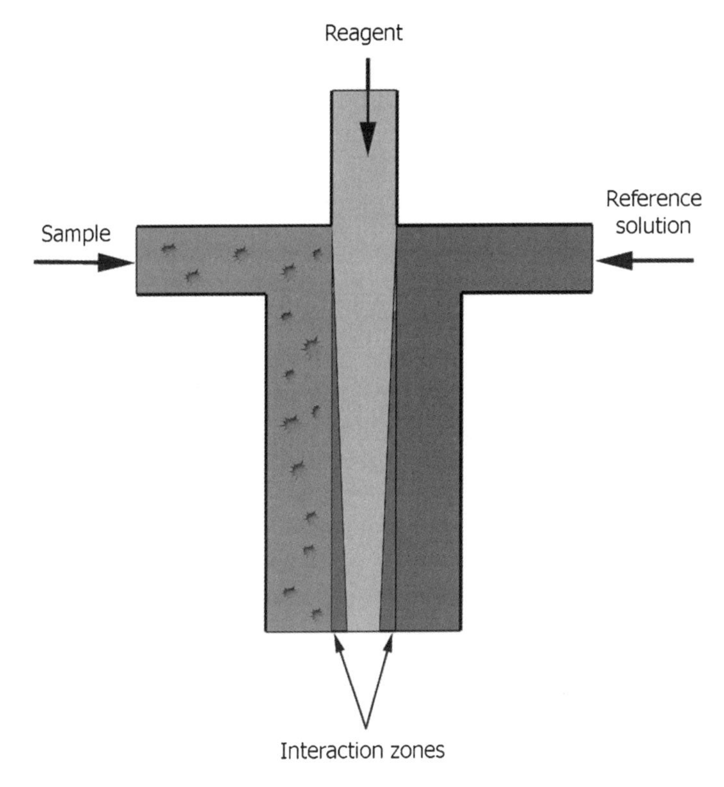

Figure 5.17 T-sensor design.[202]

Heating and cooling. In one example, localized heating and cooling are controlled by positioning the endothermic (evaporation of acetone) and exothermic (dissolution of concentrated sulfuric acid in water) processes in microreaction chambers near the reactant flow interface.[201]

Detectors. Detection in noninstrumental microfluidics must rely on human senses and therefore can only be based on colorimetry. A hydrostatically driven T-sensor has been proposed for sensing and viewing analytes in noninstrumental microfluidic devices.[202] It is very similar to the H-filter, but three flows instead of two are directed to the separation channel (see Figure 5.17). The three flows, which are introduced next to each other into a common channel, are the sample solution (*e.g.* whole blood), a receptor solution (*e.g.* an indicator solution) and a reference solution (a known analyte standard). Two interface zones are formed between the three layers of fluid. A sample is put into the top left reservoir, a reagent (*e.g.* an indicator dye) into the top middle reservoir and a reference solution with a known concentration of analyte into the right reservoir. Comparison of the intensity and position of the two diffusion interaction zones allows a semi-quantitative analyte determination. The ratio of a property of the two interface zones (*e.g.* colour, absorbance or fluorescence intensity) is a function of the concentration of the analyte and is largely free of cross-sensitivities to other sample components and instrument parameters. The principle is demonstrated by the determination of human albumin and ionized calcium in whole blood and serum.[203]

As Weigl *et al.* point out,[199] significant obstacles to the successful development of noninstrumented microfluidic assays remain. Solutions exist for sample preparation and even, if needed, target amplification, without the use of instrumentation, but is difficult to imagine how very low signal intensities could be seen without electronic signal amplification. This may limit the applicability of these assays where extremely high sensitivity is required. Successful integration of paper-based microfluidics and electrochemical detection has been demonstrated.[204] One solution could be the use of consumer electronics-based analytics, which has been described in Chapter 4.4.3. It is tempting to envision the application of CSPT, for example, to noninstrumental microfluidics devices with T-sensor output. No reports of such applications are available, so there is no idea what the detection limits of such devices could be, but the approach seems to be reasonable.

5.2.3.3 Droplet and Digital Microfluidics: from Microflows to Microdrops. Various "toolboxes" are currently proposed for microfluidics.[205] The first and most important is the "microflow" toolbox (see Figure 5.18, top left). It is now a mature approach, notwithstanding concerns about

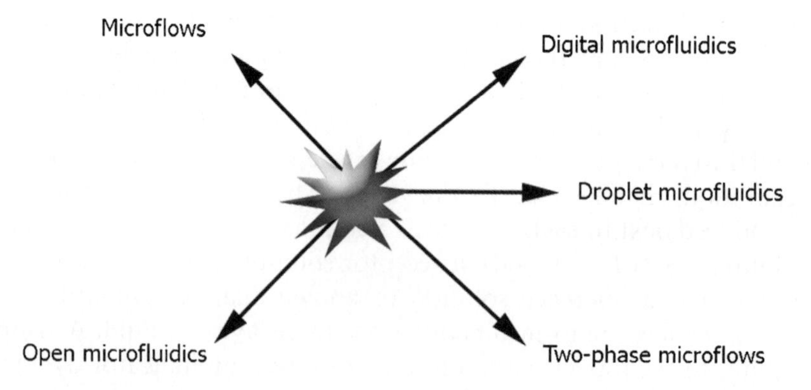

Figure 5.18 The position of droplet microfluidics among different domains (toolboxes) of microfluidics.[205]

"world-to-chip" interfacing. Digital microfluidics (see Figure 5.18, top right) is a new toolbox. More recently, even newer concepts have been developed to complement the possibilities of microfluidics: the droplet microfluidics and two-phase microflow toolboxes (see schematics in Figure 5.18, bottom right). "Open" microfluidics, *i.e.* microflows in partially opened channels where the flow is stabilized by capillarity forces in the absence of a solid boundary, is at the crossroads between single-phase and two-phase microflows (see Figure 5.18, bottom left). This is the least advanced method in microfluidics today. Droplet and digital microfluidics are emerging concepts in analytical chemistry. They are still at the development stage, and their possibilities have not all been investigated, but it seems to be a promising area from the standpoint of the twelve principles of Green Chemistry. Digital and droplet microfluidics both have in common the use of microdrops and an associated reduction in volume. Droplet microfluidics involves two- or multi-phase microflows, in which droplets are transported by a carrier flow, whereas digital microfluidics is inspired by microplate systems into which a manipulation robot is incorporated. In such systems, the microdrops are moved and treated individually as digital entities on a planar surface.

Droplet/digital microfluidics has great potential in some areas of analytical chemistry. A striking example from the standpoint of Green Analytical Chemistry is screening – searching for a specific component in a sample. The time spent on screening is a critical issue. If the screening is performed in microplates comprising one tenth of 10 μL wells (the conventional approach) and is performed for 100 000 components, the proess can take approximately four months and require up to 10 L of reagent.[205] Using microflow systems, the same screening is reduced to a week and the quantities of sample are reduced 100-fold. Droplet/digital microfluidics promises to push screening time to the extreme. The use of droplet microfluidics now reduces the time to approximately 20 minutes. As one can see, the considerable advantage of digital or droplet microfluidics is the dramatic reduction in volume. Volumes are likely to be reduced to 50 nl or less in the near future. There is another advantage of using microdroplets – the reduction of contact with solid walls. A large part of the surface of the liquid in contact with the solid wall is replaced by a liquid–liquid or liquid–gas interface, and the solid contact surface to liquid volume ratio is reduced compared with that of microflow systems, hence, adherence and adhesion problems are reduced.

Digital and droplet microfluidics have in common the use of microdrops and an associated reduction in volume (see Figure 5.19). However, they differ considerably. As we saw above, in droplet microfluidics,

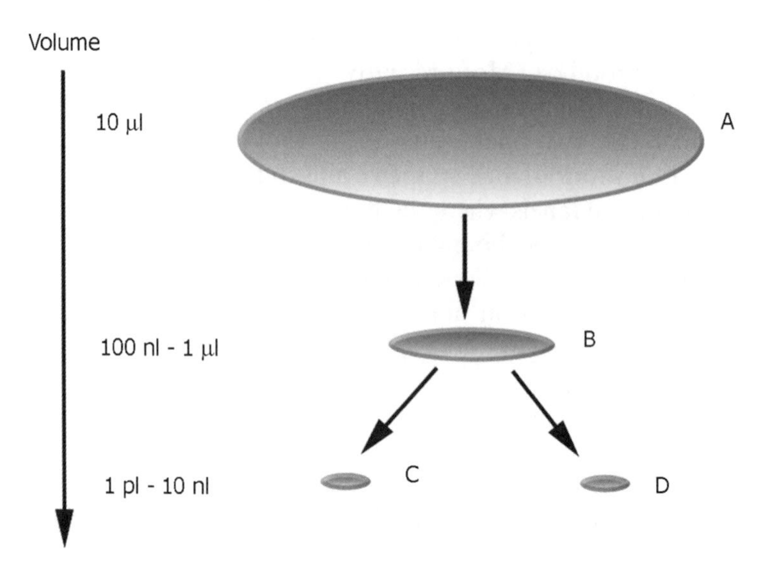

Figure 5.19 Scheme of the different scales of fluidic Microsystems.[202] Conventional (microplate) systems (A), microfluidic systems (B), digital microfluidic systems – parallel (C) and droplet microfluidic systems – serial (D).

droplets are transported by a carrier flow (liquid transport and valves are actuated by mechanical, electrical, magnetic or acoustic forces), whereas digital microfluidics is based on a totally different approach, inspired by microplate systems. In such systems the micro-drops are moved and treated individually by the phenomenon known as electrowetting on dielectric (EWOD). Digital and droplet microfluidic systems are in fact complementary. Digital microfluidics can process extremely small volumes of liquids and perform operations and manipulation in parallel and with great accuracy, whereas droplet microfluidics is well adapted to perform operations in series, such as screening or encapsulation.

Droplet actuation in digital microfluidics is radically different from that of microflow and droplet microfluidic systems (in which pressure is employed), and deserves mention. Although many phenomena such as thermocapillarity,[206] surface acoustic waves,[207] dielectrophoresis,[208] charged droplet coulombic repulsion/attraction,[209] and gravity[210] have been tested, electrowetting[211,212] is the most commonly used microdroplet actuation technique. In 1875, Gabriel Lippmann showed that the capillary depression of mercury in contact with an electrolyte solution could be controlled by the application of voltage.[213] In addition to formulating the basic theory for electrocapillarity, he also proposed several practical applications that made use of the phenomenon.

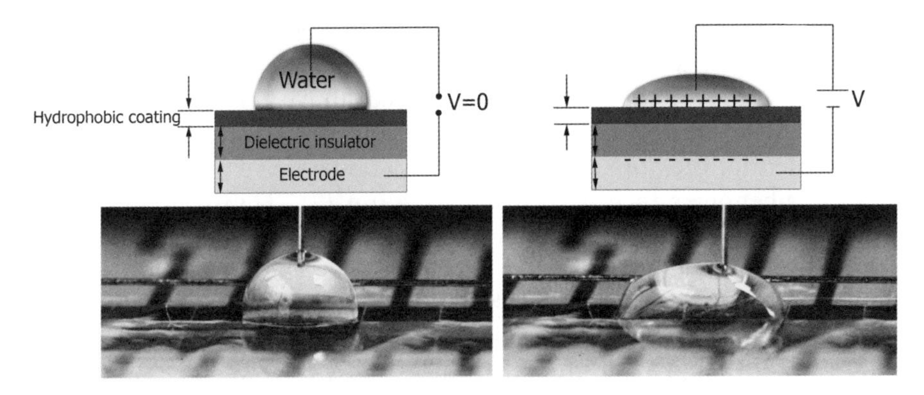

Figure 5.20 (Top left): a water droplet placed on a hydrophobic surface with a high contact angle. (Note: the two surface coatings are not drawn to scale). (Top right): EWOD phenomenon. (Bottom): experimental demonstration of EWOD – (left): no voltage applied and (right): voltage applied to droplet. (Substrate was a printed circuit covered with Saran™ food wrap; photo from the authors' laboratory).

Unfortunately, the application of the technique to aqueous electrolytes is severely restricted because of electrolysis. In the early 1990s, however, Berge realized that the problem of electrolysis could be overcome by the introduction of a thin insulating layer to separate the conductive liquid from the metal electrode.[214,215] This concept has become known as electrowetting on dielectric (EWOD) and involves applying voltage to modify the wetting behaviour of a liquid in contact with hydrophobic, insulated electrodes. Applying voltage between the drop and the electrode changes the distribution of charges due to the dielectric insulator, and significantly decreases the contact angle (see Figure 5.20). The effect is most pronounced if the insulator layer is covered with material that has a contact angle greater than 150°. Such a surface is called super-hydrophobic. A popular super-hydrophobic material is amorphous Teflon AF, from DuPont.

The change in contact angle over the electrodes can be described using the Lippmann–Young equation:

$$\cos\theta = \cos\theta_0 + \frac{\varepsilon_0\varepsilon}{2d\gamma_{LV}}U^2 \quad (7)$$

where θ is the contact angle over the active electrode; θ_0 is the equilibrium (or Young's) contact angle in the absence of an applied voltage; ε_0 is the permittivity of free space; ε is the dielectric constant of the insulating layer; d is the thickness of the insulating layer; γ_{LV} is the surface tension between the liquid and the vapour; and U is the voltage applied to the electrode. The parabolic variation of $\cos\theta$ in the Lippmann–Young equation is only applicable when the applied voltage lies below a threshold value. If the voltage is increased above this threshold, then contact angle saturation starts to occur and $\cos\theta$ eventually becomes independent of the applied voltage. Several interesting properties of the EWOD phenomenon follow from Equation (7). The insulating layer must be rather thin (the measurable effect occurs if the thin layer of dielectric is less than tens of μm and the dielectic constant of the insulator is sufficiently high). Also, because of quadratic behaviour, alternating voltage can perform as well as direct current.

A typical digital microfluidics device that uses EWOD for droplet actuation consists of a planar substrate covered with a set of (planar) electrodes patterned according to the required application. To move a microlitre-sized droplet over the electrode to the desired position, voltage is applied to the neighboring electrodes. This causes distortion of the previously semispherical droplet, and an interfacial tension gradient of the liquid inside the droplet causes it to follow the direction of the applied voltage gradient (see Figure 5.21). By applying voltage to a series of adjacent electrodes that can be turned on or off, one can manipulate droplets. The advantage of this approach is that each droplet can be individually transported, split, merged and stored simply by the application of a suitable sequence of voltages to the control electrodes.

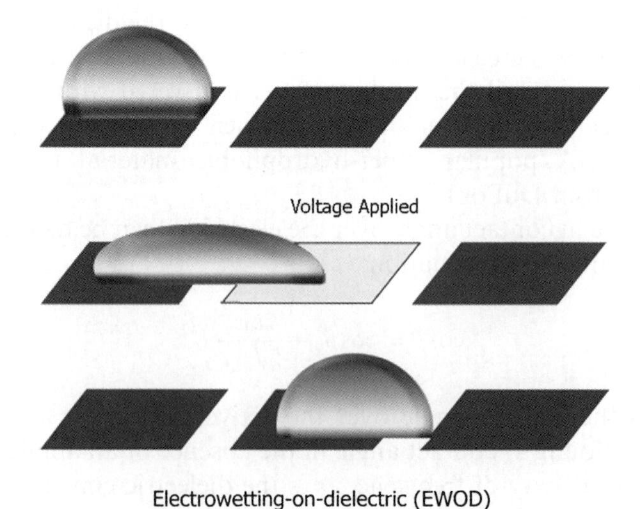

Figure 5.21 Droplet actuation mechanism by EWOD.

A closed system is used to prevent droplets from evaporating. In closed structures (coplanar design), droplets are immersed in a silicon oil film and sandwiched between two parallel plates: the bottom plate is the chip surface, which houses the addressable electrode array, and the top is a continuous ground plate (see Figure 5.22).[216] The top plate is sometimes replaced by a grounding wire system.[217] In another coplanar design, both the activation electrode and the electrodes that ground the droplet are located on the bottom of the dielectric surface.[205] The top plate is not required for this type of coplanar device, but it is included because manufacturers or users can customize the passive top plate with specific chemistry or structures appropriate for each application. Therefore, the top plate can be a disposable component.

A digital-microfluidic platform has many attractive properties for practical application. Fair *et al.* list several advantages of digital microfluidics.[218]It:

- has no moving parts. All operations take place between two plates under direct electrical control without the use of pumps or valves;
- requires no channels. The gap is simply filled with liquid; channels exist only in a virtual sense and can be instantly reconfigured by software;
- controls many droplets independently, because the electrowetting force is localized at the surface;
- uses no ohmic current. Although capacitive currents exist, the device blocks direct current, minimizing sample heating and electrochemical reactions;
- works with a wide variety of liquids, *i.e.* most electrolyte solutions;
- uses close to 100% of the sample or reagent, by wasting no fluid for priming channels or filling reservoirs;
- is compatible with microscopy. Glass substrates and transparent indium–tin–oxide (ITO) electrodes make the chip compatible with observation under a microscope;

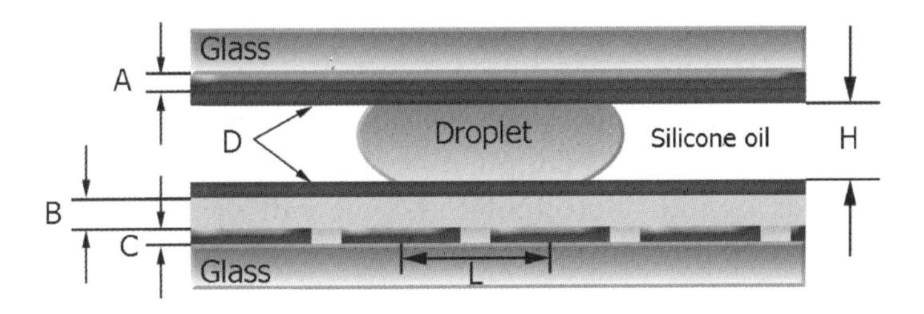

Figure 5.22 A coplanar design: indium tin oxide (A), parylene (B), indium tin oxide (C) and Teflon AF (D).[212]

- is extremely energy efficient – uses nanowatts to microwatts of power per transfer;
- achieves high droplet speeds – up to about 25 cm s^{-1};
- uses droplet-based protocols functionally equivalent to bench-scale wet chemistry. Thus, users can simply scale-down, automate, and integrate established assays and protocols;
- permits maximum operational flexibility. Direct computer control of each step allows conditional execution steps.

As shown in this list, many of the entries are relevant to the principles of Green Chemistry. However, the main disadvantage of digital microfluidics systems is that they are difficult to produce. Electrical engineers are more suited to the task than analytical chemists. The preparation of digital microfluidic systems requires various skills that analytical chemists usually do not possess, such as a facility with clean room technologies for patterning the electrodes or preparing thin (μm and nm size) films using various technologies (*e.g.* spin coating) not available in ordinary analytical laboratories. Also, the chemicals, especially hydrophobic Teflon AF, are expensive. The difficulties associated with the production of digital microfluidic systems could become an obstacle to the development of the whole field. To address this issue, Abdelgawad and Wheeler have demonstrated that digital microfluidic devices can be produced from consumer products.[219] The authors hope that their fabrication method will put digital microfluidics within the reach of any laboratory with minimal resources. According to this method, devices are formed from copper substrates used for the preparation of printed circuits in the electronics industry, or from gold compact disks. The authors used a marker to make masks to replace photolithography. This method is capable of fabricating devices with inter-electrode gaps as small as 50 μm. The gaps between electrodes were produced by a single stroke of a razor blade. Saran™ food wrap (a polyethylene film) and commercial water repellents were used as dielectric and hydrophobic coatings, respectively, to replace commonly used and more expensive materials such as Parylene-C and Teflon-AF. Devices formed by the new method enabled single- and two-plate actuation of droplets with volumes of 1–10 μL. The devices were successfully tested for droplet manipulation, merging and splitting.

Even simpler digital microfluidic systems can be envisioned. Yoon and You used commercially available glass slides covered with superhydrophobic thin film and demonstrated that droplets on a superhydrophobic surface can be actuated by the drag of a metal wire immersed into the droplet.[220] In this way, the movement of a droplet can be controlled by a programmable xyz-stage and no electrode patterning is necessary. Although no serious applications have been reported, the number of possibilities is enormous, *e.g.* integrating sample preparation with separation and detection. One can picture a droplet as a sample container and, moreover, as an optical component of the detecting instrumentation (due to its almost spherical form on the super-hydrophobical surface). The advantages of this system over conventional microfluidics or microwell plate assays include: minimized biofouling and repeated use of a platform, possible manipulation of nanoliter droplets, and re-programmability with a computer interface.

The list of digital microfluidics applications is still short. A good example[218] of the possibilities of digital microfluidics is DNA sequencing-by-synthesis (also known also as pyrosequencing).[221] The sequence of the chemical reactions involved in pyrosequencing is shown in Figure 5.23 After incorporating a nucleotide – in this case, deoxyadenosine triphosphate (dATP) – a washing step removes excess substrate. Of the two strings shown, one is arbitrary DNA, and the other is complementary to the first string (ATP is adenosine triphosphate and PPi is inorganic pyrophosphate). The process of pyrosequencing involves enzymatic extension by polymerase through the iterative (one by one) addition of labeled nucleotides. The process begins with the addition of a known nucleotide to the DNA (or RNA) strand of interest. The reaction to incorporate a nucleotide is carried out by DNA polymerase. Upon incorporation of the nucleotide, inorganic pyrophosphate (PPi) is released. The enzyme ATP sulfurylase converts the PPi to adenosine triphosphate (ATP), which then provides the energy for the enzyme lucerifase to oxidize luciferin. One of the byproducts of this final oxidation reaction is light, generated by the reaction at approximately 560 nm. Figure 5.23(a) illustrates this sequence. The light can be detected by various photodetectors. Since the order in which the nucleotide addition occurs is known, it is easy to determine the unknown strand's sequence by the formation of its complementary strand. The entire pyrosequencing process takes 3 to 4 seconds per nucleotide. Figure 5.23(b) shows a preliminary digital-microfluidic pyrosequencing platform.[218] The chip has reservoirs to contain nucleotides and enzymes. The pyrosequencing protocol is easy to integrate in such a system. The on-chip chain of events is as follows:

- the reservoir dispenses a droplet containing nucleotides of a single type and polymerase, and the droplet goes to site S, which contains immobilized single-stranded DNA;
- the droplet incubates, allowing time for polymerization;

(a)

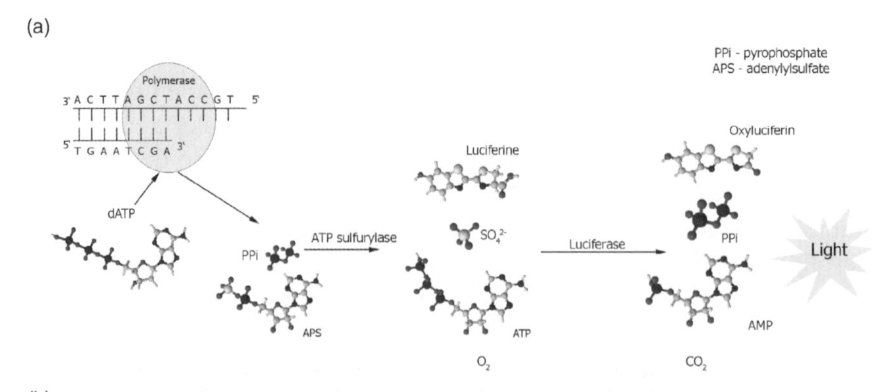

(b)

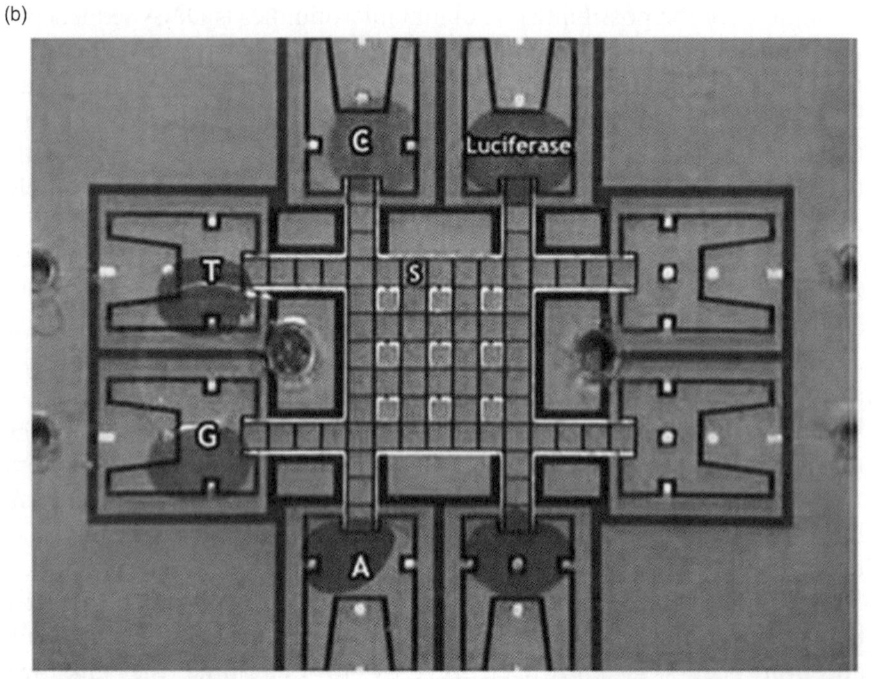

Figure 5.23 (a): sequence of reactions involved in solid-phase pyrosequencing. (b): Advanced Liquid Logic's prototype pyrosequencing chip, with on-chip reservoirs for dispensing reagents and nucleotides A, T, C and G. The DNA is attached to sample location S.[218] (Reproduced with kind permission of Springer Science + Business Media).

- the droplet leaves site S and reacts with luciferase;
- the droplet containing lucerifase moves to a detector array (not shown in Figure 5.23).

While the detector array detects a light signal from the droplet, the next nucleotide-containing droplet enters site S, so that DNA synthesis occurs simultaneously with detection. This protocol allows optimized synthesis reactions, and as synthesis is physically separated from detection, the system potentially has no inherent limitation on its read length. Unlike traditional pyrosequencing, this approach does not limit the system to detecting light before the next nucleotide is added. The rate-limiting step in the scheme is the conversion of PPi to ATP-by-ATP sufurylase. This reaction takes approximately 1.5 seconds, and the incorporation of a base takes less than 0.5 seconds. Luciferase completes the light generation in less than 0.2 seconds. In the current solid-phase pyrosequencing protocol, all these reactions must take place before the next nucleotide can be introduced.

For the authors, one of the chief design issues for performing pyrosequencing on an electrowetting chip platform has been immobilizing DNA through surface attachment. Because the platform's top plate is glass, this feature makes the attachment of DNA molecules to the top plate appealing, since it conveniently enables chip reuse and provides new possibilities for surface chemistry. The challenge is to attach DNA to hydrophobic Teflon. A method used by Fair *et al.*[223] for DNA immobilization requires patterning the Teflon film with streptavidin to create small access windows to the treated glass surface. Streptavidin binds covalently to the Teflon surface, ensuring that the protein remains attached after droplet transport. Furthermore, biotinylated DNA is guaranteed to bind well to streptavidin because the latter has an extremely high affinity for biotin.

At the time of writing (early 2009), many groups worldwide are working on droplet-based electrowetting science and technology. The number of papers published each year in this specialty is growing exponentially. Diverse areas to investigate include:[218]

- viable on-chip sample preparation methods, which is currently the biggest obstacle to commercial acceptance of microfluidic technologies, including digital microfluidics
- system integration and interfacing to other laboratory formats and devices;
- reagent packaging and storage;
- maintaining control of the temperature of the chips during field operation;
- scalable, compatible detector technology.

This list indicates that the number and variety of analyses being performed on-chip have increased, as well as the need to perform multiple-sample manipulations. It is worth repeating that the main challenge for both microfluidics and digital microfluidics is "world-to-chip" interfacing. It is often desirable to isolate components that produce a signal of interest, so that they can be detected. Currently, mass separation methods, such as capillary electrophoresis, are not part of the digital microfluidic toolkit, and the integration of separation methods presents a significant challenge. There have been some attempts to implement molecular separation on a digital microfluidic platform.[222,223] The quest to make analytical methods greener could motivate efforts to develop digital microfluidic devices, to put the field on a solid footing, and to find the unique niche that this technology can fill.

5.3 NEW VERSUS REFURBISHED INSTRUMENTATION: THE LIFE CYCLE OF AN INSTRUMENT

Recycling is important to a green view of the world. The seventh principle of Green Chemistry concerns the use of renewable feedstock, and states: "a raw material or feedstock should be renewable rather than depleting whenever technically and economically practicable".[224] Although Keith, Gron and Young considered this principle irrelevant to Green Analytical Chemistry,[225] this opinion could be debated if we consider the reuse of HPLC eluents, for example. Moreover, the seventh principal could be extended to the instrumentation used in analytical chemistry. In the context of Green Analytical Chemistry, this means, among other things, extending the life cycle of an analytical instrument. This position is further supported by a principle of Green Engineering that concerns designing for commercial "afterlife" – products, processes and systems should be designed for functionality in a commercial "afterlife". To reduce waste, components that remain functional and valuable can be recovered for reuse and/or reconfiguration. The next-generation design of products, processes and systems must take into account the functional use of the valuable properties of recovered components.

Considering the prices of modern instruments like mass, NMR or optical spectrometers, this objective seems to be obvious. However, it is not popular among practitioners and manufacturers. Whereas the attitude of the latter is understandable, the desire of the analyst to replace instrumentation every second year is hardly logical, even if the laboratory budget can accommodate it. Take gas chromatography, for example. The technique has been mature for 30 years. Chromatographs made in the seventies still perform well, and new models do not improve the quality of analysis. There is some irony in the situation: countries that produced gas chromatographs such as Czechoslovakia or the Soviet Union have disappeared, but the machines they produced are still being used in Eastern Europe.

The appearance of new instruments on the market can be welcomed if this is a reflection of a scientific breakthrough. An excellent example is the development of soft ionization methods in mass spectroscopy: MALDI and electrospray. The old chromatographs mentioned above require interfacing with personal computers and relevant software – a reflection of the information technological revolution that has occurred in the last two decades. The instrument markets are saturated. Nevertheless, a plethora of new models of standard scientific instruments appears on the market every year even though no new scientific discoveries have been implemented. Whether this is a manifestation of consumerism in analytical chemistry is outside the scope of this book. Nevertheless, Google searches such as "new *vs.* refurbished/reused chromatograph" or "new *vs.* refurbished spectrometer" do not produce any hits, indicating that this topic is not of much concern to the analytical community. This is understandable – most researchers find reassurance in the knowledge that they are the first ones to use the new equipment.

Nevertheless, the market for used analytical equipment has evolved in the last few years, with several websites offering used instruments for sale.[226] The list covers almost all types of analytical instrumentation. The website lists more than 30 companies which sell refurbished chromatographic instruments, including: gas chromatographs, ion chromatographs, reconditioned liquid chromatographs, used spectrometers (atomic absorption, inductively coupled plasma, UV-Vis and mass) electrophoresis instruments, laboratory balances, laboratory mixers, microscopes, DNA analyzers, centrifuges, fume hoods, pH meters, biotech laboratory fermenters, microplates, incubators, laboratory ovens, autoclaves, PCR thermal cyclers, DNA sequencers, flow cytometers, biological safety cabinets, gamma counters, scanning electron microscopes, elemental analyzers and used laboratory furniture. One such company is Analytical Instruments Recycle (CO, USA)[227] who states that its primary goal is "to reduce the cost of laboratory operations". Why purchase used equipment? The obvious answer is cost. Often, these instruments can be purchased for one half to two thirds of their original price. Whereas cost is the issue for most laboratories, the important aspect from the Green Analytical Chemistry perspective is the much smaller carbon footprint of refurbished instruments. Potential problems to be aware of when purchasing refurbished instruments are listed by one of the vendors of reused instruments:[228]

- Is the supplier reputable?
- Make sure that the company offers a minimum of a six-month warranty.
- Condition: unless you can personally inspect the product before you buy (and you know what to look for), you rely completely on the seller to accurately describe the instrument.
- Misrepresentation: incorrect information can be the result of ignorance or deception.
- Financial risk: this requires a level of trust on both sides – the seller does not want to send the goods without assurance of being paid, and the buyer does not want to release payment until he has received and inspected the merchandise.
- Trial: higher quality dealers will offer you a minimum of five days to accept your purchase.
- Wear/abuse: be sure to check the unit carefully for any damage – inside and out – before your acceptance period expires. Of course, normal wear and tear is to be expected, but signs of abuse are reasons to reconsider your equipment supplier.
- Assistance: companies cannot and should not be expected to provide technical support and expertise for instruments not acquired from their company.

The proponents of Green Analytical Chemistry can take comfort from the increasing popularity of refurbished instrumentation. A growing number of smaller laboratories and the increasing cost of scientific instrumentation cause one to question whether a new model of analytical instrument is required every year, as instrument manufacturing companies attempt to convince us. If a method is sufficiently mature, could instrument recycling be an option?

REFERENCES

1. W. Wardencki and J. Namieśnik, *Polish J. Environ. Studies*, 2002, **11**, 185.
2. J. Namiesnik and W. Wardencki, *J. High Resolut. Chromatogr*, 2000, **23**, 297.
3. M. Harper, *J. Chromatogr., A*, 2000, **885**, 129.

4. R. Sacks, H. Smith and M. Nowak, *Anal. Chem.*, 1998, **70**, 29A.
5. C. A. Cramers, H. -G. Jannssen, M. M. Van Deurse and P. A Leclerc, *J. Chromatogr., A*, 1999, **856**, 315.
6. Technical note: *Microfluidic Flow Control*; http://www.eksigent.com/
7. 1200 Series HPLC-Chip/MS System; http://www.chem.agilent.com/en-US/products/instruments/lc/1200serieshplc-chipmssystem/pages/default.aspx
8. T. Greibrokk and T. Andersen, *J. Chromatogr., A*, 2003, **1000**, 743.
9. P. Jandera, L. Blomberg and E. Lundanes, *J. Sep. Sci.*, 2004, **27**, 1402.
10. B. A. Jones, *J. Liq. Chromatogr. Relat. Technol.*, 2004, **27**, 1331.
11. C. R. Zhu, D. M. Goodall and S. A. C. Wren, *LC-GC Eur.*, 2004, **17**, 530.
12. C. R. Zhu, D. M. Goodall and S. A. C. Wren, *LC-GC N. Am.*, 2005, **23**, 54.
13. *J. Sep. Sci.* 2001, **24**
14. P. Sandra, D. Felix, S. M. B. Jones and J. Clark, *Talk at Pittcon*, 2005.
15. G. Vanhoenacker and P. Sandra, *J. Sep. Sci.*, 2006, **29**, 1822.
16. G. Vanhoenacker and P. Sandra, *Anal. Bioanal. Chem.*, 2008, **390**, 245.
17. H. Chen and Cs. Horváth, *Anal. Methods Instrum.*, 1993, **1**, 213.
18. F. D. Antia and Cs. Horvath, *J. Chromatogr.*, 1988, **435**, 1.
19. B. Yan, J. Zhao, J. S. Brown, J. Blackwell and P. W. Carr, *Anal. Chem.*, 2000, **72**, 1253.
20. G. Vanhoenacker and P. Sandra, *J. Chromatogr. A*, 2005, **1082**, 193.
21. Y. Yang, A. D. Jones and C. D. Eaton, *Anal. Chem.*, 1999, **71**, 3808.
22. J. Clark, W. Felix, B. Jones, S. Marin and N. Porter, in *Proceedings of the 57th Pittsburgh Conference on Analytical Chemistry and Applied Spectroscopy*, March 12–17, Orlando, FL, USA, 2006, pp. 1170–1172
23. R. M. Smith, R.J. Burgess, O. Chientavorn and J. R. Bone, *LC-GC N. Am.*, 1999, **17**, 938–945.
24. J. Coym and J. G. Dorsey, *J. Chromatogr., A*, 2004, **1035**, 23.
25. T. S. Kephart and P. K. Dasgupta, *Talanta*, 2002, **56**, 977.
26. T. Kondo and Y. Yang, *Anal. Chim. Acta*, 2003, **494**, 157.
27. R. M. Smith and R.J. Burgess, *Anal. Comm.*, 1996, **33**, 327.
28. R. M. Smith and R. J. Burgess, *J. Chromatogr., A*, 1997, **785**, 49.

29. R. J. Burgess and R. M Smith, in *the Proceedings of the 19th ISCC*, May 18–22, Wintergreen, USA, 1997, pp. 414.
30. B. A. Ingelse, H.-G. Janssen and C. A. Cramers, *J. High Resolut. Chromatogr.*, 1998, **21**, 613.
31. H. E. W. Hooijschuur, C. E. Kientz and U. A. T. Brinkman, *J. High Resolut. Chromatogr.*, 2000, **23**, 309.
32. D. Guillarme, S. Heinisch, J. Y. Gauvrit, P. Lanteri and J. L. Rocca, *J. Chromatogr., A*, 2005, **1078**, 22.
33. Y. Yang, T. Kondo and T. J. Kennedy, *J. Chromatogr. Sci.*, 2005, **43**, 518.
34. J. Thompson and P. W. Carr, *Anal. Chem.*, 2002, **74**, 1017.
35. S. M. Fields, C. Q. Ye, D. D. Zhang, B. R. Branch, X. J. Zhang and N. Okafo, *J. Chromatogr., A*, 2001, **913**, 197.
36. M. M. Sanagi, H. H. See, W. A. W. Ibrahim and A. Abu Naim, *J. Chromatogr., A*, 2004, **1059**, 95.
37. R. G. Wolcott, J. W. Dolan, L. R. Snyder, S. R. Bakalyar, M. A. Arnold and J. A. Nichols, *J. Chromatogr., A*, 2000, **869**, 211.
38. http://www.selerity.com; High Temperature HPLC Brochure
39. P. L. Zhu, J. W. Dolan, L. R. Snyder, D. W. Hill, L. Van Heukelem and T. J. Waeghe, *J. Chromatogr., A*, 1996, **756**, 51.
40. J. Kobayashi, A. Kikuchi, K. Sakai and T. Okano, *Anal. Chem.*, 2001, **73**, 2027.
41. H. Kanazawa, *J. Sep. Sci.*, 2007, **30**, 1646.
42. A. Kikuchi and T. Okano, *Prog. Polymer Sci.*, 2002, **27**, 165.
43. F. Lynen, J. M. D. Heijl, F. E. Du Prez, R. Brown, R. Szucs and P. Sandra, *Chromatographia*, 2007, **66**, 143.
44. W. R. Melander, B. K. Chen and C.s. Horváth, *J. Chromatogr.*, 1985, **318**, 1.
45. S. Goga, S. Heinisch, E. Lesellier, J. L. Rocca and A. Tchapla, *Chromatographia*, 2000, **51**, 536.
46. R. G. Wolcott, J. W. Dolan and L. R. Snyder, *J. Chromatogr., A*, 2000, **869**, 3.
47. J. Li, *Anal. Chim. Acta*, 1998, **369**, 21.
48. S. Heinisch and J.-L. Rocca, *J. Chromatogr., A*, 2009, **1216**, 642.
49. J. Nawrocki, C. Dunlap, A. McCormick and P. W. Carr, *J. Chromatogr., A*, 2004, **1028**, 1.
50. J. Nawrocki, C. Dunlap, J. Li, J. Zhao, C. V. McNeff, A. McCormick and P. W. Carr, *J. Chromatogr., A*, 2004, **1028**, 31.
51. E. Destandau and E. Lesellier, *Chromatographia*, 2008, **68**, 985.
52. Q. Jiang, H. Qiu, X. Wang, X. Liu and S. G. Zhang, *J. Liq. Chromatogr. Relat. Technol.*, 2008, **31**, 1448.

53. F. Tang, L. Tao, X. Luo, L. Ding, M. Guo, L. Nie and S. Yao, *J. Chromatogr., A*, 2006, **1125**, 182.
54. P. R. Griffiths, *Anal. Chem.*, 1988, **60**, 593A–597A.
55. L. Toribio, M. J. del Nozal, J. L. Bernal, C. Alonso and J. J. Jimenez, *J. Sep. Sci.*, 2006, **29**, 373–1378.
56. Y. Hsieh and F. Li, *Am. Pharm. Rev.*, 2007, **10**, 10–14.
57. L. T. Taylor, *Anal. Chem.*, 2008, **80**, 4285.
58. http://www.laboratoryequipment.com/article-the-greening-of-chromatography.aspx#
59. Angelo DePalma Genetic Engineering & Biotechnology News, Vol. 29, No. 10: http://www.genengnews.com
60. G. van der Vorst, H. van Langenhove, F. de Paep, W. Aelterman, J. Dingenen and J. Dewulf, *Green Chem.*, 2009, **11**, 1007.
61. P. Perrot, *A to Z of Thermodynamics*, Oxford University Press, 1998.
62. R. M. Smith, O. Chienthavorn, I. D. Wilson, B. Wright and S. D. Taylor, *Anal. Chem.*, 1999, **71**, 4443.
63. R. M. Smith, O. Chienthavorn, S. Saha, I. D. Wilson, B. Wright and S. D. Taylor, *J. Chromatogr., A*, 2000, **886**, 289.
64. D. Louden, A. Handley, S. Taylor, I. Sinclair, E. Lenz and I. D. Wilson, *Analyst*, 2001, **126**, 1625.
65. R. M. Smith, R. J. Burgess, O. Chienthavorn and R. Stuttard, *LC–GC Int.*, 1999, **1**, 30.
66. M. Albert, G. Cretier, D. Guillarme, S. Heinisch and J. L. Rocca, *J. Sep. Sci.*, 2005, **28**, 1803.
67. A. Loeser and P. Drumm, *J. Sep. Sci.*, 2006, **29**, 2847.
68. L. R. Snyder, *J. Chromatogr. Sci.*, 1970, **8**, 692.
69. T. Andersen, P. Molander, R. Trones, D. R. Hegna and T. Greibrokk, *J. Chromatogr., A*, 2001, **918**, 221.
70. J. W. Lee and P. C. Wankat, *Ind. Eng. Chem. Res.*, 2008, **47**, 9601.
71. J. W. Lee and P. C. Wankat, *Ind. Eng. Chem. Res.*, 2009, **48**, 7724.
72. A. de Villiers, F. Lestremau, R. Szucs, S. Gelebart, F. David and P. Sandra, *J. Chromatogr., A*, 2006, **1127**, 60.
73. M. E. Swartz, *Separation Science Redefined*, May 2005; http//www.chromatographyonline.com
74. H. Poppe, *J. Chromatogr., A*, 1997, **778**, 3.
75. M. E. Swartz, Contemporary Liquid Chromatographic Systems for Method Development, In *HPLC Method Development for Pharmaceuticals*, Ed. S. Ahuja and H. Rasmussen, Elsevier, Academic Press, Amsterdam, 2007, p. 1145–187.
76. T. Wu, C. Wang, X. Wang, H. Xiao, Q. Ma and Q. Zhang, *Chromatographia*, 2008, **68**, 803.

77. S. Chen and A. Kord, *J. Chromatogr., A*, 2009, **1216**, 6204.
78. F. Gritti and G. Guiochon, *Anal. Chem.*, 2008, **80**, 5009.
79. Anon,"Reducing Acetonitrile Usage", *Chromatography online*, Jan 9, 2009
80. J. Castro-Perez, R. Plumb, J. H. Granger, I. Beattie, K. Joncour and A. Wright, *Rapid Commun. Mass Spectrom.*, 2005, **19**, 843.
81. I. D. Wilson, J. K. Nicholson, J. Castro-Perez, J. H. Granger, K. A. Johnson, B. W. Smith and R. S. Plumb, *J. Proteome Res.*, 2005, **4**, 591.
82. M. Mezcua, A. Agüera, J. L. Lliberi, M. A Corés, B. Bagó and A. R. Fernández-Alba, *J. Chromatogr., A.*, 2006, **1109**, 222.
83. S. A. C. Wren and P. Tchelitcheffa, *J. Chromatogr., A*, 2006, **1119**, 140.
84. L. G. Apollonioa, D. J. Piancab, I. R. Whittallb, W. A. Mahera and J. M. Kyda, *J. Chromatogr., B: Biomed. Appl.*, 2006, **836**, 111.
85. J. Z. Min, T. Toyo'oka, T. Kurihara, M. Kato, T. Fukushima and S. Inagaki, *Biomed. Chromatogr.*, 2007, **21**, 852.
86. S. J. Bruce, P. Jonsson, H. Antti, O. Cloarec, J. Trygg, S. L. Marklund and T. Moritz, *Anal. Biochem.*, 2008, **372**, 237.
87. G. F. Wang, Y. S. Hsieh, K. C. Cheng, R. A. Morrison, S. Venkatraman, F. G. Njoroge, L. Heimark and W. A. Korfmacher, *J. Chromatogr., B: Biomed. Appl.*, 2007, **852**, 92.
88. H. Licea-Perez, S. Wang, C. L. Bowen and E. Yang, *J. Chromatogr., B: Biomed. Appl.*, 2007, **852**, 69.
89. X. Y. Wang, T. Zhao, X. F. Gao, M. Dan, M. M. Zhou and W. Jia, *Anal. Chim. Acta*, 2007, **594**, 265.
90. O. Y. Al-Dirbashi, T. Santa, K. Al-Qahtani, M. Al-Amoudi and M. S. Rashed, *Rapid Commun. Mass Spectrom.*, 2007, **21**, 1984.
91. R. G. Kay, B. Gregory, P. B. Grace and S. Pleasance, *Rapid Commun. Mass Spectrom.*, 2007, **21**, 2585.
92. L. Y. Chen, F. Qin, Y. Y. Ma and F. M. Li, *J. Chromatogr., B: Biomed. Appl.*, 2007, **855**, 255.
93. N. Rehmann, P. Hess and M. A Quilliam, *Rapid Commun. Mass Spectrom.*, 2008, **22**, 549.
94. M. Kalovidouris, S. Michalea, N. Robola, M. Koutsopoulou and I. Panderi, *Rapid Commun. Mass Spectrom.*, 2006, **20**, 2939.
95. A. Kaufmann, P. Butcher, K. Maden and M. Widmer, *Anal. Chim. Acta*, 2007, **586**, 13.
96. Y. Y. Ma, F. Qin, X. H. Sun, X. M. Lu and F. M. Li, *J. Pharm. Biomed. Anal.*, 2007, **43**, 1540.
97. E. Fux, D. McMillan, T. Bire and P. Hess, *J. Chromatogr., A*, 2007, **1157**, 273.

98. D. O'Connor and R. Mortishire-Smith, *Anal. Bioanal. Chem.*, 2006, **385**, 114.
99. A. Fekete, M. Frommberger, M. Rothballer, X. J. Li, M. Englmann, J. Fekete, A. Hartmann, L. Eberl and P. Schmitt-Kopplin, *Anal. Bioanal. Chem.*, 2007, **387**, 455.
100. H. Kawanishi, T. Toyo'oka, K. Ito, M. Maeda, T. I. Hamada, T. Fukushima, M. Kato and S. Inagaki, *J. Chromatogr., A*, 2006, **1132**, 148.
101. G. J. Dear, A. D. James and S. Sarda, *Rapid Commun. Mass Spectrom.*, 2006, **20**, 1351.
102. X. Q. Li, Z. L. Xiong, X. X. Ying, L. C. Cui, W. L. Zhu and F. M. Li, *Anal. Chim. Acta*, 2006, **580**, 170.
103. R. P. Li, L. L. Dong and J. X. Huang, *Anal. Chim. Acta*, 2005, **546**, 167.
104. R. S. Plumb, K. A. Johnson, P. Rainville, J. P. Shockcor, R. Williams, J. H. Granger and I. D. Wilson, *Rapid Commun. Mass Spectrom.*, 2006, **20**, 2800.
105. E. C. Y. Chan, S. L. Yap, A. J. Lau, P. C. Leow, D. F. Toh and H. L. Koh, *Rapid Commun. Mass Spectrom.*, 2007, **21**, 519.
106. K. Yu, D. Little, R. Plumb and B. Smith, *Rapid Commun. Mass Spectrom.*, 2006, **20**, 544.
107. T. Yamashita, Y. Dohta, T. Nakamura and T. Fukami, *J. Chromatogr., A*, 2008, **1182**, 72.
108. L. S. New, S. Saha, M. M. K. Ong, U. A. Boelsterli and E. C. Y. Chan, *Rapid Commun. Mass Spectrom.*, 2007, **21**, 982.
109. K. Yu, L. Di, E. Kerns, S. Q. Li, P. Alden and R. S. Plumb, *Rapid Commun. Mass Spectrom.*, 2007, **21**, 893.
110. K. Y. Li, Y. G. Zhou, H. Y. Ren, F. Wang, B. K. Zhang and H. D. Li, *J. Chromatogr., B: Biomed. Appl.,* 2007, **850**, 581.
111. L. G. Apollonio, I. R. Whittall, D. J. Pianca, J. M. Kyd and W. A. Maher, *Rapid Commun. Mass Spectrom.*, 2006, **20**, 2259.
112. G. F. Wang, Y. Hsieh, X. M. Cui, K. C. Cheng and W. A. Korfmacher, *Rapid Commun. Mass Spectrom.*, 2006, **20**, 2215.
113. Y. Pico, M. Farre, C. Soler and D. Barcelo, *J. Chromatogr., A*, 2007, **1176**, 123.
114. Y. Zhang, J. J. Jiao, Z. X. Cai, Y. Zhang and Y. P. Ren, *J. Chromatogr., A*, 2007, **1142**, 194.
115. J. Wang and D. Leung, *Rapid Commun. Mass Spectrom.*, 2007, **21**, 3213.
116. C. C. Leandro, P. Hancock, R. J. Fussell and B. J. Keely, *J. Chromatogr., A*, 2007, **1144**, 161.

117. L. Bendahl, S. H. Hansen, B. Gammelgaard, S. Sturup and C. Nielsen, *J. Pharm. Biomed. Anal.*, 2006, **40**, 648.
118. M. Ventura, D. Guillen, I. Anaya, F. Broto-Puig, J. L. Lliberia, M. Agut and L. Comellas, *Rapid Commun. Mass Spectrom.*, 2006, **20**, 3199.
119. E. Barcelo-Barrachina, E. Moyano, M. T. Galceran, J. L. Lliberia, B. Bago and M. A. Cortes, *J. Chromatogr., A*, 2006, **1125**, 195.
120. T. Kovalczuk, M. Jech, J. Poustka and J. Hajslova, *Anal. Chim. Acta*, 2006, **577**, 8.
121. Y. P. Ren, Y. Zhang, S. L. Shao, Z. X. Cai, L. Feng, H. F. Pan and Z. G. Wang, *J. Chromatog., A*, 2007, **1143**, 48.
122. C. C. Leandro, P. Hancock, R. J. Fussell and B. J. Keely, *J. Chromatogr., A*, 2006, **1103**, 94.
123. M. R. Boleda, M. T. Galceran and F. Ventura, *J. Chromatogr., A*, 2007, **1175**, 38.
124. M. Farre, M. Gros, B. Hernandez, M. Petrovic, P. Hancock and D. Barcelo, *Rapid Commun. Mass Spectrom.*, 2008, **22**, 41.
125. B. Kasprzyk-Hordem, R. M. Dinsdale and A. J. Guwy, *J. Chromatogr., A*, 2007, **1161**, 132.
126. L. G. Apollonio, D.J. Pianca, I. R. Whittall, W. A. Maher and J. M. Kyd, *J. Chromatogr., B: Biomed. Appl.*, 2006, **836**, 111.
127. M. Mezcua, A. Aguera, J. L. Lliberia, M. A. Cortes, B. Bago and A. R. Fernandez-Alba, *J. Chromatogr., A*, 2006, **1109**, 222.
128. M. Petrovic, M. Gros and D. Barcelo, *J. Chromatogr., A*, 2006, **1124**, 68.
129. Y. Wu, J. R. Engen and W. B. Hobbins, *J. Am. Soc. Mass Spectrom.*, 2006, **17**, 163.
130. M. Farre, M. Kuster, R. Brix, F. Rubio, M. J. L. de Alda and D. Barcelo, *J. Chromatogr., A*, 2007, **1160**, 166.
131. D. O'Connor, R. Mortishire-Smith, D. Morrison, A. Davies and M. Dominguez, *Rapid Commun. Mass Spectrom.*, 2006, **20**, 851.
132. J. Castro-Perez, R. Plumb, J. H. Granger, L. Beattie, K. Joncour and A. Wright, *Rapid Commun. Mass Spectrom.*, 2005, **19**, 843.
133. R. Plumb, J. Castro-Perez, J. Granger, I. Beattie, K. Joncour and A. Wright, *Rapid Commun. Mass Spectrom.*, 2004, **18**, 2331.
134. I. D. Wilson, J. K. Nicholson, J. Castro-Perez, J. H. Granger, K. A. Johnson, B. W. Smith and R. S. Plumb, *J. Proteome Res.*, 2005, **4**, 591.
135. A. Tullo, *Chem. Eng. News*, 2008, **86**, 27.
136. Sigma-Aldrich Technical Report: *Approaches to Lessening the Impact of the Acetonitrile Shortage on Your Reversed-Phase HPLC Separations, 2009*; http//:www.sigmaaldrich.com

137. M. Harding, *J. Pharm. Pharmacol.*, 2008, **60**, A67 (Suppl. 1 Meeting Abstr. 167).
138. J. Kelsey, *The Column*, 2009, **5**, 15.
139. C. J. Welch, T. Brkovic, W. Schafer and X. Gong, *Green Chem.*, 2009, **11**, 1232.
140. F. F. Reuss, *Mem. Soc. Imp. Naturalistes Moscou*, 1809, **2**, 327.
141. A. V. Pertsov and E. A. Zaitseva (Baum), in *III International Conference on Colloid Chemistry and Physicochemical Mechanics*, Moscow, Russia, 24–28th of June, 2008.
142. F. Kohlrausch, *Ann. Phys. Chem.(Leipzig)*., 1897, **62**, 209.
143. A. W. K. Tiselius, *The Moving-Boundary Method of Studying the Electrophoresis of Proteins*, PhD thesis, University of Uppsala, Sweden, 1930.
144. S. Hjertén, *Chromatogr. Rev.*, 1967, **9**, 122.
145. F. M. Everaerts and W. M. L. Hoving-Keulemans, *Sci. Tools.*, 1970, **17**, 25.
146. R. Virtanen, *Acta Polytech. Scand.*, 1974, **123**, 1.
147. J. W. Jorgenson and K. D. Lukacs, *Anal. Chem.*, 1981, **53**, 1298.
148. Performing search with SciFinder®™ Scholar™ program (ACS) on the keyword: "capillary electrophoresis" resulted in about 30 000 hits.
149. A. Jouyban and E. Kenndler, *Electrophoresis*, 2006, **27**, 992.
150. S. Terabe, K. Otsuka, K. Ichikawa, A. Tsuchiya and T. Ando, *Anal. Chem.*, 1984, **56**, 111.
151. A. J. Zemann, E. Schnell, D. Volgger and G. K. Bonn, *Anal. Chem.*, 1998, **70**, 563.
152. J. A. F. da Silva and C. L. do Lago, *Anal. Chem.*, 1998, **70**, 4339.
153. A. Seiman, M. Vaher and M. Kaljurand, *J. Chromatogr., A*, 2008, **1189**, 266.
154. L. Xu, P. C. Hauser and H. K. Lee, *J. Chromatogr., A*, 2009, **1216**, 5911.
155. A. Schuchert-Shi and P. C. Hauser, *Anal. Biochem*, 2009, **387**, 202.
156. W. Pormsila, S. Krahenbuhl and P. C. Hauser, *Anal. Chim. Acta*, 2009, **636**, 224.
157. T. Kappes, P. Schnierle and P. C. Hauser, *Anal. Chim. Acta*, 1999, **393**, 77.
158. P. Kuban, H. T. A. Nguyen, M. Macka, P. R. Haddad and P. C. Hauser, *Electroanalysis*, 2007, **19**, 2059.
159. A. Seiman, M. Jaanus, M. Vaher and M. Kaljurand, *Electrophoresis*, 2009, **30**, 507.
160. T. Knjazeva, M. Kulp and M. Kaljurand, *Electrophoresis*, 2009, **30**, 424.

161. J. Tanyanyiwa, S. Leuthardt and P. C. Hauser, *J. Chromatogr., A*, 2002, **978**, 205.
162. J. P. Hutchinson, C. J. Evenhuis, C. Johns, A. A. Kazarian, M. C. Breadmore, M. Macka, E. F. Hilder, R. M. Guijt, G. W. Dicinoski and P. R. Haddad, *Anal. Chem.*, 2007, **79**, 7005.
163. J. P. Hutchinson, C. Johns, M. C. Breadmore, E. F. Hilder, R. M. Guijt, C. Lennard, G. Dicinoski and P. R. Haddad, *Electrophoresis*, 2008, **29**, 4593.
164. G. V. Kaigala, M. Behnam, C. Bliss, M. Khorasani, S. Ho, J. N. McMullin, D. G. Elliott and C. J. Backhouse, *IET Nanobiotechnol.*, 2009, **3**, 1.
165. G. V. Kaigala, V. N. Hoang, A. Stickel, J. Lauzon, D. Manage, L. M. Pilarski and C. J. Backhouse, *Analyst*, 2008, **133**, 331.
166. A. Manz, N. Graber and H. M. Widmer, *Sens. Actuators, B*, 1990, **1**, 244.
167. D. R. Reyes, D. Iossifidis, P.-A. Auroux and A. Manz, *Anal. Chem.*, 2002, **74**, 2623.
168. P.-A. Auroux, D. Iossifidis, D. R. Reyes and A. Manz, *Anal. Chem.*, 2002, **74**, 2637.
169. S. A. Soper, S. M. Ford, S. Qi, R. L. McCarley, K. Kelly and M. C. Murphy, *Anal. Chem.*, 2000, **72**, 642.
170. T. D. Boone, Z. H. Fan, H. H. Hooper, A. J. Ricco, H. Tan and S. J. Williams, *Anal. Chem.*, 2002, **74**, 78.
171. keyworld "microfluidics" returns more than 4000 hits in the ISI Web of Science database search.
172. C. K. Fredrickson and Z. H. Fan, *Lab Chip*, 2004, **4**, 526.
173. Y. Xia and G. M. Whitesides, *Angew. Chem., Int. Ed. Engl.*, 1998, **37**, 551.
174. J. Wang, *Talanta*, 2002, **56**, 223.
175. G. Jesson, G. Kylberg and P. Andersson, in *Micro Total Analysis Systems 2003*, ed. M. A. Nothrup, K. F. Jensen and D. J. Harrison, pp. 155–158
176. V. Nittis, R. Fortt, C. H. Legge and A. J. de Mello, *Lab Chip*, 2001, **1**, 128.
177. A. Puntambekar and C. H. Ahn, *J. Micromech. Microeng.*, 2002, **12**, 35.
178. C. Gonzalez, S. D. Collins and R. L. Smith, *Sens. Actuators, B*, 1998, **49**, 40.
179. H. Chen, D. Acharya, A. Gajraj and J.-C. Melners, *Anal. Chem.*, 2003, **75**, 5287.
180. J. M. Ramsey, *Nat. Biotechnol*, 1999, **17**, 1061.

181. S. Attiya, A. B. Jemere, T. Tang, G. Fitzpatrick, K. Seiler, N. Chiem and D. J. Harrison, *Electrophoresis*, 2001, **22**, 318.
182. N. H. Bings, C. Wang, C. D. Skinner, C. L. Colyer, P. Thibault and D. J. Harrison, *Anal. Chem.*, 1999, **71**, 3292.
183. Z. Yang and R. Maeda, *Electrophoresis*, 2002, **23**, 3474.
184. J. Liu, C. Hansen and R. Q. Stephen, *Anal. Chem.*, 2003, **75**, 4718.
185. Y. Wang, W.-Y. Lin, K. Liu, R. J. Lin, M. Selke, H. C. Kolb, N. Zhang, X.-Z. Zhao, M. E. Phelps, C. K. F. Shen, K. F. Faull and H.-R. Tseng, *Lab Chip*, 2009, **9**, 2281.
186. J.-H. Wang and Z.-L. Fang, *Chin. J. Anal. Chem*, 2004, **32**, 1401.
187. M. Kulp, M. Vaher and M. Kaljurand, *J. Chromatogr., A*, 2005, **1100**, 126.
188. Y. H. Lin, G. B. Lee, C. W. Li, G.-R. Huang and S. H. Chen, *J. Chromatogr., A*, 2001, **937**, 115.
189. H. Liu and P. K. Dasgupta, *Anal. Chem.*, 1997, **69**, 1211.
190. C.-G. Fu and Z.-L. Fang, *Anal. Chim. Acta*, 2000, **422**, 71.
191. X.-J. Huang, Q.-S. Pu and Z.-L. Fang, *Analyst*, 2001, **126**, 281.
192. X.-D. Cao, Q. Fang and Z.-L. Fang, *Anal. Chim. Acta*, 2004, **513**, 473.
193. S. L. Wang, X. J. Huang, Z.-L. Fang and P. K. Dasgupta, *Anal. Chem.*, 2001, **73**, 4545.
194. S. Attiya, A. Jemere, T. Tang, G. Fitzpatrick, K. Seiler, N. Chiem and D. J. Harrison, *Electrophoresis*, 2001, **22**, 318.
195. K. Ohno, K. Tachikawa and A. Manz, *Electrophoresis*, 2008, **29**, 4443.
196. R. Mukhopadhyay, *Anal. Chem.*, 2009, **81**, 4172.
197. H. Becker, *Lab Chip*, 2009, **9**, 2119.
198. A. W. Martinez, S. T. Phillips and G. M. Whitesides, *Proc. Natl. Acad. Sci. U. S. A.*, 2008, **105**, 19606.
199. B. Weigl, G. Domingo, P. LaBarre and J. Gerlach, *Lab Chip*, 2008, **8**, 1999.
200. L. Yao, B. Liu, T. Chen, S. Liu and T. Zuo, *Biomed. Microdevices.*, 2005, **7**, 253.
201. R. M. Guijt, A. Dodge, G. W. van Dedem, N. F. de Rooij and E. Verpoorte, *Lab Chip*, 2003, **3**, 1.
202. B. H. Weigl and P. Yager, *Science*, 1999, **283**, 346.
203. A. E. Kamholz, B. H. Weigl, B. A. Finlayson and P. Yager, *Anal. Chem.*, 1999, **71**, 5340.
204. W. Dungchai, O. Chailapakul and C. S. Henry, *Anal. Chem.*, 2009, **81**, 5821.
205. J. Berthier, in *Microdrops and Digital Microfluidics*, ed. W. Andrew, Applied Sci. Publishers, NY, 2008, p. 2.

206. A. A. Darhuber, J. P. Valentino, S. M. Troian and S. Wagner, *J. Microelectromech. Syst.*, 2003, **12**, 873.
207. A. Renaudin, P. Tabourier, V. Zhang, J. C. Camart and C. Druon, *Sens. Actuators, B*, 2006, **113**, 389.
208. J. A. Schwartz, J. V. Vykoukal and P. R. C. Gascoyne, *Lab Chip*, 2004, **4**, 11.
209. Y.-M. Jung and I. S. Kang, *Biomicrofluidics*, 2009, **3**, 022402-1.
210. T. Gilet, D. Terwagne and N. Vandewalle, *Appl. Phys. Lett.*, 2009, **95**, 014106-1.
211. M. G. Pollack, A. D. Shenderov and R. B. Fair, *Lab Chip*, 2002, **2**, 96.
212. R. B. Fair, *Microfluidics Nanofluidics*, 2007, **3**, 245.
213. G. Lippmann, *Ann. Chim. Phys.*, 1875, **5**, 494.
214. B. Berge, *C. R. Acad. Sci. Paris, Ser. II*, 1993, **317**, 157.
215. R. I. Shamai, D. Andelman, B. Berge and R. Hayes, *Soft Matter*, 2008, **4**, 38.
216. M. G. Pollack, R. B. Fair and A. D. Shenderov, *Appl. Phys. Lett.*, 2000, **77**, 1725.
217. P. Dubois, G. Marchand, Y. Fouillet, J. Berthier, T. Douki, F. Hassine, S. Gmouh and M. Vaultier, *Anal. Chem.*, 2006, **78**, 4909.
218. R. B. Fair, A. Khlystov, T. D. Tailor, V. Ivanov, R. D. Evans, V. Srinivasan, V. K. Pamula, M. G. Pollack, P. B. Griffin and J. Zhou, *IEEE Design Test Computers*, 2007, **1**, 10–24.
219. M. Abdelgawad and A. R. Wheeler, *Microfluidics Nanofluidics*, 2008, **4**, 349.
220. J.-Y. Yoon and D. J You, *J. Biol. Eng.*, 2008, **2**, 15.
221. A. A. Gharizadeh, B. Gustafsson, A. C. Sterky, F. Nyren, P. Uhlen and M. Lundeberg, *J. Anal. Biochem.*, 2000, **280**, 103.
222. M. Abdelgawad, M. W. L. Watson and A. R. Wheeler, *Lab Chip*, 2009, **9**, 1046.
223. J. Gorbatsova, M. Jaanus and M. Kaljurand, *Anal. Chem.*, 2009, **81**, 8590.
224. P. T. Anastas and J. C. Warner, *Green Chemistry Theory and Practice*, Oxford University Press, New York, 1998.
225. L. H. Keith, L. U. Gron and J. L. Young, *Chem. Rev.*, 2007, **107**, 2695.
226. see for example. "*Used Equipment Repair, Dealer Distributors/ Analytical Chemistry, Chromatography, Spectroscopy*"; http:// www.medibix.com/CompanySearch.jsp?cs_choice = c&clt_choice = t&treepath = 12392&stype = i
227. http:/www.aironline.com/company/overview
228. D. Gleb, *New vs. Refurbished. What to Consider When Purchasing Test Equipment*; http://www.testsolu.com/html/articles.php

CHAPTER 6

Greening Analytical Chemistry by Improving Signal Acquisition and Processing

Surprisingly, chemometrics is not yet recognized as the ultimate green approach to chemical analysis. Its relevance to green analytical chemistry remains to be proved. A rare exception is the paper by Armenta, Garrigues and de la Guardia,[1] who wrote that "the evolution of the available instrumentation and mathematical data treatments (chemometrics) has allowed the development of solvent-free methodologies based on direct measurements of solid or liquid samples without any chemical sample pre-treatment". The examples they give include methodologies based on direct measurement and chemometric data treatment of vibrational spectrometric data – near infrared (NIR),[2] mid-infrared (mid-IR) or Raman spectrometry, fluorescence, UV–Vis spectroscopy and nuclear magnetic resonance (NMR). They note that the main advantage of these methods is their avoidance of sample pretreatment, thus reducing the use of solvents and reagents, and also the analysis time. Chemometrics can significantly improve the performance of these methods. For example, the use of transmittance measurements in NIR spectroscopy and partial least squares calibration has proved to be a powerful tool for the determination of the peroxide index in edible oil without using solvents or derivatisation reagents.[3]

Green Analytical Chemistry
By Mihkel Koel and Mihkel Kaljurand

Published by the Royal Society of Chemistry, www.rsc.org

6.1 CHEMOMETRICS AND PROCESSING ANALYTICAL RESULTS

A definition of chemometrics is provided by Otto:[4] "chemometrics is the chemical discipline that uses mathematical and statistical methods to design or select optimal measurement procedures and experiments, and to provide maximum chemical information by analyzing chemical data". Chemometrics may be one route to the ultimate green analytical chemistry since it enables results of analyses to be obtained *via* calculations only, thus reducing the amount of actual measurement. Even if one feels that "reducing the amount of measurement" is overstated, chemometrics frequently enables results to be obtained by using much simpler measurement processes. A good example is recording individual concentrations and spectra of mixtures using excitation–emission spectroscopy with a chemometric technique known as PARAFAC, which will be explained below.

Chemometrics is used to solve problems involving large amounts of data. Within process-analysis and monitoring, chemical analysis, spectroscopy, molecular modelling, sensory analysis and many other fields, large data-tables are produced that need to be analyzed and visualized in order to properly understand the problem. The most popular methods for exploring data, and building calibration; regression and classification models, are principal component analysis (PCA), partial least squares regression (PLS), and several variants of multivariate classification. Chemometrics consists of various subfields, such as descriptive and inference statistics, signal processing, experimental design, modelling, optimization, pattern recognition, classification, artificial intelligence methods, image processing, and system theory.

The essence of chemometrics and one of the most popular chemometric techniques is principal component/factor analysis which is a method of exploratory data analysis, *i.e.* examining relationships between samples (patients, food samples, organisms, chromatographic columns, spectra) and between variables (compound concentrations, spectral peaks, chromatographic peak areas, elemental compositions). The initial data are usually sample measurements presented in a table, where rows represent samples (*e.g.* patients, animals) and columns represent different properties (elemental concentrations, retention times, *etc.*). For example, if x_{ij} is a j-th characteristic of i-th sample, the data table A will be subjected to PCA as follows:[1]

[1] We assume that the reader has a basic knowledge of matrix operations. A good introduction to that field is a paper by B. M. Wise and N. B. Gallagher, An Introduction to Linear Algebra, *Crit. Rev. Anal. Chem.*, 1998, **28**, 1.

$$A = \begin{bmatrix} x_{1,1} & x_{1,2} & \cdots & x_{1,n} \\ x_{2,1} & x_{2,2} & \cdots & x_{2,n} \\ \vdots & \vdots & \ddots & \vdots \\ x_{m,2} & x_{m,2} & \cdots & x_{m,n} \end{bmatrix}$$

where m is the number of samples and n is the number of performed measurements. PCA attempts to decompose the initial data table, (*i.e.* the data matrix in the terminology of linear algebra) into the product of two matrices – T (matrix of scores) and P (a matrix of loads) as $A = TP^T$. The upper index denotes a matrix transpose. The aim of decomposition is to reduce the number of columns in T and P, to k, for example, so that it is much less than the number of initial measurements, *i.e.* $k \ll n$. If this can be done, it means that the different columns (*i.e.* measurements) are strongly correlated and that the data structure is best described not by n measurements, but by a much smaller number of "latent" factors equal to k. Thus, with information about the independent latent variables, the data can be greatly reduced. However, the chemical nature of those variables is not disclosed by the PCA procedure, and further mathematical operations are necessary to convert the abstract factors into physically meaningful variables. Because PCA is straightforward, mathematical operations can be performed easily with available software. The latter operation – factor analysis – is a heuristic technique that requires chemical intuition. Nevertheless, it is in the spirit of green analytical chemistry because no experimentation with chemicals is involved.

It can be easily demonstrated how qualitative and quantitative analysis is possible using factor analysis of HPLC–diode array detector data (DAD). The result of HPLC–DAD measurement is a chromatospectrogram, which is a table of the recorded absorbances of different compounds of various concentrations measured on different wavelengths. After the factor analysis procedure, this table is decomposed into chromatographic peaks of compounds that have eluted together from the column and are each associated with corresponding UV spectrum. Computation algorithms that perform such decompositions are rather elaborate and, as mentioned previously, chemical intuition is involved in many steps of factor analysis, rendering the procedure subjective and sometimes precluding a unique result. Even worse, no result might be forthcoming at all. On the other hand, if decomposition is successful, performing separations with different columns and mobile phases can be avoided and much of the analysis process can be done *in silico*. There are a large number of applications of mathematical resolution of chromatographic peaks. The ISI Web of Science yields more than 300 hits on the keywords "curve resolution" and "factor analysis". Despite the problem of multiple (in fact, an infinite number of) solutions, the resolution algorithms are available in commercial software.

A popular topic in PCA applications is the grouping and classification of datasets that can display complex data as groupings on a simple plot. PCA finds the greatest variation in the data set and plots the data in a way that displays this variation. In this approach, only two or three factors, at most, that are responsible for the greatest variation of the data are taken into account. If the principal components of the data set are found, the samples are represented as points on the plane of the coordinates, which are determined, for example, by the two most significant (in terms of the variability of the data) factors (see Figure 6.1).

The structure underlying the data reveals itself by visual inspection of the points because PCA distinguishes similarities and differences immediately. Thousands of PCA applications have been published and we will not describe them here. The Web of Science reveals 2500 hits on the keyword "principal component analysis". Applications cover all aspects of analytical chemistry: biochemical, food, forensic, environmental analysis, *etc*. Various chemometric techniques have also been used to reduce analysis times in multicomponent systems.[5] In this research, several algorithms were compared for the selection of optimal UV absorption wavelengths for analysis and prediction of reactant and product concentrations involved in the enzymatic synthesis of ampicillin. The results compared favourably with HPLC data.

6.1.1 EEM Spectroscopy and Chemometrics

More than other methods, chemometrics has played a special role in the development of EEM, as demonstrated by numerous authors. Chemometritians have proved that collecting 2D fluorescence spectra of various samples significantly increases the selectivity of the method. In many cases, it has become possible to obtain information on individual components in a sample – both spectral (identity) and quantitative (concentrations) – without separating the sample into components. Chemometrics in EEM spectroscopy is usually called multi-way analysis. EEM fluorescence spectroscopy, in conjunction with multi-way spectral deconvolution, can be used to simultaneously determine multiple analytes in the presence of unknown, uncalibrated interferents.

The algorithms are too complex to be described here. However, the basic concept of multi-way analysis can be presented. As we saw in

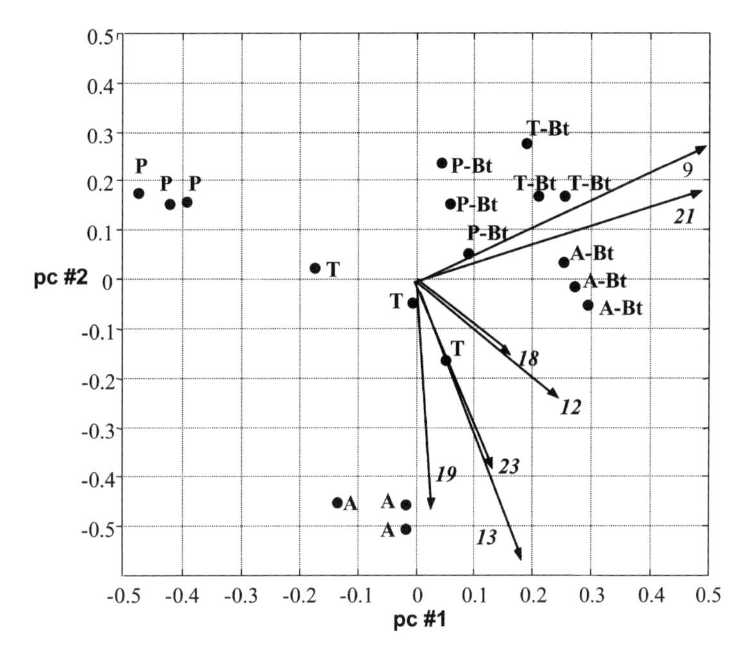

Figure 6.1 CE-MS electropherograms of maize varieties in the first principal component coordinates (accounting for 70% variability). "Aristis", "R33P66" and "Tietar" are the names of the varieties, and "Aristis-Bt", "R33P66-Bt" and "Tietar-Bt" correspond to their respective transgenic varieties. Vectors are loadings that indicate the metabolites responsible for the distinction of wild and transgenic varieties.[5] (Reproduced with kind permission of the American Chemical Society).

Chapter 4.4.2, in EEM spectroscopy, a total fluorescence spectrum for a sample is obtained by varying the excitation photons wavelengths λ_i and measuring the fluorescence of a sample at certain other wavelengths λ_j to collect the resultant i by j data matrix with a CCD camera. Each of the ith rows in the EEM spectrum is the emission spectrum at the ith excitation wavelength. Each of the j columns in the EEM spectrum is the excitation spectrum at the jth emission wavelength. If the sample consists of several, for example, n compounds, then it can easily be shown[6] that the so-called bilinearity property of EEM data holds: if EEM data is represented as a matrix D, then it can be represented as a multiplication of excitation X_k and emission M_k profiles (as vectors) of individual components as follows:

$$D = X_1 M_1^T + X_2 M_2^T + \dots + X_n M_n^T + E \tag{1}$$

where T means the transpose of a matrix as is common in linear algebraic notations, and E accounts for the part of the data that can not be represented by excitation/emission profiles (noise, scattering, *etc.*). Note that bilinearity is a feature of many methods of analytical chemistry, especially in handling data from hyphenated instruments. HPLC with diode array detector (DAD) is a textbook example where X represents instant concentrations of components eluting into DAD and M is their absorption spectra. If the excitation/emission profiles of the individual components are known, then Equation (1) allows the EEM to be predicted easily. This direct task is rarely of interest, however, and what is usually measured is D, in the hope that X_k and M_k can be estimated. As was previously mentioned, the solution is not unique: within a given noise intensity, an infinite number of spectra will suit Equation (1) in terms of least squares. Geometrically, D can be represented as a data plane.

Stacking multiple EEMs from various samples with different concentrations or time series from the same (but varying) concentrations of the components creates a data "cube" that can be mathematically broken down into a set of tri-linear components. Equation (1) becomes the following equation of three vectors, with the help of a less familiar matrix operation, as follows:

$$\hat{D} = C_1 \otimes (X_1 M_1^T) + C_2 \otimes (X_1 M_2^T) + \ldots + C_n \otimes (X_n M_n^T) + E \qquad (2)$$

where the symbol "$\otimes$" means "outer product", and element of $\hat{D}$, $\hat{D}_{i,j,k}$, is the fluorescence intensity of sample k at the excitation wavelength i and the emission wavelength j, columns X, M, and C are the estimates of the pure excitation, emission, and concentration profiles, respectively, and E is the residual error matrix from fitting the model to the collected data.[2] PARAFAC is an algorithm for applying the model in Equation (2) to collections of EEM data.[7] It is an iterative, least squares type of algorithm that can be used to solve the above equation by simultaneously extracting multiple pure spectral profiles from multi-way data even in the presence of unknown, uncalibrated interferents.[7] The inherent excitation (X) and emission (M) profiles and the resolved relative concentrations (C) of the n components in the samples can be recorded, because mathematical proofs show that this decomposition of $\hat{D}$ to X, M and C is unique except for a scaling factor.[8] Therefore, provided that no more than the correct number of components is extracted, the profiles are determined uniquely up to permutations and scaling. This allows for resolving pure chemical spectra.

[2] The data structure in Equation (2) deserves a few words. C_i, X_i, and M_i are vectors characterizing component i in the sample. The way the product $X_i M_i^T$ is written suggest that it is a matrix and that the outer product of this matrix with C_i, vector, C_i, $\otimes(X_i^T M_i)$ results in a tensor, which can be imagined as a set of matrices arranged in the form of a cube.

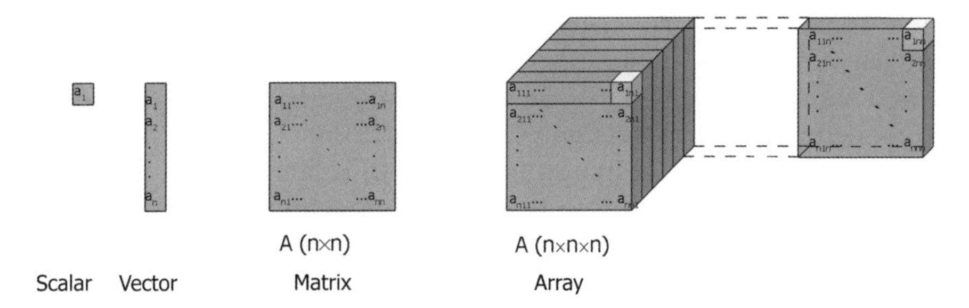

Figure 6.2 Representation of data structures with the corresponding order in parentheses. A-object, n-number of rows, columns and layers.

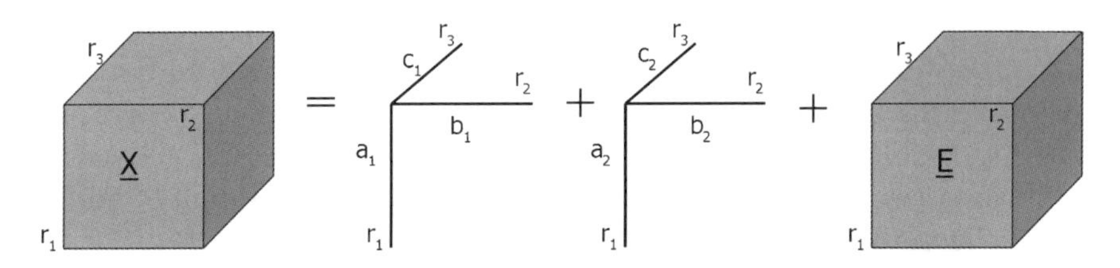

Figure 6.3 Geometrical explanation of PARAFAC decomposition. r_1, r_2, r_3 - number of rows, columns and layers; a_i, b_i, c_i - denote independent variables (*e.g*, concentrations, wavelengths *etc.*)

When algebraic notations are too abstract, geometrical images may help clarify the concepts related to multivariate analysis. In Figure 6.2, possible data structures and their dimensions, encountered in science, are represented. A single measurement (number) is represented as a dot, a collection of dots is a line (vector, *e.g.* a chromatogram), a collection of vectors is a square in a plane (a matrix, *e.g.* spectrochromatogram, EEM spectrum), and, finally, three-way data is a cube. Principal component analysis and PARAFAC can thus be represented as the decomposition of an initial array into the sum of properly multiplied component vectors, *i.e.* vectors are expressed in multiplication terms in such a way that the result is not a common dot product but an outer product which results in a data plane, a matrix for PCA, or a data cube for PARAFAC (see Figure 6.3).

An example of PARAFAC was a study[9] that investigated whether fluorescence spectroscopy of serum samples in combination with multivariate data analysis could be used to discriminate healthy females from breast cancer patients with solitary and multiple metastases, respectively. Serum samples were obtained from 39 females: 13 healthy females (controls) and 26 clinically diagnosed patients with either solitary metastases (11 patients) or multiple metastases (15 patients). The fluorescence spectroscopic results were compared with results based on three tumour markers, cancer antigen (CA 15-3), carcinoembryonic antigen (CEA), and tissue polypeptide antigen (TPA). Seven was the lowest number of errors obtained using multivariate analysis on the biomarkers. Furthermore, fluorescence spectroscopy made it possible to discover sample sub-groupings: females with solitary and multiple metastases could be divided into two subgroups according to the spectral patterns of the samples.

6.2 QUANTITATIVE STRUCTURE–PROPERTY RELATIONSHIP (QSPR)

Closely related to the above mathematical methods is another computational chemistry method: quantitative structure–activity relationship and quantitative structure–property relationship (QSAR/QSPR). They are processes by which chemical structure is quantitatively correlated with a well-defined property of a molecule, such as biological activity, or with some property of the chemical under study (solubility, boiling point, chemical reactivity, *etc.*).

The basic idea of QSPR is rather simple: molecular structure is characterized by a set of quantitative "descriptors" (such as molecular weight, volume, number of C, O, N, *etc.* atoms, functional group counts, HOMO energy, electrophilicity, dipole moment, and ionization potential). Hundreds of descriptors exist.[10] The property of interest for a particular molecule is presented as a linear sum of the descriptors of that molecule. Thus, $x_{1k}, x_{2k}, \ldots$ are the descriptor values and y_k is the property (*e.g.* biological activity) value for a particular k-th molecule. A QSAR model generally takes the form of a linear equation

$$y_k = const + a_1 x_{1k} + a_2 x_{2k} + a_3 x_{3k} + \ldots + a_n x_{nk} \quad (3)$$

The descriptors x_{1k} through x_{nk} are computed for each molecule in the series and the coefficients a_1 through a_n are calculated by least squares, and the empirical data entered into Equation (3). The derived regression equation can then be used to predict the properties of new or hypothetical molecules from their structure alone, reducing the need for experimental synthesis and measurement. The most appropriate descriptors are automatically selected from a very large pool. The latest generation of computational tools makes it relatively simple to prescreen large libraries of hypothetical candidate compounds for environmentally-significant properties, using a regular Windows-based computer. The maximum number of descriptors that one should use in the model should be reasonable, otherwise over-fitting error may appear in the results – the correlation is good but the results are meaningless. The rule of thumb is that the number of descriptors used in a model should be at least five times less than the number of compounds, the experimentally measured properties of which are used for developing the model.

QSPR has been applied in environmental chemistry. Pre-screening for environmentally-significant properties through computational chemistry has been proposed by Liang and Gallagher.[11] They demonstrated that by using computational chemistry, (specifically, QSPR and other computational methods based on statistical thermodynamics), it is possible to pre-screen for environmentally-significant properties of candidate compounds and products, particularly before they are synthesized. There are various approaches to the prediction of environmentally-significant properties (such as aquatic fate, persistence, atmospheric fate, atmospheric lifetimes of halocarbons, soil remediation rates and the design of ionic liquids for greener processes), that use readily available tools based on quantum chemistry, statistical thermodynamics and related methods. Using a computer to predict properties of compounds can save significant laboratory time and expenses. Besides saving research and development costs, the associated advantages of shortening time-to-market and minimizing environmental risk offer the greatest potential for long-term profitability. It is well known that many environmentally-significant properties are closely related to physical and chemical properties that can be predicted from their molecular structure by calculations using various algorithms.

The current risk paradigm calls for individual consideration and evaluation of each separate environmental pollutant, but this does not accurately reflect the cumulative impact of anthropogenic chemicals. Because of the enormous amount of possible pollutants and toxic industrial chemicals, interest in QSPR/QSAR in environmental analytical chemistry is considerable. In one recent study,[12] previously validated QSPRs were used to estimate simultaneously the baseline toxicity and atmospheric persistence of approximately 50 000 compounds. It was found that the baseline toxicity was not significantly altered by halogen substitution, but a distinct increase occurred in environmental persistence with increased halogenation. Halogenated compounds as a group obtained a high ranking in this data set, with well-known pollutants at the top: DDT metabolites and derivatives, polychlorinated biphenyls, diphenyl ethers and dibenzofurans, chlorinated paraffins, chlorinated benzenes and derivatives, hydrochlorofluorocarbons and dichlorononylphenol. Environmentally-friendly chemicals that received the lowest ranking are nearly all hydroxylated and water-soluble. Virtual screening can assist Green Chemistry to design safe and degradable products and to enable assessment of efficiency in chemical risk management.

Over the last few decades, the QSAR discipline has emerged as a major scientific methodology to aid in our understanding of the behaviour of chemicals in the environment, as well as their potential adverse effects. This interest is reflected in the fact that QSAR workshops and conferences have developed into highly anticipated international forums. At a recent conference,[13] many applications of QSPR/QSAR in environmental analytical chemistry were reported, such as the prediction of:

- physico–chemical properties of polybrominated diphenyl ethers;
- rate constants of OH radical reactions with organic pollutants;
- metabolic biotransformation rates for fish;
- environmental toxicity against *T. pyriformis*;
- stereospecificity of representative chiral organophosphorus pesticides;
- melting points for substituted benzenes;
- regioselectivity in cytochrome p450 mediated xenobiotic metabolism;
- acute toxicity of chemicals to fish, algae and daphnia;
- night-time degradability of volatile organic compounds by NO_3 radicals;
- endocrine disruption potencies of brominated flame retardants;
- aqueous solubility of military compounds.

The vapour pressure of a chemical is one of the most important parameters for determining its toxicity in the atmosphere; thus, the knowledge of melting points is important for the design and application of green solvents. Ionic liquids – a class of compounds popular in Green Chemistry studies – have low vapour pressure. A melting point at or below ambient temperature is an essential property of ionic liquids that are being considered as non-volatile replacement solvents. The flexibility of ionic liquid design is both an advantage and a disadvantage. The melting points of two kinds of room-temperature ionic liquids, imidazolium tetrafluoroborates ($[C_nC_mIm][BF_4]$) and imidazolium hexafluorophosphates ($[C_nC_mIm][PF_6]$), were examined using the QSPR approach.[14] The descriptors were first selected by studying the optimized geometries of four ILs: $[C_2C_1Im][BF_4]$, $[C_4C_1Im][BF_4]$, $[C_2C_1Im][PF_6]$ and $[C_4C_1Im][PF_6]$. Electrostatic, quantum mechanical and topological descriptors were considered effective to describe the melting points of ionic liquids. A three-parameter model with a squared correlation coefficient, R^2, of 0.9047 was developed for 16 kinds of imidazolium tetrafluoroborates, and a six-parameter equation with an R^2 of 0.9207 was formulated for 25 kinds of imidazolium hexafluorophosphates. The proposed models were used to predict the melting points of ILs with similar structural features. In another study[15] that attempted to develop predictive tools for the determination of new ionic liquid solvents, QSPR models for the melting points of 126 structurally diverse pyridinium bromides were developed. Six- and two-descriptor equations with squared correlation coefficients, R^2, of 0.788 and 0.713, respectively, were reported for the melting temperatures. The models illustrate the importance of information content indices, total entropy, and the average nucleophilic reactivity index of a nitrogen atom. Other researchers have also used QSPR to study the melting points of ionic liquids. A correlation with $R^2 = 0.790$ was created for a set of 75 tetraalkyl-ammonium bromides, and two correlations were created with $R^2 = 0.716$ and $R^2 = 0.766$ for a set of 34 (*n*-hydroxyalkyl)-trialkyl-ammonium bromides.[16] The descriptors used in the correlations were analyzed to determine structural features that lower the melting point, and melting points were predicted for salts that incorporate these features. In another article, a recursive neural network algorithm was used to predict the melting points of several pyridinium-based ionic liquids.[17] The algorithmic descriptions of chemical compounds are depicted as trees and their root systems. It takes a direct approach to the quantitative structure–property relationship (QSPR) of ILs, and avoids the use of dedicated molecular descriptors. The model was applied to a set of 126 pyridinium bromides, and was validated by splitting the dataset into a disjoint training set (100 compounds) and test set (26 compounds).

Despite the obvious and significant economic and environmental advantages of QSPR/QSAR over experimentation in chemistry, the availability of so many different descriptors merits a word of warning. In recent years, increased attention has been directed to and concern has been expressed regarding the pitfalls of QSAR.[18] With a large array of possible descriptors, each of which can be coupled with many statistical methods, the number of equivalent solutions that fit the experimental data is typically substantial. Each of these equivalent solutions, however, represents a hypothesis regarding the underlying physical or biological phenomenon. It may be that each solution encodes the same basic hypothesis in only subtly different ways. However, it may also be that many of the hypotheses are meaningfully different from each other and a chemist may repeatedly arrive at wrong conclusions. With the nearly infinite number of molecular descriptors coupled with incredibly flexible machine learning algorithms, this can be expected. Indeed, the very feature that makes QSPR/QSAR appealing – that it is possible to identify a few critical molecular properties from a nearly infinite pool of detailed possibilities – makes the method difficult to control.[18] Predictions regarding future compounds could be disappointing if the wrong model were selected from statistically nearly equivalent models. This means that QSPR/QSAR results must be confirmed by carefully designed experiments for testing hypotheses.

6.3 IMPROVING THE PERFORMANCE OF FLOW TECHNIQUES BY COMPUTERIZED MULTIPLE-INPUT

If chemometrics, as it is commonly understood, implements computers for *post factum* data analysis of experiments, this means that the information is already present in the data. Computers can also contribute to the process of producing information by controlling experiments. The Fourier transform spectroscopic methods, either NMR[19] or infrared,[20] are well known examples. Although these techniques have little relevance to Green Analytical Chemistry, their less well-known counterparts in chromatography and correlation (multiplex) chromatography do. In correlation chromatography, improving performance is possible because the experiment is performed under computer control, with a slight modification of the sampling part of the instrument. The performance of chromatography can be improved in two ways:

- The detection limit can be improved without any "chemical" manipulation of the sample (such as extraction or concentration) and without increasing the time of the experiment. The latter is of special interest for chromatographic modes where eluent consumption is an issue.
- Sample analysis throughput can be improved.

The opportunity to improve measurement performance by simply reorganizing the experiment without any modification of the experimental hardware was recognized as early as 1935 by Yates, who developed weighting design.[21] Instead of weighting individual masses sequentially, he proposed to load all masses of interest simultaneously in both pans of the beam balance, using rules carefully formulated according to the theory of experimental design. The number of measurements remains equal to the traditional weighting design, but the precision of weighting increases. In correlation chromatography, the sample is introduced at short intervals according to a pseudo random sequence, resulting in many overlapping chromatograms. If the sequence is properly selected, the chromatograms can be deconvolved into a single-injection chromatogram with reduced baseline noise.

Robert Annino[22] and Henri Smit[23] initiated experiments with correlation chromatography (CC) in the middle of the last century and the results of 1980s research are summarized in a book.[24] Unfortunately, the application of CC has been very limited since then. Recently however, there has been renewed interest in applying the multiplex method in capillary electrophoresis [25,26] and in liquid chromatography (see Figure 6.4).[27] The latter research illustrates well how the "greenness" of HPLC can be improved through random sampling: a seven-fold improvement in the detection limit was achieved compared with that of a single injection within approximately the same time and eluent consumption needed to run a single-injection chromatogram (Figure 6.4). The disadvantage of correlation chromatography seems to be the impossibility of performing gradient elution; stationarity and stability are the main drawbacks of the method.

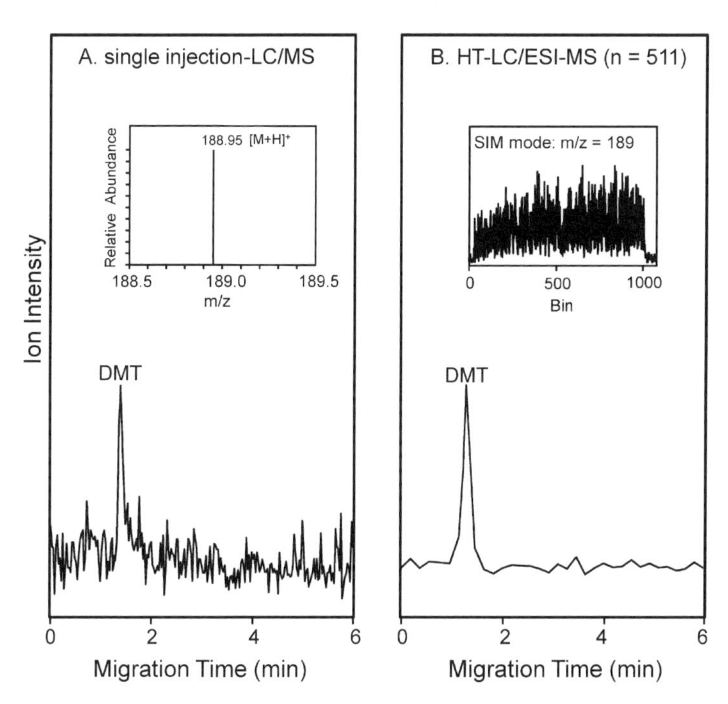

Figure 6.4 (A) A typical LC-MS chromatogram of *N,N*-dimethyltryptamine (DMT) obtained by single injection based on the SIM mode (ion peak at $m/z = 189$ was selected for monitoring); the concentration of DMT was $1.0\,\mu g\,mL^{-1}$. (B) HT-LC-MS (order of matrix, 511) chromatogram of DMT. Inset, the raw data of TIC (by selecting the ion at $m/z = 189$; Bin = 3 sec) chromatogram before inverse Hadamard transformation.[27] (Reproduced with kind permission of the American Chemical Society).

Smit *et al.* recognized quite early the opportunity to use correlation chromatography to decrease analysis time.[28] Sampling was performed with high frequency but from many sources. Using different sequences for different sample sources, it is possible to deconvolve single-injection chromatograms from different sources simultaneously, which improves analysis throughput significantly. This approach was recently employed by Trapp, who used a multiplexing sampler to rapidly inject samples onto a separation column according to a pseudo-random sequence. The pseudo-random sequence is divided into sub-sequences assigned to individual reactor channels encoding individual samples with similar analyte composition but different concentrations.[29] The technique developed by Trapp increases the information content and decreases the overall time for the chromatographic analysis. This type of time-resolved quantification of analytes in complex mixtures could be useful for real-time analysis in kinetic and mechanistic studies conducted in parallelized chemical reactors. Ultra-fast separation techniques can be envisaged.

Despite the efforts of some enthusiasts, the influence of multiplex techniques in separation science has remained marginal. Since nothing relevant to correlation chromatography (software and hardware) is commercially available, implementation of the method requires a good knowledge of mathematics and engineering in order to modify a chromatograph sampler and reprogram its computer. It is not surprising that less exotic (and less green) chemical approaches for improving signals have been adopted by chromatographers.

REFERENCES

1. S. Armenta, S. Garrigues and M. de la Guardia, *Trends Anal. Chem.*, 2008, **27**, 497.
2. S. Armenta, S. Garrigues and M. de la Guardia, *Vib. Spectrosc.*, 2007, **44**, 273.
3. M. H. Moh, Y. B. C. Man, F. R. Van De Voort and W. J. W. Abdullah, *J. Am. Oil Chem. Soc.*, 1999, **76**, 19.
4. M. Otto, *Chemometrics*, Wiley-VCH Verlag GmbH & Co. KGaA, Weinheim, 2007.
5. M. P. A. Ribeiro, T. F. Pádua, O. D. Leite, R. L. C. Giordano and R. C. Giordano, *Chemometrics Intell. Lab. Syst.*, 2008, **90**, 169.
6. F. -T. Chau, Y. -Z. Liang, J. Gao and X. -G. Shao, *Chemometrics: From Basics to Wavelet Transform (Chemical Analysis: A Series of Monographs on Analytical Chemistry and its Applications)*, Wiley-Interscience, 2004, p. 263.

7. R. Bro, *Chemometrics Intell. Lab. Sys.*, 1997, **38**, 149.
8. H. -L. Wu, R. -Q. Yu and K. Oguma, *Anal. Sci.*, 2001, **17**(Supplement), i483.
9. L. Nørgaard, G. Söletormos, N. Harrit, M. Albrechtsen, O. Olsen, D. Nielsen, K. Kampmann and R. Bro, *J. Chemometrics*, 2007, **21**, 451.
10. M. Karelson, *Molecular Descriptors in QSAR/QSPR*, Wiley-Interscience, 2000.
11. C. K. Liang and D. A. Gallagher, *J. Chem. Inf. Comput. Sci.*, 1998, **38**, 321.
12. T. Oberg, *Environ. Toxicol. Chem.*, 2006, **25**, 1178.
13. *The 13th International Workshop on Quantitative Structure-Activity Relationships (QSARs) in the Environmental Sciences*, June 8–12 2008, Syracuse, New York, USA
14. N. Sun, X. Z. He, X. P. Zhang, X. M. Lu, H. Y. He and S. J. Zhang, *Fluid Phase Equilib.*, 2006, **246**, 137.
15. A. R. Katritzky, A. Lomaka, R. Petrukhin, R. Jain, M. Karelson, A. E. Visser and R. D. Rogers, *J. Chem. Inf. Comput. Sci.*, 2001, **36**, 785–804.
16. D. M. Eike, J. F. Brennecke and E. J. Maginn, *Green Chem.*, 2003, **5**, 323.
17. R. Bini, C. Chiappe, C. Duce, A. Micheli, R. Solaro, A. Starita and M. R. Tiné, *Green Chem.*, 2008, **10**, 306.
18. S. R. Johnson, *J. Chem. Inf. Modeling*, 2008, **48**, 25.
19. J. D. Roberts, *ABCs of FT-NMR*, University Science Books, 2000.
20. P. R. Griffiths and J. A. De Haseth, *Fourier Transform Infrared Spectrometry*, 2nd edn IM Publications, 2007.
21. F. Yates, *J. Royal Statist. Soc. Suppl.*, 1935, **2**, 181.
22. R. Annino and E. L. Bullock, *Anal. Chem.*, 1973, **45**, 1221.
23. H. C. Smit, *Chromatographia*, 1970, **3**, 515.
24. M. Kaljurand and E. Küllik, *Computerized Multiple Input Chromatography*, Ellis Horwood, Chichester, 1989.
25. T. Kaneta, Y. Yamaguchi and T. Imasaka, *Anal. Chem.*, 1999, **71**, 5444.
26. A. Seiman, M. Kaljurand and A. Ebber, *Anal. Chim. Acta*, 2007, **589**, 71.
27. C.-H. Lin, T. Kaneta, H.-M. Chen, W.-X. Chen, H.-W. Chang and J.-T. Liu, *Anal. Chem.*, 2008, **80**, 5755.
28. H. C. Smit, C. Mars and J. C. Kraak, *Anal. Chim. Acta*, 1986, **181**, 37.
29. O. Trapp, *Angew. Chem. Int. Ed.*, 2007, **46**, 5609.

CHAPTER 7

Conclusions

In the preceeding pages, many different methods are described that meet various greenness criteria. The criteria may differ, but the overall greenness of a method relates to its ecological footprint. From the description of the methods, greenness can be manifested in four different ways:

- Instrumental methods such as X-ray fluorescence, laser ablation and sensors are green *per se*. These methods use instruments that can be portable, used on-site, do not require sample treatment, and give almost immediate results.
- Methods that must be modified to make them green. This is mainly relevant to separation methods and especially to HPLC. High-temperature and micro-column HPLC meet the requirements of green analytical techniques. Electroseparation techniques like capillary electrophoresis and separations on microchips also belong in this category.
- Non-instrumental analytical techniques and methods based on the use of consumer electronics. Very few applications have been reported, so these methods are outside the mainstream of analytical chemistry. They could find their place in the developing world, however, if an important application were found. Home pregnancy tests are an example of a success story of such methods and could inspire the development of further non-instrumental methods.
- Application of chemometrics to data produced by well-known and well-developed methods.

Green Analytical Chemistry
By Mihkel Koel and Mihkel Kaljurand

Published by the Royal Society of Chemistry, www.rsc.org

An important question is how widely those green methods and procedures are applicable. Is it possible that a non-instrumental method based on paper microfluidics, for example, could replace a well-established but expensive instrumental method? Who would welcome such a development? Nobody yet knows. Different methods are in competition and most of the green methods might not appear useful. Comparative studies are completely lacking and may never be done. In this book, by describing various emerging, less well-known alternatives to mainstream analytical methods, we have attempted to demonstrate that green analytical chemistry is possible. Its adoption depends largely on political or ecological factors that are beyond the purview of analytical chemistry. The only way that green methods will be widely accepted is through economic pressure (such as we witnessed with the acetonitrile crisis in 2009), or pressure from society for a cleaner and safer environment.

The development of analytical chemistry continues at a steady rate and every new discovery in chemistry, physics, molecular biology and material science finds an application in analytical chemistry as well. Because it is concerned with the analysis, collection, classification, manipulation, storage, retrieval and dissemination of information, analytical chemistry can be considered a branch of interdisciplinary information science. It deals with the acquisition and processing of information about chemicals in our environment. Like any other branch of information science, it requires only the means of transmitting information and therefore consumes no more than the resources needed to accomplish that purpose. The essence of green analytical chemistry could well be in achieving that aim.

Subject Index

Note: Figures are indicated by *italic page numbers*, Tables by **bold page numbers**